T0343400

Genetic Manipulation of the Nervous System

NEUROSCIENCE PERSPECTIVES

Titles in this series
Roger Horton and Cornelius Katona (eds), Biological Aspects of Affective Disorders
Trevor Stone (ed), Adenosine in the Nervous System
Judith Pratt (ed), The Biological Bases of Drug Tolerance and Dependence
Michel Hamon (ed), Central and Peripheral 5-HT_3 Receptors
John Waddington (ed), D_1 : D_2 Dopamine Receptor Interactions
Thomas Barnes (ed), Antipsychotic Drugs and Their Side-Effects
Eva Giesen-Crouse (ed), Peripheral Benzodiazepine Receptors
David Nicholson (ed), Anti-Dementia Agents
Alan Harvey (ed), Natural and Synthetic Neurotoxins
John Wood (ed), Capsaicin in the Study of Pain
Yossef Itzhak (ed), Sigma Receptors
John Mayer/Ian Brown (eds), Heat Shock Proteins in the Nervous System
Chris Carter (ed), Neuropharmacology of Polyamines
Roger Pertwee (ed), Cannabinoid Receptors

Genetic Manipulation of the Nervous System

edited by

D. S. Latchman
Department of Molecular Pathology,
University College London Medical School
London, UK

ACADEMIC PRESS
Harcourt Brace & Company, Publishers
London San Diego New York
Boston Sydney Tokyo Toronto

ACADEMIC PRESS LIMITED
24/28 Oval Road,
London NW1 7DX

United States Edition published by
ACADEMIC PRESS INC.
San Diego, CA 92101

This book is printed on acid free paper

A catalogue record for this book
is available from the British Library

ISBN 0-12-437165-5

Typeset by Servis Filmsetting Ltd, Manchester, UK
Printed and bound in the United Kingdom
Transfered to Digital Printing, 2011

Contents

Contributors

Manish Aghi Molecular Neurogenetics Unit, Neurology Service, Massachusetts General Hospital and Neuroscience Program Harvard Medical School, Boston, MA 02129, USA

Arturo Alvarez-Buylla The Rockefeller University, 1230 York Avenue, New York, NY 10021, USA

D. Bain Molecular Medicine Unit, Department of Medicine, University of Manchester School of Medicine, Stopford Building, Manchester M13 9PT, UK

Frederick M. Boyce Department of Neurology, Harvard Medical School and Massachusetts General Hospital, Charlestown, MA 02129, USA

Xandra O. Breakefield Molecular Neurogenetics Unit, Neurology Service, Massachusetts General Hospital and Neuroscience Program Harvard Medical School, Boston, MA 02129, USA

M.H. Buc-Caron C 9923 CNRS, Laboratoire de Génétique Moléculaire de la Neurotransmission et des Processus Dégénératifs, 1 Avenue de la Terrasse, 91198 Gif-sur-Yvette, France

Maria G. Castro Molecular Medicine Unit, Department of Medicine, University of Manchester School of Medicine, Stopford Building, Manchester M13 9PT, UK

Robert S. Coffin Department of Molecular Pathology, University College London Medical School, The Windeyer Building, 46 Cleveland Street, London W1P 6DB, UK

Lawrence T. Feldman Departments of Microbiology and Immunology, UCLA School of Medicine, Los Angeles, CA 90024, USA

A.R. Fooks Molecular Pathology Section, Research Division, Centre for Applied Microbiology and Research, Porton Down, Salisbury SP4 0JG, UK

Fred H. Gage Department of Neurosciences, University of California, San Diego, La Jolla, CA 92093, USA

Alfred I. Geller Division of Endocrinology, Children's Hospital Boston, MA 02115, and Program in Neuroscience, Harvard Medical School, Boston, MA 02129, USA

Markus Heilig Department of Clinical Neuroscience, Magnus Huss Clinic, Karolinska Hospital, S-17176 Stockholm, Sweden

E.L. Hooghe-Peters Department of Pharmacology, Neuroendocrinology Unit, Faculty of Medicine and Pharmacy, Free University of Brussels, Laarbeeklaan 103, B1090 Brussels, Belgium

P. Horellou C 9923 CNRS, Laboratoire de Génétique Moléculaire de la Neurotransmission et des Processus Dégénératifs, 1 Avenue de la Terrasse, 91198 Gif-sur-Yvette, France

Takashi Imaoka Department of Neurological Surgery, Okayama University Medical School, Shikata-Cho 2-5-1, Okayama 700, Japan

Makote Ishibashi Institute for Immunology and Department of Anatomy, Kyoto University Faculty of Medicine, Sakyo-Ku, Kyoto 606, Japan

Ryoichiro Kageyama Institute for Immunology, Kyoto University Faculty of Medicine, Sakyo-Ku, Kyoto 606, Japan

Edward M. Kaye Molecular Neurogenetics Unit, Neurology Service, Massachusetts General Hospital and Neuroscience Program Harvard Medical School, Boston, MA 02129, USA

David S. Latchman Department of Molecular Pathology, University College London Medical School, The Windeyer Building, 46 Cleveland Street, London W1P 6DB, UK

Carlos Lois Department of Biology, 68–380 MIT, 77 Massachusetts Avenue, Cambridge, MA 02139, USA

Pedro R. Lowenstein Molecular Medicine Unit, Department of Medicine, University of Manchester School of Medicine, Stopford Building, Manchester M13 9PT, UK

J. Mallet C 9923 CNRS, Laboratoire de Génétique Moléculaire de la Neurotransmission et des Processus Dégénératifs, 1 Avenue de la Terrasse, 91198 Gif-sur-Yvette, France

Koki Moriyoshi Institute for Immunology, Kyoto University Faculty of Medicine, Sakyo-Ku, Kyoto 606, Japan

Rachael L. Neve Department of Genetics, Harvard Medical School and McLean Hospital, Belmont MA 02178, USA

Karen L. O'Malley Department of Anatomy and Neurobiology, Washington University School of Medicine, St Louis, MO 63110, USA

Mary Lou Oster-Granite Division of Biomedical Sciences, University of California, Riverside, CA 92521, USA

Peter A. Pechan Molecular Neurogenetics Unit, Neurology Service, Massachusetts General Hospital and Neuroscience Program Harvard Medical School, Boston, MA 02129, USA

F. Revah C 9923 CNRS, Laboratoire de Génétique Moléculaire de la Neurotransmission et des Processus Dégénératifs, 1 Avenue de la Terrasse, 91198 Gif-sur-Yvette, France

J.J. Robert C 9923 CNRS, Laboratoire de Génétique Moléculaire de la Neurotransmission et des Processus Dégénératifs, 1 Avenue de la Terrasse, 91198 Gif-sur-Yvette, France

Karl-Hermann Schlingensiepen Max-Planck-Institut für Biophysikalische Chemie, Göttingen, Germany

F. Sebaté C 9923 CNRS, Laboratoire de Génétique Moléculaire de la Neurotransmission et des Processus Dégénératifs, 1 Avenue de la Terrasse, 91198 Gif-sur-Yvette, France

Miguel Sena-Esteves Molecular Neurogenetics Unit, Neurology Service, Massachusetts General Hospital, and Neuroscience Program Harvard Medical School, Boston, MA 02129, USA

Marie-Claude Senat Department of Neurosciences, University of California, San

Diego, La Jolla, CA 92093, USA, and INSERM U161, 2 Rue D'Alésia, 75014 Paris, France

H. David Shine Departments of Neurosurgery and Cell Biology, Baylor College of Medicine, Houston, TX 77020, USA

Steven T. Suhr Department of Neurosciences, University of California, San Diego, La Jolla, CA 92093, USA

L. Tenenbaum Department of Pharmacology, Neuroendocrinology Unit, Faculty of Medicine and Pharmacy, Free University of Brussels, Laarbeeklaan 103, B1090 Brussels, Belgium

Masaaki Tsuda Department of Microbiology, Faculty of Pharmaceutical Sciences, Okayama University, Tsushima-Naka 1-1-1, Okayama 700, Japan

Gavin W.G. Wilkinson Department of Medicine University of Wales College of Medicine, Heath Park, Cardiff CF4 4XN, UK

Savio L.C. Woo Department of Cell Biology and The Howard Hughes Medical Institute Baylor College of Medicine, Houston, TX 77030, USA

Series Preface

The neurosciences are one of the most diverse and rapidly changing areas in the biological sphere. The need to understand the workings of the nervous system pervades a vast array of different research areas. By definition research in the neurosciences encompasses anatomy, pathology, biochemistry, physiology, pharmacology, molecular biology, genetics and therapeutics. Ultimately, we are striving to determine how the human brain functions under normal circumstances and perhaps more importantly how function changes in organic disease and in altered states of mind. The key to many of these illnesses will unlock one of the major therapeutic challenges remaining in this era.

The difficulty lies in the vastness of the subject matter. However I try, I find it almost impossible to keep abreast of the changes occurring in my immediate sphere of interest, let alone those exciting advances being made in other areas. The array of journals dealing with neurosciences is enormous and the flow of reprints needed to keep me updated is daunting. Inevitably piles of papers accumulate on my desk and in my briefcase. Many lie there unread until sufficient time has passed for their content to be overtaken by yet more of the ever rising tide of publications.

There are various approaches that one can take to deal with this problem. There is the blinkered approach in which you pretend that literature outside your area does not exist. There is the ignore it totally option. Indeed, one colleague of mine has ceased to read the literature in the belief that, if there is a publication of critical importance to his research, someone will tell him about it. I am not that brave and instead I arrived at what I thought was the ideal solution. I started to read critical reviews of areas of current interest. But I soon came unstuck as I realized that, for many subjects of importance to the neurosciences, such authoritative works did not exist.

Instead of simply moaning that the literature was incomplete, I conceived the idea of *Neuroscience Perspectives*. For purely selfish reasons I wanted to have available a series of individually edited monographs dealing in depth with issues of current interest to those working in the neuroscience area. Already a number of volumes have been published which have been well received and the series is thriving with books on a range of topics in preparation or in production. Each volume is designed to bring a multidisciplinary approach to the subject matter by pursuing the topic from the laboratory to the clinic. The editors of the individual volumes are producing balanced critiques of each topic to provide the reader with an up-to-date, clear and comprehensive view of the state of the art.

As with all ventures of this kind, I am simply the individual who initiates a chain of events leading to the production of the series. In reality, it is key individuals at Academic Press who are really responsible for the success of *Neuroscience Perspectives*. In particular, Dr Carey Chapman and Leona Daw have the uneviable task of recruiting editors and authors and keeping the ship on an even keel.

Finally, I hope that *Neuroscience Perspectives* will continue to be enjoyed by my colleagues in the neurosciences. Already the series is being read, understood and enjoyed by a wide audience and it is fast becoming a reference series in the field.

Peter Jenner

Preface

The ability to deliver specific genes to individual cells within an intact animal or human offers enormous possibilities both for scientific analysis of the gene's function and for therapeutic intervention in genetic diseases. The enormous complexity of the nervous system and the many diseases in which it is involved indicate that such *in vivo* gene delivery is likely to have enormous implications in this system. Unfortunately, however, the application of *in vivo* gene delivery to the nervous system has lagged behind that of other systems both because of its complexity and because the retroviral vectors used in other tissues do not infect non-dividing cells such as terminally differentiated neurons.

The aim of this work is to consider the approaches which have been developed to overcome these difficulties allowing the *in vivo* delivery of specific genes in a safe and effective manner. In view of the inability of retroviruses to infect non-dividing neurons, many laboratories have attempted to use other viruses as vectors for gene delivery to such cells. These approaches, including the use of adenovirus, adeno-associated virus and herpes simplex virus, are discussed in the first section of this book. In contrast, others have taken advantage of the safety and efficiency of retroviral vectors to use these for particular approaches to *in vivo* gene delivery in the nervous system. Such approaches are discussed in the second section of the book, and include the use of retroviral vectors to deliver genes to dividing neuronal precursor cells and to specifically target dividing cancer cells in the brain, as well as to engineer non-neuronal cells in culture prior to transplant into the brain.

As well as these viral-based methods, it is evidently also possible to deliver genes to the nervous system by non-virally mediated means, and the methods to do this are discussed in the final section of the book. These include, for example, the creation of transgenic animals in which all the cells have been genetically modified. As well as studying these animals, they can also serve as a source of material for transplantation of genetically modified cells into the brain. Further approaches of this type include the direct injection of plasmic DNA and of antisense oligonucleotides into the brain.

All these approaches have proved of advantage in particular situations and have aided an understanding of the nervous system as well as opening up therapeutic possibilities. Overall therefore it is hoped that this book will provide an indication of the range of techniques which are available for the genetic manipulation of the nervous system together with the advantages and disadvantages of each. Finally I would like to thank Professor Peter Jenner who suggested that I should edit a volume in the Neuroscience Perspectives series as well as Carey Chapman, Susan Lord and the staff at Academic Press for their efficiency in producing it.

D.S. Latchman

CHAPTER 1

GENETIC MANIPULATION OF THE NERVOUS SYSTEM: AN OVERVIEW

David S. Latchman

Department of Molecular Pathology,
University College London Medical School,
The Windeyer Building, 46 Cleveland Street, London W1P 6DB, UK

Table of Contents

1.1 Introduction

The ability to express specific genes in particular cell types *in vivo* evidently has numerous applications. Thus, in scientific terms, it is possible by this means to test the effects of enhancing or eliminating the expression of a particular gene product allowing an assessment of its functional role. In addition however, the ability to deliver a gene to specific cells *in vivo* in which it is inactive due, for example, to genetic disease, has obvious therapeutic applications and such so-called 'gene therapy' approaches have aroused considerable interest (for review see Miller, 1992; Latchman, 1994a).

Both the scientific and therapeutic aspects of gene delivery have considerable applications in the nervous system. An overview of these applications is provided in this chapter and many are discussed in succeeding chapters. The development of gene delivery to the nervous system for these purposes has suffered from considerable difficulties however, principally due to the non-dividing nature of neuronal cells and the enormous complexity of the brain. This introductory chapter therefore also outlines the approaches which have been used to overcome these problems, many of which are discussed in detail in later chapters.

Genetic Manipulation of the Nervous System
ISBN 0-12-437165-5

1.2 Applications of *in vivo* gene delivery to the nervous system

1.2.1 Experimental applications

As in any other cell type, the ability to over-express a particular gene product can aid in elucidating its function in neuronal cells. This can be achieved either by introducing the gene of interest into the germ line to produce transgenic mice (for review see Evans, 1994) or by directly injecting it into the brain in a suitable delivery system, such as a viral vector (see e.g. Ho *et al.*, 1993). In both of these cases it will be necessary to target expression of the gene either to all neuronal cells or to specific neuronal cell types by using an appropriate promoter either derived from the gene of interest itself or from another gene with the appropriate neuronal-specific expression pattern. Thus, if this is not done, the germ line transgene may be expressed in a wide variety of inappropriate cell types whilst even when the gene is injected directly into the brain it may be expressed by non-neuronal cells.

These considerations indicate that the successful over-expression of a particular gene in neuronal cells *in vivo* requires the characterization of neuronal-specific promoters. Indeed *in vivo* gene delivery to produce transgenic animals has also been used as a means of characterizing such promoters. Thus in this approach, various fragments of the gene regulatory region are linked to a reporter gene such as that encoding ß-galactosidase, and following injection into the fertilized egg, the pattern of reporter gene activity in the developing animal allows an assessment of the activity of the promoter fragment under test (see e.g. Forss-Petter *et al.*, 1990).

Interestingly, in many neuronally expressed genes, much larger promoter fragments are required to direct the correct pattern of expression *in vivo* compared to those required following transfection of cultured neuronal cells *in vitro*. Thus, for example, in the case of the tyrosine hydroxylase gene a construct containing only 272 bases of upstream promoter sequence is specifically active in neuronal cells lines when introduced by transfection (Yoon and Chikaraishi, 1992). In contrast, this construct was inactive when tested in transgenic animals *in vivo*, with the correct pattern of promoter activity, mimicking endogenous tyrosine hydroxylase expression, being produced only with 4800 bases of upstream promoter sequence (Banerjee *et al.*, 1992). Hence the features which allow a promoter to work *in vivo* may not, in many cases, be readily assayable in cultured cells *in vitro*. For this reason, the development of effective means of gene delivery to neuronal cells *in vivo* needs to be paralleled by the use of *in vivo* methods to characterize neuronal promoters in order to allow the identification of suitable means of specifically expressing the gene of interest in the desired cell type.

Such means of targetting expression to specific cell types when over-expressing a specific gene product are of less importance when attempting to eliminate the activity of a particular gene product. Such elimination of a functional gene product can be achieved either by homologous recombination in the fertilized egg to produce a so-called 'knock out' mouse (for review see Bronson and Smithies, 1994) or by introducing anti-sense oligonucleotides or constructs which bind to the mRNA and promote its degradation and/or inhibit its translation (for review see Wagner, 1994 and Chapter 15, this volume). Thus, in both these cases it is assumed that gene inactivation will occur in all cells to which the exogenous construct is delivered, allowing an assessment of the overall role of the inactivated gene product. In this situation therefore, cellular specificity is provided by the fact that the targetted gene is only expressed in specific cell types and its inactivation will therefore have an effect only in those cells. In contrast, an over-expressed gene can clearly produce effects not only in cell types which normally express it but also in those which do not normally do so.

Even in cases of gene inactivation however, it may be of advantage to use a neuronal-specific promoter to target expression of the inactivating construct to a specific cell type. Thus in 'knock out' experiments, the inactivation of gene expression in a range of cell types may have lethal effects preventing the study of individual cell types. This renders it of interest to attempt to 'knock out' the gene only in neuronal cells or in a subset of neurons. Interestingly such a cell type-specific 'knock out' has recently been achieved in the case of the DNA polymerase ß gene whose inactivation in all cells is evidently lethal. In this case (Gu *et al.*, 1994) the gene was deleted only in T lymphocytes by using a T cell-specific promoter to activate the gene encoding the Cre recombinase. Since expression of this recombinase is necessary for the gene rearrangement which inactivates the DNA polymerase gene, this occurs only in T cells allowing its effect in these cells to be analysed in viable animals.

Thus further work on the characterization of neuronal-specific promoters using *in vivo* gene delivery will not only be of interest in terms of the promoters themselves but will also provide a means of targetting over- or under-expression of individual genes to specific cell types *in vivo*.

As well as such experiments at the level of the individual gene, *in vivo* gene delivery can also be used at the level of the individual cell to monitor the fate of that cell or its interconnections with other cells. Thus, for example, injections of retroviral vectors expressing the marker ß-galactosidase gene into the developing embryo results in the labelling of dividing neuronal precursor cells. The progress of these cells can then be identified at a subsequent stage of development simply by staining appropriate sections for ß-galactosidase activity (see e.g. Price *et al.*, 1987). Similarly, precursor cells which have been genetically marked either by retroviral injection *in vivo* or by being obtained from transgenic animals can be injected into an intact animal and the fate and progress of the cell monitored (Snyder *et al.*, 1992). For further discussion of such neuronal stem cells, see Chapter 13.

Approaches of this type can also be used in the adult brain as a means of mapping neuronal connections. Thus, in this case, the brain is injected with the herpes virus, pseudorabies which has been modified to express ß-galactosidase. As this virus spreads from cell to cell via synaptic connections, the pattern of ß-galactosidase activity which is observed following injection at different sites can be used to map such connections (see e.g. Card *et al.*, 1991).

Hence *in vivo* gene delivery has a wide range of experimental applications ranging from mapping cellular connections and cell lineages via the mapping of promoter activity to the functional assessment of the effects of altering the expression of specific genes. The breadth of these experimental applications is paralleled by the breadth of potential therapeutic applications which are discussed in the next section.

1.2.2 Therapeutic applications

The most obvious means of using *in vivo* gene delivery would be to introduce into the brain a functional copy of a specific gene which was defective in a specific neurological disease. As the genes whose inactivation underlies many neurological disorders are progressively identified (for review see Freidmann, 1994), such an approach has obviously attracted increasing interest. It should be noted, however, that in many neurological dseases where the genetic basis has been identified, such as Huntington's chorea (Huntington's Disease Collaborative Research Group, 1993), the mutant allele produces the disease in a dominant manner even in the presence of the normal allele. It is by no means clear therefore that the over-expression of the wild type allele in such situations would have functional effects and it may be necessary to inactivate the mutant allele or preferably to replace it with a wild type allele.

Despite these caveats, the feasibility of this approach to characterized genetic diseases of the nervous system has been demonstrated in animal models. Thus, for example, Wolfe *et al.* (1992) used a herpes simplex virus (HSV) vector expressing the ß-glucuronidase gene to successfully express this enzyme in the neuronal cells of animals which lack it, paralleling the absence of this enzyme in the human genetic disease mucopolysaccharidosis VII (Sly disease).

Obviously, however, such an approach cannot be used in those neurological diseases where the genetic defect has not been identified or where the disease is not caused by a single genetic defeat. Nonetheless, *in vivo* gene delivery can be of therapeutic benefit in these situations also. Thus, in Parkinson's disease, the primary defect is caused by the loss of dopaminergic neurons. Many early studies showed, however, that in both animal models of this disease and human patients the symptoms could be relieved by transplantation of fetal neurons or adrenal medulla cells which naturally express tyrosine hydroxylase, the rate-limiting step in the dopaminergic pathway (Backland *et al.*, 1985).

Unfortunately, the utility of this approach was hampered by difficulties in obtaining fetal material suitable for use in transplantation. These studies did however establish the principle that the delivery of the tyrosine hydroxylase gene would be of therapeutic benefit in this disease. Indeed this has now been demonstrated directly, either by transplanting non-neuronal cells such as fibroblasts or muscle cells which have been genetically modified to express tyrosine hydroxylase, or by direct injection of viral vectors expressing this gene (see below and Friedmann, 1994, for review). Hence *in vivo* gene delivery can have therapeutic applications even in situations where the primary genetic defect is unknown, if it can minimize the biochemical consequences of the disease.

A similar application may occur in Alzheimer's disease where the primary defect is due to the death of cholinergic neurons. As these neurons are dependent on nerve growth factor (NGF) for survival (Vantini *et al.*, 1989), the *in vivo* delivery of the gene encoding NGF might have therapeutic benefit in delaying or halting the progressive cell death characteristic of the disease. Thus, as in the case of tyrosine hyroxylase, early studies showing that injection of NGF itself could prevent neuronal death in the surgically damaged brain *in vivo* (Williams *et al.*, 1986) have been supplemented by studies showing that a similar effect can be achieved by transplanting cells genetically modified with the gene encoding this factor (Rosenberg *et al.*, 1988).

Indeed the observations of Rosenberg *et al.* (1988) that the endogenous neurons in animals given NGF-expressing transplants not only survived but sprouted axons offers hope that it may be possible for such methods both to enhance survival and to improve or modify the functioning of specific neurons. This would have major implications, for example in the area of spinal injury since many mammalian neurons regenerate very poorly following damage. Hence the *in vivo* delivery of genes which promoted the ability to regenerate might allow improved recovery following such injury. Similarly, since the genes encoding the heat shock proteins (hsps) have been shown to protect neuronal cells from stress (Mailhos *et al.*, 1994) it is possible that the *in vivo* expression of these factors might be of benefit in allowing neurons to survive for example the ischaemic stress which occurs during repetitive strokes.

A number of therapeutic applications thus exist for *in vivo* gene delivery in the nervous system, ranging from the correction of a genetic defect via the minimization of its biochemical consequence to the improved functioning of neurons in damaged or stressed areas.

1.3 Methods of *in vivo* gene delivery to the nervous system

The many potential experimental and therapeutic benefits of *in vivo* gene delivery to the nervous system have led to much effort being expended in the development of systems to do this. Unfortunately, the retroviral vectors which are widely used in gene therapy procedures for other tissues (for review see Miller, 1992) will not infect non-dividing cells such as neurons (Miller *et al.*, 1990) and hence cannot be used for this purpose. A variety of means have therefore been devised to overcome this problem and these will be discussed in turn.

1.3.1 Germ line delivery

As discussed above, an effective means of gene delivery for functional studies is to inject the gene into the fertilized egg so ensuring that it is inherited by all cells and directing its expression to neuronal cells by using a suitable promoter. Evidently, however, such germ line gene delivery in which the introduced gene will be inherited by any offspring is unlikely to be used in human therapeutic procedures for ethical reasons (for discussion see Latchman, 1994b). However it has been used, for example, to generate animal models of human diseases such as Alzheimer's disease, and this is discussed in Chapter 12.

1.3.2 Retroviral methods

1.3.2.1 Infection of non-neuronal cells in culture followed by cellular transplantation

One means of avoiding the problem that retroviruses will not infect non-dividing neurons, is to use these viruses to transduce dividing cells in culture with the gene of interest and then to transplant these cells into the brain. As described above and in subsequent chapters (see Chapters 10 and 11), the genetic modification of different cell types, such as fibroblasts, the PC12 neuronal cell line or muscle cells, has been used in a number of different situations and shown following transplantation to have therapeutic benefit, for example in animal models of Parkinson's disease and in the surgically lesioned brain.

An elegant example of such transplantation work which takes advantage of the ability of retroviruses to infect only dividing cells is provided by the work of Culver *et al.* (1992) on the treatment of brain tumours. Thus these workers introduced fibroblast cells infected with a retroviral vector expressing the HSV thymidine kinase gene into the brains of rats with a cerebral glioma. The retrovirus produced by the transplant was only able to infect the dividing glioma cells of the tumour and not the non-dividing neuronal cells. Hence when the animals were treated with the anti-herpes drug gancyclovir which kills only cells expressing the HSV thymidase kinase, the tumour cells died, resulting in tumour regression, leaving the normal neuronal cells intact. This approach therefore offers a realistic hope of treating normally intractable tumours of this type and is now being extended to human patients (for further details see Chapter 10). It should be noted that this approach to the therapy of brain tumours is also being pursued with other viruses, and this is discussed in Chapter 4.

1.3.2.2 Infection of dividing cells in the brain by direct retroviral injection

It would evidently also be possible to directly inject dividing tumour cells in the brain with such retroviral constructs in order to deliver the thymidase kinase gene to them. Thus it is possible to use retroviruses to deliver genes to the brain where the target cells are dividing. This approach has already been discussed above in terms of its experimental use to follow the lineage of dividing progenitor cells which have been stably modified to express a marker gene. In addition it can also be used to modify the expression of a specific gene of interest in dividing progenitor cells in order to assess its functional role in their proliferation and/or differentiation (see Chapter 9).

As noted above, it is also possible to combine the retroviral infection of dividing progenitor neuronal cells with the cellular transplantation approach. Thus, if such progenitor cells are infected with a retrovirus carrying a marker gene *in vitro*, their fate can be followed following *in vivo* transplantation into the brain. In addition to being of use for such experimental studies, this approach may also have an ultimate therapeutic benefit if such progenitor cells can be induced to differentiate appropriately and replace lost or damaged neurons. This is discussed further in Chapter 13.

Table 1 Relief of apomorphine-induced rotational behaviour using virally expressed tyrosine hydroxylase

Virus	Longest positive test point	Reference
Adenovirus	2 weeks	Horellou *et al.* (1994)
Adeno-associated virus	2 months	Kaplitt *et al.* (1994)
HSV episomal vector	1 year	During *et al.* (1994)

1.3.3 DNA-based approaches

1.3.3.1 DNA viruses

An obvious approach since retroviruses cannot infect non-dividing cells is to carry out *in vivo* gene delivery by injection of other viruses which can do so. Several DNA viruses have been used in this manner, including HSV, adenovirus and adeno-associated virus (for review see Ali *et al.*, 1994). The use of each of these viruses is discussed in individual chapters of this book, i.e. adenovirus in Chapters 2 and 3, HSV in Chapters 6, 7 and 8 and adeno-associated virus in Chapter 5, and will not be extensively described here. The potential effectiveness of each of these viruses is demonstrated for example by the observed symptomatic relief obtained in animal models of Parkinson's disease when tyrosine hydroxylase is expressed via HSV (During *et al.*, 1994), adenovirus (Horellou *et al.*, 1994) or adeno-associated virus (Kaplitt *et al.*, 1994) vectors (Table 1).

1.3.3.2 Plasmid DNA

Although vectors based on DNA viruses thus appear to be able to successfully deliver genes to the nervous system, they do suffer from difficulties from the safety point of view. Thus, even when they are disabled so that they cannot replicate lytically, there still remains the possibility of recombination with endogenous latent virus within the treated individual, resulting in the formation of a potentially dangerous wild type virus. For this reason, the possibility of delivering the gene of interest simply by direct injection of a plasmid vector has been investigated. Indeed expression in neuronal and glial cells can be obtained when the foreign gene is injected in this way, particularly if the efficiency of DNA uptake by intact cells is enhanced by using cationic liposomes (see Chapter 14 for further details). Although methods of this type thus have potential and are much safer than the viral vectors, it remains to be seen whether they can be modified so as to attain the delivery efficiency provided by viral vectors.

1.4 Conclusions

This introductory chapter has described a wide variety of experimental and therapeutic benefits which would arise from *in vivo* gene delivery to the nervous system as well as a variety of methods for achieving this. Although several of the methods described are of use in a particular situation, the existence of such a wide variety of methods in current use indicates that no single method is of real efficiency in a wide variety of situations. Nonetheless, the variety of approaches outlined here and described in subsequent chapters offer hope that effective *in vivo* genetic manipulation of the nervous system for experimental and therapeutic purposes can be achieved.

References

Ali, M., Lemoine, N.R. & Ring, C.J.A. (1994) The use of DNA viruses as vectors for gene therapy. *Gene Ther.* **1**, 367–384.

Backland, E-O., Granberg, P.O., Hamberger, B., Knutsson, E., Martensson, A., Sedirall, G., Seigar, A. & Olson, L. (1985) Transplantation of adrenal medulla tissue to striatum in Parkinsonism. *J. Neurosurg.* **62**, 169–173.

Banerjee, S.A., Hoppe, P., Brillian, M. & Chikariashi, D.M. (1992) 5′ flanking sequences of the rat tyrosine hydroxylase gene target accurate tissue-specific, developmental and trans-synaptic expression in transgenic mice. *J. Neurosci.* **12**, 4460–4467.

Bronson, S.K. & Smithies, O. (1994) Altering mice by homologous recombination using embryonic stem cells. *J. Biol. Chem.* **269**, 27155–27158.

Card, J.P., Wheatley, M.E., Robbins, A.K., Moore, R.Y. & Enqinst, L.W. (1991) Two α-herpesvirus strains are transported differentially in the rodent visual system. *Neuron* **6**, 957–969.

Culver, K.W., Ram, Z., Wallbridge, S., Ishi, H., Oldfield, E.H., Blaese, R.M. (1992) *In vivo* gene transfer with retroviral vector–producer cells for treatment of experimental brain tumours. *Science* **256**, 1550–1552.

During, M.J., Naegel, J.R., O'Malley, K.L. & Geller, A.I. (1994) Long-term behavioural recovery in Parkinsonian rates by a HSV vector expressing tyrosine hydroxylase. *Science* **266**, 1399–1403.

Evans, M.J. (1994) Transgenic technologies. In *Encyclopedia of Molecular Biology* (ed. Kendrew, J.), pp. 1092–1093. Oxford, Blackwell Science.

Forss-Petter, S., Danielson, P.E., Catsicas, S., Bettenbergm, E., Price, J., Nerenberg, M. & Sutcliffe, J.G. (1990) Transgenic mice expressing β-galactosidase in mature neurons under neuron-specific enolase promoter control. *Neuron* **5**, 187–197.

Friedmann, T. (1994) Gene therapy for neurological disorders. *Trends Genet.* **10**, 210–214.

Gu, H., Martin, J.D., Orban, P.C., Mossman, H. & Rajewsky, K. (1994) Deletion of a DNA polymerase β gene segment in T cells using cell type specific gene targetting. *Science* **265**, 103–106.

Ho, D.Y., Mocarski, E.S. & Sapolsky, R.M. (1993) Altering central nervous system physiology with a defective herpes simplex virus vector expressing the glucose transporter gene. *Proc. Natl. Acad. Sci. USA* **90**, 3655–3659.

Horellou, P., Vigne, E., Castel, M.-N., Barneoud, P., Colin, P., Perricaudet, M., Delaere, P. & Mallet, J. (1994) Direct intracerebral gene transfer of an adenoviral vector expressing tyrosine hydroxylase in a rat model of Parkinson's disease. *NeuroReport* **6**, 49–53.

Huntington's Disease Collaborative Research Group (1993) A novel gene containing a trinucleotide repeat that is expanded and unstable on Huntington's disease chromosomes. *Cell* **72**, 971–983.

Kaplitt, M.G., Leone, P., Samulski, R.J., Xiao, X., Pfaff, D.W., O'Malley, K.L. & During, M.J. (1994) Long-term gene expression and phenotypic correction using adeno-associated virus vectors in the mammalian brain. *Nature Genet.* **8**, 148–154.

Latchman, D.S. (ed.) (1994a) *From Genetics to Gene Therapy.* Oxford, Bios Scientific Publishers.

Latchman, D.S. (1994b) Germ line gene therapy? *Gene Ther.* **1**, 277–279.

Mailhos, C., Howard, M.K. & Latchman, D.S. (1994) Heat shock proteins hsp90 and hsp70 protect neuronal cells from thermal stress but not from programmed cell death. *J. Neurochem.* **63**, 1787–1795.

Miller, A.D. (1992) Human gene therapy comes of age. *Nature* **357**, 455–460.

Miller, D.G., Adam, M.A. & Miller, A.D. (1990) Gene transfer by retrovirus vectors occurs only in cells that are actively replicating at the time of infection. *Mol. Cell Biol.* **10**, 4239–4242.

Price, J., Turner, D. & Cepko, C. (1987) Lineage analysis in the vertebrate nervous system by retrovirus-mediated gene transfer. *Proc. Natl. Acad. Sci. USA* **84**, 156–160.

Rosenberg, M.B., Friedmann, T., Robertson, R.C., Tuszynski, M., Wolff, J.A., Breakfield, X.O. & Gage, F.H. (1988) Grafting genetically modified cells to the damaged brain: restorative effects of NGF expression. *Science* **242**, 1575–1578.

Snyder, E.Y., Deitcher, D.L., Walsh, C., Arnold-Aldea, S., Hartweig, E.A. & Cepko, C.L. (1992) Multipotent neural cell lines can engraft and participate in development of mouse cerebellum. *Cell* **68**, 33–51.

Vantini, G., Schiavo, N.D., Martino, A., Polato, P., Triban, C., Callegaro, L., Toffano, G. & Leon, A. (1989) Evidence for a physiological role of nerve growth factor in the central nervous system of neonatal rats. *Neuron* **3**, 267–273.

Wagner, R.W. (1994) Gene inhibition using antisense oligodeoxynucleotides. *Nature* **372**, 333–335.

Williams, L.R., Varon, S., Peterson, G.F., Wicktorin, K., Bjocklund, A. & Gage, F.H. (1986) Continuous infusion of nerve growth factor prevents basal forebrain neuronal death after fimbra fornix transfection. *Proc. Natl. Acad. Sci. USA* **83**, 9231–9235.

Wolfe, J.H., Deshmane, S.L. & Fraser, N.W. (1992) Herpes virus vector gene transfer and expression of β-glucuronidase in the central nervous system of MPS VII mice. *Nature Genet.* **1**, 379–384.

Yoon, S.O. & Chikaraishi, D.M. (1992) Tissue-specific transcription of the rat tyrosine hydroxylase gene requires synergy between an AP-1 motif and an overlapping E box-containing dyad. *Neuron* **9**, 55–67.

CHAPTER 2

NON-NEUROTROPIC ADENOVIRUS: A VECTOR FOR GENE TRANSFER TO THE BRAIN AND POSSIBLE GENE THERAPY OF NEUROLOGICAL DISORDERS

Pedro R. Lowenstein*, Gavin W. G. Wilkinson†, M. G. Castro*, A. F. Shering*, A. R. Fooks‡ and D. Bain*

*Molecular Medicine Unit, Department of Medicine, University of Manchester School of Medicine, Stopford Building, Manchester M13 9PT, †Department of Medicine, University of Wales College of Medicine, Heath Park, Cardiff CF4 4XN, and ‡Molecular Pathology Section, Research Division, Centre for Applied Microbiology and Research, Porton Down, Salisbury SP4 0JG, UK

Table of Contents

2.1 Introduction

Historically, it has been virtually impossible to transfer genes into brain cells *in vivo* either to study or manipulate molecular processes. In addition to the physical barriers which protect the brain, the most commonly used gene delivery system (retroviral vectors) requires cells to divide to integrate the transgenes into their genomes; thus

Genetic Manipulation of the Nervous System
ISBN 0-12-437165-5

they can only be used to transduce the dividing cells of the nervous system, e.g. astrocytes and oligodendrocytes (Fischer and Gage, 1993). Since neurons in the adult brain do not divide retroviruses are of limited use to neurobiologists wanting to manipulate the molecular basis of neuronal physiology. Microinjection can be reliably used *in vitro*, but not effectively to transduce cells of the brain *in vivo* (Allsopp *et al.*, 1993; Mulderry *et al.*, 1993). However, novel vector systems have now emerged to provide a highly effective technology with which to explore the molecular basis of neuronal function, as well as the prospects of *in vivo* gene transfer of DNA into neurons and other brain cells to achieve a therapeutic benefit, i.e. neurological gene therapy. This review will concentrate on the potential application of replication-deficient adenovirus vectors as a vehicle for gene delivery.

A significant feature of adenovirus as a potential vector for DNA delivery in human gene therapy protocols has been that a live (replication-competent) vaccine has been safely administered to many million human beings (US military recruits) over several decades to provide protection against natural adenovirus infections (Rubin and Rorke, 1994). More recently, similar replication-competent adenovirus recombinants encoding defined immunogens have also been used in human vaccine trials. While most adenovirus vaccine development has been based on exploiting replication-competent systems, replication-deficient adenovirus vectors have recently also proved to be highly effective agents with which to generate both humoral and cell-mediated immune responses to expressed transgenes. These vectors have definite attributes as agents for immunization. There is therefore considerable experience in using live adenovirus isolates, replication-competent adenovirus recombinants, and replication-deficient recombinant adenovirus as immunizing agents (Top *et al.*, 1971a,b; Schwartz *et al.*, 1974; Morin *et al.*, 1987; Prevec *et al.*, 1989, 1991; Eliot *et al.*, 1990; Jacobs *et al.*, 1992; Gallichan *et al.*, 1993; Rubin and Rorke, 1994; Wilkinson, 1994; Wilkinson and Borysiewicz, 1995). Furthermore, human gene therapy programmes have been launched by a number of groups in the USA and continental Europe, resulting in the application of replication-deficient adenovirus recombinants in the treatment of patients suffering cystic fibrosis; cells in the nose and the lungs have been successfully transduced with recombinants expressing the cystic fibrosis trasnsmembrane regulator (CFTR; Rosenfeld *et al.*, 1992, 1994; Engelhardt *et al.*, 1993; Mastrangeli *et al.*, 1993; Simon *et al.*, 1993; Zabner *et al.*, 1993a,b, 1994; Boucher *et al.*, 1994; Bout *et al.*, 1994a,b; Brody *et al.*, 1994b; Mittereder *et al.*, 1994; Welsh *et al.*, 1994; Wilson *et al.*, 1994; Yang *et al.*, 1994; Yei *et al.*, 1994b). Thus, a wealth of data continues to accumulate concerning the deliberate therapeutic administration of adenovirus and the response of the human patient.

Interest in exploiting the new capabilities offered by adenoviruses as gene therapy vectors has been keen. The key features which make adenovirus potentially a credible vector for neurological gene therapy are that: (1) they can be administered *in vivo*; (2) they can transduce many different differentiated cell types, including post-mitotic neurons; and (3) expression in the target cell can be restricted to the transgene only (reviewed in Wilkinson *et al.*, 1994). So far Ad recombinants have been used to transduce cells of the lung (see above for CFTR transfer; Gilardi *et al.*, 1990; Rosenfeld *et al.*, 1991; Yei *et al.*, 1994), liver (Herz and Gerard, 1993; Ishibashi *et al.*, 1993; Engelhardt *et al.*, 1994; Hayashi *et al.*, 1994; Kozarsky *et al.*, 1994), arteries and other blood vessels (Lemarchand *et al.*, 1992), joints (Roessler *et al.*, 1993), bone marrow cells and various differentiated circulating cells of the immune system (Haddada *et al.*, 1993), heart (Stratford-Perricaudet *et al.*, 1992; Kass-Eisler *et al.*, 1993), skeletal muscle (Stratford-Perricaudet *et al.*, 1990, 1992; Quantin *et al.*, 1991; Acsadi *et al.*, 1994a,b) brain (Akli *et al.*, 1993; Bajocchi *et al.*, 1993; Davidson *et al.*, 1993; Le Gal La Salle *et al.*, 1993), and spinal cord (Lisovoski *et al.*, 1994), and to release factor IX into the bloodstream (Smith *et al.*, 1993). The widespread potential applications of adenovirus systems has promoted vector development, in particular efficient strategies have been devised for the insertion of transgenes into recombinant adenovirus. Inserts of up to 8 kbp are feasible in replication-deficient recombinants, which can be grown without helper virus on trans-complementing 293 cells (Graham *et al.*, 1977; Berkner, 1988; Wilkinson *et al.*, 1994).

Recent reviews have provided detailed protocols and methodologies involved in constructing and utilizing Ad vectors for gene transfer to cells of the brain both *in vitro* and *in vivo* (Lowenstein and Enquist, 1995). In this chapter, we will consequently concentrate on progress that has been made in the application of such vectors within the context of their exploitation for neurobiology and discuss potential future developments in this field.

2.2 The vector system

Adenovirus binds to an as yet unidentified receptor, through the fibre protein of the viral capsid, and is internalized through an interaction with integrins on the cell surface receptors, in particular $\alpha_v\beta_3$ and $\alpha_v\beta_5$ (Wickham *et al.*, 1993). Entry into the cell is by receptor-mediated endocytosis (Wickham *et al.*, 1994), after which the viral nucleocapsid is released from endosomes into the cytoplasm, and transported to the nucleus (Greber *et al.*, 1993). The double-stranded DNA genome (~36 kb; Figure 1) is then released into the nucleus of the cell. Upon infection of a permissive cell, the replication cycle can be divided into three phases of gene expression: immediate early (E1 genes), early (E2–E4 genes) and finally the late events, which follows the onset of viral DNA replication. The very first region of the genome to become active is the left-hand region of the left strand, E1, which encodes a series of transcriptional activators which will promote progression into the early phase of gene expression. These early phases of virus replication are associated with host cell cycle progression to S1, onset of viral DNA replication as well as the production of gene products associated with evasion of specific and non-specific host anti-virus defences. Five late transcriptionally active regions are found at the right of the left strand, and are produced from a single major late promoter by alternative splicing (L1, L2, L3, L4, L5). Expression of late viral genes becomes activated after the start of viral DNA replication and will

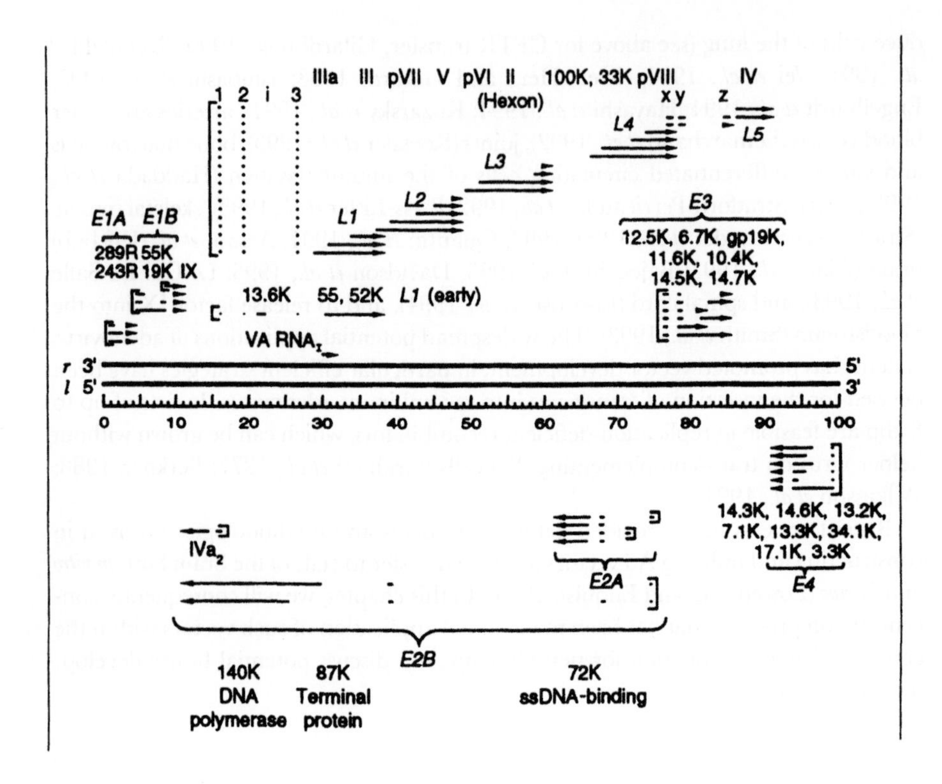

Figure 1 Map of the adenovirus genome. This figure represents the genome of adenovirus type 2, a linear double-stranded DNA comprising 36 000 bp; r and l refers to rightward and leftward transcription of each of the strands. Split arrows indicate splicing events of the mRNAs; exons are shown. Transcriptional units are indicated by E1A, E1B, E2A, E2B, E3, L1–L5. The other names (e.g. 55K, II, or 72K ssDNA-binding) refer to individual adenovirus proteins. [This figure is slightly modified from Wold *et al.* (1994).]

eventually lead to the assembly of progeny virions, in excess of 10 000 new virus particles per infected cell (Shenk, 1995).

Most available recombinant adenoviruses for use as vectors for gene transfer are rendered replication defective through deletions in the E1 gene region. These vectors, which have been termed 'first generation' adenovirus vectors, may also carry deletions of varying sizes in the non-essential E3 region, to allow the cloning of larger inserts into the viral genome (Gilardi *et al.*, 1990; Kozarsky and Wilson, 1993; Brody *et al.*, 1994a,b; Wilkinson *et al.*, 1994). Although this region is non-essential for adenovirus growth *in vitro*, it encodes genes associated with avoiding host immune response and consequently such deletions are not neutral, even in an Ad E1 vector. Replication-deficient adenovirus E1 recombinants are commonly grown on the trans-complementing 293 cells which express an integrated copy of the Ad5 E1 gene as a helper function (Graham *et al.*, 1977).

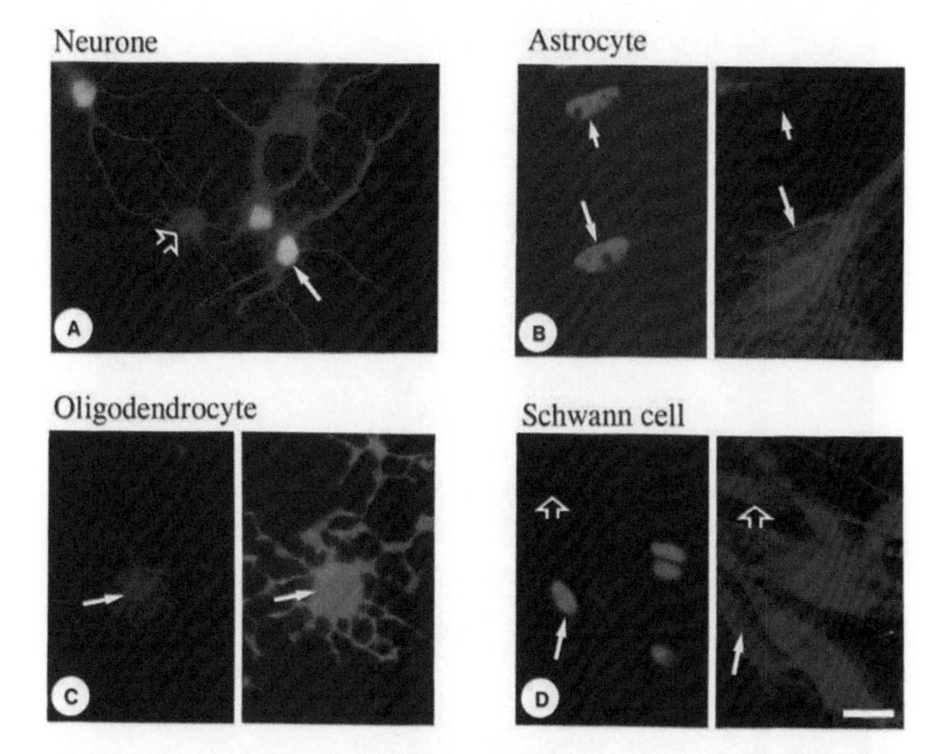

Plate 1 Expression of transgenes in identified cells from the central and peripheral nervous system in culture. All cells were infected at a multiplicity of infection of 10 with a recombinant adenovirus expressing β-galactosidase under the control of an RSV-LTR promoter linked to a nuclear localization signal (kindly provided by M. Perricaudet) (A, B, D), or with wild type adenovirus type 5 at the same multiplicity (C) (kindly provided by K. Leppard), and appropriate antigens detected by double immunofluoresence labelling. A shows a double exposure of infected neocortical neurons (white arrow) double immunostained with specific neuronal anti-Map2 antibodies (red, Sigma), and anti-β-galactosidase antibodies (yellow in this double exposed print, kindly provided by J. Price) – a white arrow outline indicates a non-transduced neuron; B shows the expression of nuclear β-galactosidase (green) in GFAP (red, Sigma) immunoreactive cortical astrocytes (long white arrow); a shorter white arrow indicates a GFAP-negative cell (probably a fibroblast) expressing nuclear β-galactosidase; C indicates a myelin basic protein (MBP) immunoreactive oligodendrocyte (green, specific MBP antibodies provided by P. Morgan), immunoreactive for E1a products (red; antibodies provided by E. Harlow) expressed after infection of cells with wild type adenovirus type 5 (white arrows indicate the position of the mainly nuclear E1a immunoreactivity); D illustrates a culture of peripheral rat nerve Schwann cells immunoreacted with anti-NGF receptor antibodies (red; antibodies kindly provided by R. Mirsky), some of which express nuclear β-galactosidase after infection with recombinant adenovirus (green; a white arrow indicates a transduced and a white arrow outline a non-transduced Schwann cell). Scale bar shown in D equals 50 μm.

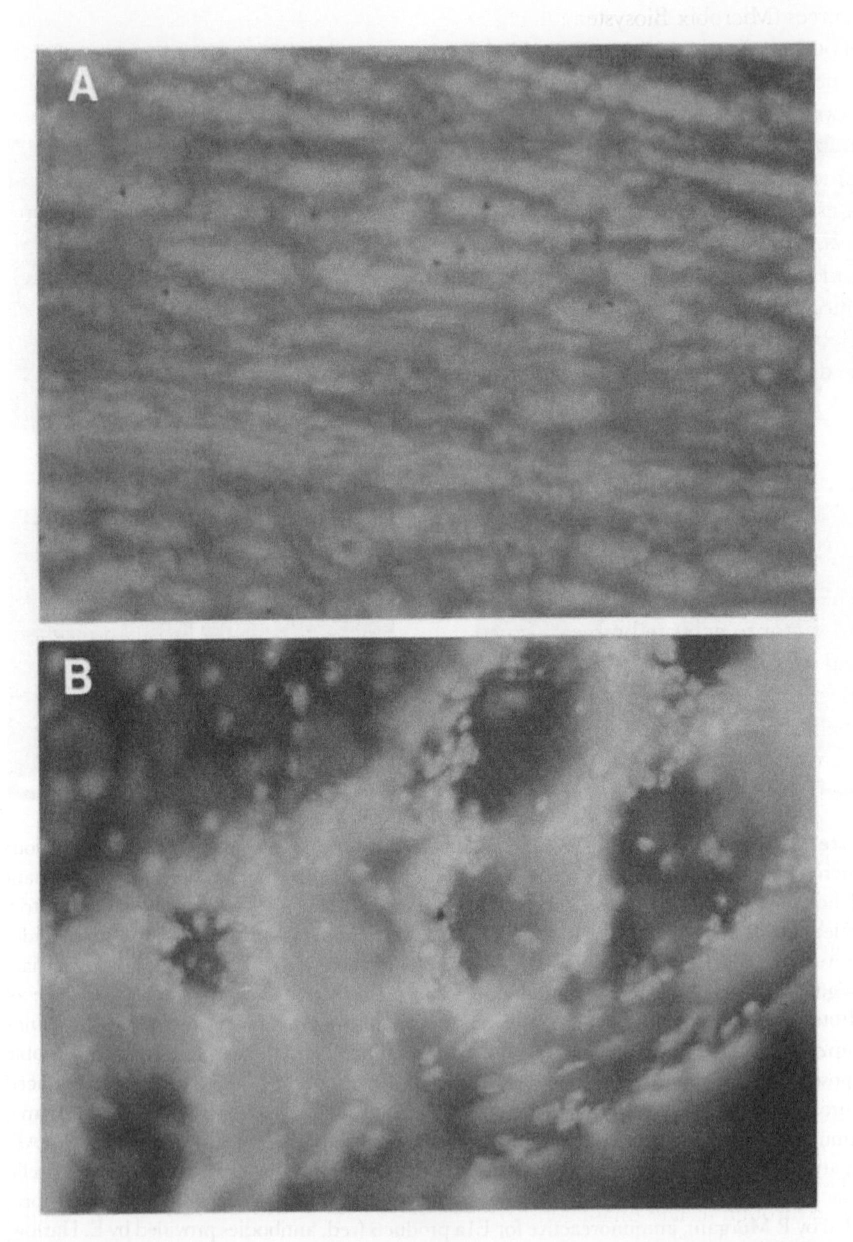

Plate 2 Photomicrographs of transgenic neurons (blue cells) that have migrated into the olfactory bulb after homotopic transplantation of SVZ of the lateral ventricle. (A) Many transgenic cells that emigrated from such a graft reach the granular layer of the olfactory bulb where they differentiate into neurons. (B) Fewer cells reach the glomerular layer where they differentiate into periglomerular neurons. The photomicrograph shows one transgenic neuron between two glomeruli.

Efficient systems to construct adenovirus recombinants are now available both from a number of different laboratories, and more recently also from commercial sources (Microbix Biosystems Inc., Toronto, Canada). These allow the construction of both replication-competent and replication-deficient recombinant adenoviruses for gene transfer (Prevec *et al.*, 1989), using conventional and well-established DNA cloning and transfection procedures (Sambrook *et al.*, 1989). Most published reports utilize adenovirus recombinants derived from either serotypes 5 or 2 (see Table 2), although there are some 47 human serotypes identified and many more animal isolates, some of which have also been used within the context of gene transfer (Cotten *et al.*, 1993). In Ad $E1^-$ vectors, transgenes are usually inserted into the E1 region under the control of exogenous viral promoters (e.g. RSV-LTR, IE-hCMV). Cell-specific promoters can also be used effectively to target expression to specific cell types (Babiss *et al.*, 1986; Besserau *et al.*, 1994; Grubb *et al.*, 1994), although this has yet to be demonstrated in the brain.

2.3 Developments in the use of adenovirus recombinants as vectors

The deletion of the Ad E1 gene region results in a failure of the replication-deficient Ad vector to activate both early and late phase transcription from the viral genome and consequently expression only of a transgene under the control of a constitutive or specifically active promoter is achieved in the target cell. While this genetic barrier to adenovirus gene expression is extremely efficient, it is not absolute. Even limited breakthrough to Ad gene expression can be problematical; it may either by itself affect the physiology of the target cell or induce a cellular immune response (Engelhardt *et al.*, 1994; Wold *et al.*, 1994; Yang *et al.*, 1994a,b). The deletion of both the Ela and Elb regions in currently used vectors clearly provides a better block than simply an Ela deletion (Imperiale *et al.*, 1984; Spergel and Chen-Kiang, 1991). Nevertheless, extremely high input multiplicities of infection (in excess of 100 infectious units per cell) can result in breakthrough to low level expression of the Ad genome (Figure 2; Engelhardt *et al.*, 1994; Kass-Eisler *et al.*, 1994; Yang *et al.*, 1994a,b). Here we also show that the addition of forskolin to cultures can increase the breakthrough to late gene expression. Forskolin results in enhanced levels of intracellular calcium and activation of the transcription factor cAMP responsive element binding protein (CREB or ATF), a factor which is known to bind and stimulate transcription from the Ad early promoter (Muchardt *et al.*, 1990). It is therefore not surprising that forskolin treatment of cells infected with a replication-deficient Ad recombinant can enhance breakthrough to late phase gene expression. Both a high input multiplicity of infection (around the site of virus inoculation or instillation) and elevated calcium levels are likely to occur in transduced cells following *in vivo* gene transfer. Both these phenomena are very likely to coincide in neurons surrounding the site of adenovirus direct injection into the brain.

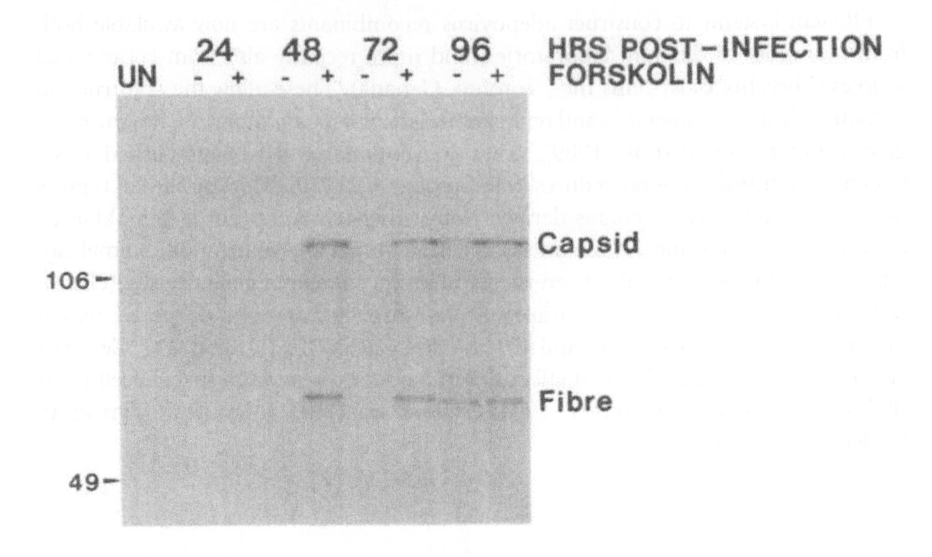

Figure 2 Breakthrough of expression of late adenovirus proteins. Human fibroblasts were infected with a replication-deficient E1 adenovirus recombinant encoding the measles virus nucleocapsid protein (Rad 68) at a high virus input multiplicity of infection (100 pfu/cell) and cell extracts were prepared at the times indicated. Samples were electrophoresed using SDS-PAGE and transferred to a nitrocellulose membrane. The Western transfer was probed with human polyclonal adenovirus antiserum as primary antibody and peroxidase conjugated anti-human ECL (Amersham) reagents were used to identify the expression of adenovirus antigens. The expression of two major late gene products can be detected at 96 h post-infection in the absence of forskolin, but after only 48 h post-infection in its presence.

The induction of an immune response to elements of the Ad vector is now perceived as a significant limitation of the vector in long-term gene therapy protocols. Although gene transduction is possible in seropositive animal models and in human clinical trials, repeated exposure to viral antigens can be expected to reduce the efficiency of gene transfer. This problem can to some extent be circumvented by utilizing different Ad serotypes as vectors. Breakthrough to early phase transcription was demonstrated by the identification of a humoral immune response to early phase proteins. In a mouse model breakthrough has been strongly correlated with the induction of a $CD8^+$ cytotoxic-T-lymphocyte (CTL) response to adenovirus proteins which is responsible both for an inflammatory response and an elimination of target cells expressing a transgene (Yang *et al.*, 1994a,b). Such experiments have been performed using extremely high doses of input virus. If lower doses of adenovirus recombinant are inoculated into a mouse, major histocompatibility complex-I (MHC-I) responses specific for a viral transgene in the absence of a detectable response to the vector can be induced (unpublished results). However, high input doses of recombinant may be essential for efficient *in vivo* gene transfer. A further modification to current vectors is therefore of interest, i.e. the supplementation of deletions in E1⁻, E3⁻, with an additional conditional mutation in E2a, the so-called 'second generation vectors' (Engelhardt *et al.*, 1994; Yang *et al.*, 1994b). The E2a region plays a role in activating late phase gene expression. Recombinants carrying this additional mutation have proved to be less reactogenic and permit prolonged expression of the transgene (Engelhardt *et al.*, 1994; Yang *et al.*, 1994).

It is important to realize that many experiments are performed in non-permissive or semi-permissive species (Ginsberg *et al.*, 1991; Ross and Ziff, 1992), e.g. mice, rats and non-human primates. In such systems, Ad E1⁻ recombinants are unable to replicate due to a 'species block'. It would thus be useful to test recombinants in a more relevant species, such as the 'permissive' cotton rat (*Sigmoidon hisipidus*), to determine whether cell-specific factors in addition to species-specific factors regulate the capacity of adenovirus to replicate in particular target organs (Oualikene *et al.*, 1994).

Many researchers see the current cloning capacity of ~7–8 kb as a limitation towards the more general applicability of adenoviral vectors. There is a stringent limitation exerted by the adenovirus particle, which will package only an additional 3% (~1 kb) in addition to the full length viral genome (Shenk, 1995). Thus, a number of groups are working towards expanding the cloning capacity by designing defective helper-dependent adenoviruses. In these, only the original packaging signals and origins of replication would be maintained within the recombinant for gene transfer. This could theoretically increase the cloning capacity to around 35 kb. Such helper-dependent vectors can, so far, only be grown in the presence of a helper virus required to provide in *trans* the necessary functions to replicate and package the 'gene transfer recombinant' into virions. This system is essentially analogous to herpes simplex virus 1 (HSV1) amplicons and shares some of their limitations (Lowenstein *et al.*, 1994, 1995a), e.g. low titres and the presence of variable amounts of helper virus in viral stocks. Using such vectors expression of *lacZ* has been demonstrated in 293 cells and a murine muscle cell line (G. Dickson, personal communication). Eventually, it should be possible to develop this system so that recombinants can be produced in the absence of helper virus to generate a safer vector with markedly reduced potential to be reactogenic.

2.4 Adenoviral recombinant vectors: applications to basic neuroscience

Although the technology for gene transfer to the brain is still in its infancy, new systems are being rapidly developed. Table 1 compares the major systems for gene transfer into neurons both *in vitro* and *in vivo*. Vectors vary dramatically in their efficiency of gene transfer, longevity of expression, size of the transgene they will accommodate and their associated toxicity. Most vectors mainly direct short-term expression of transgenes, either because of their intrinsic toxicity *in vitro* (e.g. Semliki Forest Virus (SFV), vaccinia), or because of the immune response the host will mount against the recombinant *in vivo* (e.g. adenovirus).

Table 1 A comparison between adenovirus and other gene transfer methods

	Vectors								
	Adenovirus	HSV1/r	HSV1/a	AAV	Retrovirus	Vaccinia	SFV	Microinjection	Transfection
Size (kb)	36	152	5–30	4.68	7–10	185	11.51	10–30	10–30
Cloning capacity	7.5	10–30	5–25	4.5	7–8	30	9–20*	10–30	10–30
Neuronal transduction									
In vivo?	Yes	Yes	Yes	Yes	Yes†	Yes	Yes	No	No
In vitro?	Yes	Yes	Yes	Yes	Yes†	Yes	Yes	Yes	Yes
Very long term gene expression	?	In DRG; hippocampus?	Yes?	Yes	Yes	No	No	Yes	Yes
Gene therapy?	Yes	Yes	Yes	Yes	Yes	No	No	No	Yes
Vaccination	Yes	?	?	?	?	Yes	Yes	No	No

*P. Liljeström, personal communication.
†Only into dividing neuronal precursors.
HSV-1/r, recombinant HSV1-based vectors; HSV-1/a, amplicon HSV1-based vectors.

In 1993 four groups independently reported *in vivo* gene transfer into several types of brain cells using adenovirus recombinants encoding β-galactosidase (Akli *et al.*, 1993; Bajocchi *et al.*, 1993; Davidson *et al.*, 1993; Le Gal La Salle *et al.*, 1993). Since this initial breakthrough, there have been approximately 20 original papers using adenovirus recombinants to transduce brain cells (see Table 2). Although different adenovirus recombinants can display a selective affinity for transducing different target cells (Acsadi *et al.*, 1994a,b; Grubb *et al.*, 1994; Besserau *et al.*, 1994), it is now clear that adenovirus recombinants can indeed express transgenes in all cells found within the central nervous system, e.g. neurons, astrocytes, oligodendrocytes, ependymal cells, fibroblasts, macrophages, endothelial blood vessel cells, retinal pigment epithelium, photoreceptors, as well as retinal neurons proper, and peripheral nerve Schwann cells either *in vivo* or *in vitro* (Plate 1 and Table 2; Akli *et al.*, 1993; Bajocchi *et al.*, 1993; Caillaud *et al.*, 1993; Davidson *et al.*, 1993; Le Gal La Salle *et al.*, 1993; Bain *et al.*, 1994; Bennet *et al.*, 1994; Jomary *et al.*, 1994; Li *et al.*, 1994; Byrnes *et al.*, 1995; Lowenstein *et al.*, 1995b,c). While initial experiments have utilized recombinant adenovirus to transfer marker proteins into the brains of both rodents and primates, recombinant vectors encoding therapeutic genes for use in the treatment of neurological diseases have now been produced (Chen *et al.*, 1994; Davidson *et al.*, 1994; Perez Cruet *et al.*, 1994 Shewach *et al.*, 1994; Lowenstein *et al.*, 1995b). Expression of transgenes after *in vivo* administration of vectors to the brain was detected up to 5–6 months post-inoculation (Table 2).

The delivery to the brain of genes of possible therapeutic value has also been attempted: recombinant vectors encoding hypoxanthine phosphoribosyl transferase (HPRT) have been constructed, which could be utilized in the treatment of the Lesch–Nyhan syndrome. These vectors have mainly been utilized to transduce HPRT in primate (Davidson *et al.*, 1994) and rodent brains (Lowenstein *et al.*, 1995b) although there is yet no published account of their capacity to complement an HPRT deficiency in animal models for the Lesch–Nyhan syndrome. A vector expressing tyrosine hydroxylase, and its therapeutic efficacy in a rat model of Parkinson's disease, has also been recently reported (Horellou *et al.*, 1994).

Since adenovirus recombinants can be used to transduce essentially all brain cells, they have also been postulated to be applicable to gene therapy protocols for the treatment of brain tumours, by utilizing them to deliver cytotoxic genes directly into the tumours. Thus, a number of groups have utilized several adenovirus recombinants, e.g. expressing HSV1 thymidine kinase, under the control of viral promoters. In experimental paradigms the use of these viruses appears to be promising (Badie *et al.*, 1994; Boviatsis *et al.*, 1994; Brody *et al.*, 1994a; Chen *et al.*, 1994; Perez Cruet *et al.*, 1994; Shewach *et al.*, 1994). Retrovirus recombinants have been used before in the treatment of brain tumours (Ram *et al.*, 1993), and replication-competent HSV1 vectors could be tried out very soon (Martuza *et al.*, 1991; Markert *et al.*, 1993). It will be of interest to compare the clinical efficacy of the three different systems as strategies for brain tumour therapy (Oldfield *et al.*, 1993).

Two groups have also made interesting observations on the use of adenovirus recombinants which should be of interest to neuroanatomists, namely the use of

Table 2 Review of publications using recombinant adenovirus vectors for gene transfer into the brain or constituent cells

Reference	Recombinant adenovirus E1$^-$ vector	Promoter	Transgene	Target *in vitro* and *in vivo*	Transduction efficiency		Longevity of expression	Toxicity	Immune response	Therapeutic efficacy/type of experiment
					Expression *in vivo*	Expression *in vitro*				
Akli *et al.* (1993)	Ad-RSV βGal Ad type 5	RSV-LTR	lacZ-nls	Brain *in vivo*; rats	Neurons, astrocytes, microglia, ependymal cells. 3×10^3–3×10^8 pfu/injection; approx. 1 infected cell/100 pfu	N/A	Up to 45 days	No major cytotoxicity at 45 days post-infection. Cytopathic effect at $\geq10^7$ pfu/injection	Not examined	Gene transfer into the rodent brain *in vivo*
Badie *et al.* (1994)	AdLacZ (provided by B. Trapnell, *Genetic Therapy Inc.*)	HCMV Major IE	lacZ	Rat C6 glioma cells	1–3 × 10^8 pfu in 10 μl into the CNS; expression in 25–30% C6 glioma tumours; expression in neurons, glia, ependyma (no double staining)	50% efficiency at MOI = 30 on C6 cells;	Expression examined after 4, 7, 14 days	Mild local tissue necrosis and cystic gliosis at higher doses attributed to volume injected; Ad*LacZ* caused 50% reduction in cell density *in vitro* and 75% reduction in tumour cell volume *in vivo*	Not examined	Delivery of marker genes to the CNS and brain tumours/ tumour therapy

Table 2 *contd*

Reference	Recombinant adenovirus E1$^-$ vector	Promoter	Transgene	Target *in vitro* and *in vivo*	Transduction efficiency		Longevity of expression	Toxicity	Immune response	Therapeutic efficacy/type of experiment
					Expression *in vivo*	Expression *in vitro*				
Bajocchi *et al.* (1993)	Ad-RSVβGal Ad-α1AT Ad type 5	RSV-LTR Ad Major late promoter	lacZ, SV40 nls α 1 anti-trypsin	Intra-ventricular or intrastriatal admini-stration; 5×10^8–10^9 pfu/animal	Throughout the ventricular system; substantia nigra and thalamus	N/A	Not examined in detail for Ad.RSVβgal; up to 6–8 days for Ad-alfa1AT	Not reported	Not examined	Gene transfer of marker enzymes and transgenes into the rodent ventricles and brain *in vivo*
Bennet *et al* (1994)	Ad-CMVlacZ; Ad5	HCMV Major IE	lacZ	Adult mouse mammalian retina; subretinal injection of 10^2–10^9 pfu/retina in 1–10 μl	Positive RPE, but not neurons after 10^7 pfu/retina	N/A	Up to 13 weeks post-injection in RPE, photo-receptors and Muller cells (glial cells)	No major cytotoxicity; short-term disruption of outer retina; no systemic toxicity	Not tested	Gene transfer into retina *in vivo*
Boviatsis *et al.* (1994)	Ad5MLP-lacZ;	Major late promoter (adenovirus)	β-galactosi-dase	Rat 9L gliosarcoma cells	Yes in tumors; 3×10^7 pfu/tumor	2% at MOI = 1; 50% at MOI = 50	5% at 6 days and 1% at 12 days post-injection	Extensive *in vitro* at MOI = 50	Not major; not examined in detail	β-Galactosi-dase expression in tumours and neurons; tumour necrosis; CNS tumour gene therapy

Table 2 *contd*

Reference	Recombinant adenovirus E1$^-$ vector	Promoter	Transgene	Target *in vitro* and *in vivo*	Transduction efficiency		Longevity of expression	Toxicity	Immune response	Therapeutic efficacy/type of experiment
					Expression *in vivo*	Expression *in vitro*				
Byrnes *et al* (1995)	AdRL Ad0 Ad5	RSV-LTR No promoter	lacZ No transgene	Brain, *in vivo*; ≈ 3 × 10^6 pfu/ 0.5 μl injection site	Intrastriatal injections; expression also seen in substantia nigra	N/A	Peak: 2–9 days; positive cells up to 60 days	See immune response	Leukocyte infiltration, mainly macrophages and T cells; MHC class I expression; microglial activation; mainly during first 15 days	Study of the immune response to the direct administra-tion of adenovirus recombinants into the brain
Caillaud *et al.* (1993	Ad-RSVβGal; Ad5	RSV-LTR	lacZ-nls	Primary cultures of neurons and glial cells	N/A	*Neurons* *pfu eff.%* 4 × 10^7 1–20 2 × 10^8 10–65 4 × 10^8 30–80 *Glial cells* *pfu eff. %* 4 × 10^7 20 2 × 10^8 100	Up to a month	At ≥ 4 × 10^8 pfu/ml	N/A	Gene transfer into neural cells *in vitro*
Chen *et al.* (1994)	Ad5 ADV/RSV-tk;	RSV-LTR	TK(HSV-1)	C6 glioma	Yes in C6 glioma tumours; 3 × 10^8 pfu/tumour	100% at MOI = 125; saturation of TK at MOI = 1000; *in vivo* = 50%	Not examined	*In vitro*: at MOI ≥ 500; *in vivo*: not major, no markers used	Experiments done in athymic nude mice	Tumour treatment/ significant tumour decrease/ long-term tumour regrowth

Table 2 *contd*

Reference	Recombinant adenovirus $E1^-$ vector	Promoter	Transgene	Target *in vitro* and *in vivo*	Transduction efficiency		Longevity of expression	Toxicity	Immune response	Therapeutic efficacy/type of experiment
					Expression *in vivo*	Expression *in vitro*				
Davidson *et al.* (1993)	Ad-CMVlacZ Ad type 5 $E1^-$; $E3^-$	HCMV Major IE	lacZ	*In vivo*: intrastriatal, mice	Some neurons in the striatum and cortex; oligodendrocytes and axons	N/A	Up to 8 weeks: 10–20% of 1 week values	Non-specific local tissue damage; no detection of adenovirus proteins; no viraemia	Not examined	Gene transfer into rodent brain *in vivo*
Davidson *et al* (1994)	Ad5-RSV-lacZ and RSV-rHPRT;	RSV-LTR	lacZ and rat HPRT	Striatum and cortex *in vivo*	Primate; 3×10^{10} pfu/injection; high in neurons and glial cells; 40% increase of HPRT	N/A	Not tested; examined after 7 days	Focal inflammation	Humoral response to RAd proteins, but not transgenes; lack of widespread CNS inflammation	Gene transfer into primate brain; immune response to adenovirus vectors and encoded transgenes
Horellou *et al.* (1994)	Ad-RSVhTH Ad-RSVβGal	RSV-LTR RSV-LTR	Human tyrosine hydroxylase (TH) lacZ-nls	Brain *in vivo*	1.5×10^8 pfu/injection site (in 9 μl; up to nine sites per striatum; transduction efficiency not quantified; most TH immunoreactive cells were astrocytes; some neurons	N/A	Animals examined for 15 days	Limited tissue necrosis; GFAP immunoreactive reactive astrogliosis	Inflammatory response; not examined in detail	Gene therapy of Parkinson's disease; significant reduction in abnormal behaviour of intrastriatal AdRSVhTH vs AdRSVβgal

Table 2 *contd*

Reference	Recombinant adenovirus E1⁻ vector	Promoter	Transgene	Target *in vitro* and *in vivo*	Transduction efficiency		Longevity of expression	Toxicity	Immune response	Therapeutic efficacy/type of experiment
					Expression *in vivo*	Expression *in vitro*				
Jomary *et al.* (1994)	Ad2/CMV lacZ-1	HCMV Major IE	lacZ-nls	Retinal cells	Pigment epithelium and ganglion cells	70% at MOI = 50	Not tested	Not tested	Not tested	Gene transfer of marker cells into retina
Le Gal La Salle *et al.* (1993)	Ad-RSVβGal Ad type 5	RSV-LTR	lacZ-nls	SCG and astrocytes in culture. Hippo-campus and substantia nigra (SN) in vivo: $3–5 \times 10^7$ pfu/injection	Expression in neurons, astrocytes and microglial cells	≈100% in SCG; ≈60% in glia	*In vivo*: 60 days	Non-specific local tissue necrosis and reactive gliosis	Not examined	Gene transfer into brain cells *in vivo* and *in vitro*
Li *et al.* (1994)	Ad-CMVβA.-ntlcZ; Ad5	HCMV Major IE plus β-actin promoter	lacZ	Subretinal injection in normal and mutant mice of 4×10^4 $–4 \times 10^7$ pfu in 0.3–0.5 μl/retina	RPE, photo-receptor cells, and little in inner nuclear layer. Higher transduction of photo-receptors in mutant or regenerating retinas.	N/A	Up to 6 weeks	Partial transitory retinal detachment; no inflammatory infiltration; no systemic toxicity	Not tested	Gene transfer into retina *in vivo*

Table 2 *contd*

Reference	Recombinant adenovirus E1⁻ vector	Promoter	Transgene	Target *in vitro* and *in vivo*	Transduction efficiency		Longevity of expression	Toxicity	Immune response	Therapeutic efficacy/type of experiment
					Expression *in vivo*	Expression *in vitro*				
Lisovoski *et al* (1994)	Ad-RSVβGal Ad type 5	RSVLTR	lacZ	Spinal cord	7×10^7 pfu/injection	N/A	Animals perfused after 3 days	Not examined	Not examined	Neuro-anatomy: cell labelling for morpho-logical studies
Lowenstein *et al*. (1995)	RAd35	HCMV Major IE	lacZ	Neocortex cultures	Not tested	Astrocytes Fibroblasts	Cells fixed after 72 h	Not examined	N/A	Cell type-specific expression in neocortical cell types *in vitro*
	RAd122	RSV-LTR	lacZ		Not tested	Neurons Schwann cells Astrocytes Fibroblasts				
	Ad-RSVβGal	RSV-LTR	lacZ-nls							
Perez-Cruet *et al*. (1994)	ADV-tk	RSV-LTR	HSV1 thymidine kinase	9L gliosar-coma cells	1.2×10^9 pfu in 6 μl	N/A	N/A	No major toxicity outside tumour cell killing	Not examined	Effective tumour eradication
	ADV-βgal	RSV-LTR	lacZ							
Ridoux *et al*. (1994a)	Ad-RSVβGal	RSV-LTR	lacZ-nls	Astroycytes *in vitro;* 2×10^7 pfu/confluent dish	Transplanta-tion into the hippo-campus, substantia nigra or caudate nucleus	70–100% of cells expressing β-galactosidase	Up to 5 months	Glial reaction and host cell invasion surrounding the transplant	Not determined	Gene transfer into astrocytes prior to intracranial transplanta-tion

Table 2 *contd*

Reference	Recombinant adenovirus E1⁻ vector	Promoter	Transgene	Target *in vitro* and *in vivo*	Transduction efficiency		Longevity of expression	Toxicity	Immune response	Therapeutic efficacy/type of experiment
					Expression *in vivo*	Expression *in vitro*				
Ridoux *et al.* (1994b)	Ad-RSVβGal	RSV-LTR	β-galactosidase+nls*	CNS, basal ganglia	Not determined; 2×10^7 pfu/injection	In striatum and subtantia nigra	Up to 30 days	Microglial proliferation and astrocytic activation after the first week, but not 2 weeks post-injection	Not determined	Pathway tracing in the CNS

Abbreviations
TK; Thymidine kinase
TH; Tyrosine hydroxylase
SCG: Superior cervical ganglia
Ad; Adenovirus
RAd; Recombinant adenovirus
RSV-LTR; Rous Sarcoma Virus – Long terminal repeat
HCMV Major IE; Human cytomegalovirus major immediate early
nls; Nuclear localization signal
α1AT; α1 anti-trypsin
CMV; Human cytomegalovirus
RPE; Retinal pigment epithelium
MLP; Major late promoter

adenovirus for pathway tracing and cell labelling. Ridoux and colleagues (1994b) observed retrograde labelling of substantia nigra neurons after injections of recombinants expressing β-galactosidase into the striatum; similar results were also observed by Byrnes *et al.* (1995). Thus, replication-defective adenovirus could be used for retrograde labelling of neuronal pathways. Interestingly, HSV1-derived recombinants can also be used to trace neuronal pathways (Ugolini *et al.*, 1989), with some recombinants being specific for anterograde transport, others for retrograde transport (Zemanick *et al.*, 1991), while certain pseudorabies recombinants have been shown to be specific markers for individual neuronal circuits (Card *et al.*, 1991, 1992). Lisovoski *et al.* (1994) have utilized adenovirus recombinants to label neurons of the spinal cord at different stages of development to study their morphological development. The staining they obtain after *in vivo* administration fills the dendritic arbours of neurons, thus allowing for their morphological and morphometric examination during spinal cord development.

Ad $E1^-$ recombinants, with or without transgenes, have also been injected into the rodent brain to examine the immune response to human adenovirus (Byrnes *et al.*, 1995). Direct intraparenchymal injection induced an inflammatory response, which appeared to be induced by the virion itself, rather than by the transgenic protein. Different rat strains showed different levels of brain inflammation after similar doses of recombinant virus. Although inflammation was observed by 12 h post-injection, in spite of a significant inflammatory response, β-galactosidase was expressed for at least 2 months. While it will be important to assess the exact role of the immune system in regulating the length of expression of transgenes after both single and multiple administrations of the virus to the brain, the paper by Byrnes *et al.* (1995) suggests that in most cases described so far, transgene expression from an adenovirus recombinant has been seen against a background immune response against the vector.

An effective way of delivering gene products to the brain, rather than by introducing it into its constituent cells, is by transplanting genetically engineered cells directly into the CNS. So far, most genetically engineered cells expressing transgenes for transplantation have been transduced with retrovirus recombinant vectors. Recently, Ridoux *et al.* (1994a) have used adenovirus to transduce primary cultures of rat astrocytes *in vitro* and then transplanted these into the CNS of host rats. Expression of transgene was detected for at least 5 months. Thus, adenovirus might constitute an alternative to retrovirus for transducing cells in preparation for transplantation into the brain. In many cases expression from retroviral vectors ceases after a few days to a few weeks; thus, it will be interesting to compare both systems *vis-à-vis* length of transgene expression.

Viral promoters have so far mostly been used within adenovirus vectors applied to gene transfer into the brain, e.g. Rous Sarcoma Virus-Long terminal repeat (RSV-LTR), Major immediate early human cytomegalovirus promoter (HCMV Major IE), adenovirus major late promoter. This is probably due to the easy availability and ubiquitous activity of such promoters. However, it is likely that neuronal-specific promoters driving transgenes within adenovirus recombinants will retain their target cell specificity, as they do for muscle-specific promoters (Besserau *et al.*, 1994), or others (Babiss *et al.*, 1986). Thus, we predict that adenovirus will provide a very good system to evaluate cell-specific activities of neuronal promoters. We have already demonstrated that even a recombinant containing a viral promoter expresses in a cell-specific way after infecting primary cultures of neocortical neurons (Table 2; Lowenstein *et al.*, 1995).

Cell-specific promoters have been used within HSV1 vectors (Andersen *et al.*, 1992). However, HSV1 recombinants might be less than ideal for this purpose, since even vectors which do not replicate still express transcriptional activators. Additionally, the transcription activator VP16 (α-TIF) is part of the virion's tegument. Many HSV1 recombinants carry a mutation/deletion in the powerful transcription factor ICP4. Such deletions block the virus from completing its lytic cycle, but VP16 will still stimulate the high level expression of four other immediate early proteins. One of these especially, ICP0, can activate the expression from several viral and cellular promoters and will be expressed at high levels from an ICP4 mutant. Thus, the effect of any of these HSV1-associated transcriptional activators on promoters cloned within a HSV1 recombinant will have to be taken into account. Although cell type-specific expression has been reported (Kaplitt *et al.*, 1994a), the contribution of the HSV1 transcriptional activators on neuron-specific promoters has not yet been evaluated in detail. A possible solution to this would be to use helper-dependent HSV1 vectors, and evaluate target cells for co-infection with helper particles (Lowenstein *et al.*, 1994b, 1995a).

Scant information is available on the interactions of adenovirus with the brain. Thus, it will be of great importance to explore this field, e.g. adenovirus entry into brain cells, transport to the nucleus, viral replication in neurons of permissive species etc. In addition, the effect of adenovirus recombinants on the electrophysiological properties of infected neurons and any long-term effects on neuronal physiology will have to be investigated.

Although expression of transgenes after the administration of adenovirus recombinants into adult brains has been seen by some groups to last up to 6 months, it has been reported before that injection of recombinants into neonate has allowed this expression to be sustained over periods up to a year (Stratford-Perricaudet *et al.*, 1990; Kass-Eisler *et al.*, 1994). Long-term modulation of the immune response, and the elucidation of its exact role in the regulation of long-term expression from adenoviral recombinants administered into the CNS, will be crucial to achieve stable expression over very long times.

2.5 Adenoviral recombinant vectors: applications to neurological gene therapy

Essentially there are two types of applications for which vectors for gene transfer into the brain could be used. Short-term expression could be used for example to express cytotoxic products to kill tumour cells, or to provide drugs to block ischaemia-induced neurotoxicity. Ideally, the area of brain tissue to be targeted should be focal, and therapeutic benefit should be predicted after short-term transgene expression (Table 3).

Table 3 Neurological gene therapy 1995: what can be achieved with adenovirus vectors?

Disease	Non-inherited		Inherited				
	Tumours	PD	HD	CMT	Leukodystrophies	Lesch–Nyhan	Frax-A
Outcome	Fatal	Fatal	Fatal	Non-fatal	Fatal	Fatal	Non-fatal
Distribution	Focal	Focal†	Diffuse	Diffuse	Diffuse	Diffuse	Diffuse
Age at manifestation	Variable	Late	Variable	Variable	Early	Early	Prenatal
Presymptomatic detection*	No	No	Yes	Yes	Yes	Yes	Yes‡
Genotype predicts disease severity	ND	ND	Yes	Yes	Yes/no§	Yes	No
Gene cloned	Yes¶	Yes**	Yes	Yes	Yes	Yes	Yes
Expression needed	Short term	Long term	Long term	Long term	Long term	Long term	Long term
Experimental gene therapy strategy	Tumour cell killing	Dopamine replacement	Reduce expression of expanded allele	Normalize PMP22 expression	BMT expressing missing genes	Express HPRT in brain	Express FMR1 in brain
Clinically relevant available therapeutic protocols	Yes	Yes	No	No	Yes	No	No
Available adenoviral vectors	Yes	Yes	No	No	No	Yes	No

HD, Huntington's disease; CMT, Charcot–Marie–Tooth disease (the commonest genetic alteration is duplication of the region of chromosome 17p11.2 containing the PMP22 gene); Frax-A, fragile X syndrome type A; PD, Parkinson's disease (idiopathic) BMT, Bone marrow transplant.
*Even if presymptomatic detection is available, it is not done routinely on every pregnancy; it is performed only if there is any suspicion due to a positive family history.
†The primary pathology is loss of dopaminergic neurons from the pars compacta of the substantia nigra, but other lesions within PD are distributed throughout the brain.
‡Molecular genetic diagnosis is possible in the fetus, but it is arguable whether this would be prior to occurrence of pathological changes.
§Genotype predicts disease severity in metachromatic and Krabbe's leukodystrophy, but not yet in adrenoleukodystrophy.
¶The genes mutated in inherited brain tumours have been cloned. However, therapeutic gene therapy is more likely to utilize the transduction of cytotoxic genes to eliminate tumour cells.
**The genes for tyrosine hydroxylase and dopa decarboxylase have been cloned. These are not however mutated in either familial or sporadic PD.
This table is modified from MacMillan and Lowenstein (1995).

Table 3 Neurological gene therapy 1995: what can be achieved with adenovirus vectors?

Disease	Non-inherited		Inherited				
	Tumours	PD	HD	CMT	Leukodystrophies	Lesch–Nyhan	Frax-A
Outcome	Fatal	Fatal	Fatal	Non-fatal	Fatal	Fatal	Non-fatal
Distribution	Focal	Focal†	Diffuse	Diffuse	Diffuse	Diffuse	Diffuse
Age at manifestation	Variable	Late	Variable	Variable	Early	Early	Prenatal
Presymptomatic detection*	No	No	Yes	Yes	Yes	Yes	Yes‡
Genotype predicts disease severity	ND	ND	Yes	Yes	Yes/no§	Yes	No
Gene cloned	Yes¶	Yes**	Yes	Yes	Yes	Yes	Yes
Expression needed	Short term	Long term	Long term	Long term	Long term	Long term	Long term
Experimental gene therapy strategy	Tumour cell killing	Dopamine replacement	Reduce expression of expanded allele	Normalize PMP22 expression	BMT expressing missing genes	Express HPRT in brain	Express FMR1 in brain
Clinically relevant available therapeutic protocols	Yes	Yes	No	No	Yes	No	No
Available adenoviral vectors	Yes	Yes	No	No	No	Yes	No

HD, Huntington's disease; CMT, Charcot–Marie–Tooth disease (the commonest genetic alteration is duplication of the region of chromosome 17p11.2 containing the PMP22 gene); Frax-A, fragile X syndrome type A; PD, Parkinson's disease (idiopathic) BMT, Bone marrow transplant.

*Even if presymptomatic detection is available, it is not done routinely on every pregnancy; it is performed only if there is any suspicion due to a positive family history.

†The primary pathology is loss of dopaminergic neurons from the pars compacta of the substantia nigra, but other lesions within PD are distributed throughout the brain.

‡Molecular genetic diagnosis is possible in the fetus, but it is arguable whether this would be prior to occurrence of pathological changes.

§Genotype predicts disease severity in metachromatic and Krabbe's leukodystrophy, but not yet in adrenoleukodystrophy.

¶The genes mutated in inherited brain tumours have been cloned. However, therapeutic gene therapy is more likely to utilize the transduction of cytotoxic genes to eliminate tumour cells.

**The genes for tyrosine hydroxylase and dopa decarboxylase have been cloned. These are not however mutated in either familial or sporadic PD.

This table is modified from MacMillan and Lowenstein (1995).

system following the injection of the adenovirus type 5 into the primate brain was recently described (Davidson *et al.*, 1994).

The possibility that latent adenovirus infection might be localized to the human brain, and that adenovirus could enter the brain through infected macrophages from the periphery, has not yet been examined in enough detail, although widespread neuronal inclusions diagnostic of adenovirus particles have been seen in some fatal encephalitis cases (Chou *et al.*, 1973). However, all these data taken together with our observations that wild-type adenovirus can express endogenous proteins in neurons, and brain glial cells, strongly suggests that adenovirus could replicate, or at least express many of its genes after infecting human neurons or glial cells and that expression might proceed for a longer time than predicted, even in the absence of encephalitis symptoms. If there is a risk of adenoviral encephalitis it will have to be carefully considered at the time adenovirus vector is directly administered into the brain. It is clear that many more studies will be required to clarify the origins of neurotoxicity after direct injections of adenovirus virions into the brain.

2.6 Summary: targeting the brain with adenovirus recombinants

The replication-deficient adenovirus vector is an extremely versatile vector for gene transfer and will soon become an invaluable reagent for neurobiologists interested in either basic science or neurological gene therapy applications. Their main advantages are:

(1) the existence of well-characterized methods to produce replication-defective, high titre, recombinants;
(2) a capacity to infect both terminally differentiated (neurons) and dividing brain cells (glia);
(3) a capacity to transduce both immature neonatal and mature neurons;
(4) the potential to achieve cell type-specific expression from particular promoters to allow targeted and regulated transgene expression;
(5) an ability to achieve multiple infections of a single target cell;
(6) transduction of brain cells both *in vitro* and *in vivo*;
(7) their use to elucidate problems of neuronal morphology and connectivity;
(8) relatively long-term high level expression *in vivo*;
(9) low incidence of brain infections caused by adenovirus;
(10) experience of adenovirus recombinants in use in human gene therapy protocols;
(11) the experience in the use of adenovirus for human immunization.

While results so far have been highly encouraging (Tables 2 and 3), adenovirus vector systems have yet to be optimized for application to somatic cell gene therapy or specifically for neurological gene transfer. Currently, progress in clinical neurological gene therapy is both dependent on and limited by the technology for gene transfer. Selection of the most appropriate vector system for a study is crucial. Ad vectors clearly have distinct advantages for efficient gene delivery both *in vitro* and *in vivo*. Vector development will undoubtably further enhance the utility of these vectors.

Acknowledgements

We would like to acknowledge the support our labs receive from The Wellcome Trust, MRC, BBSRC, Muscular Dystrophy Group, The Royal Society and The Welsh Scheme for the Development of Health and Social Research. We would also like to thank the following researchers for the very kind provision of antibodies: Ed Harlow, Jack Price, Rhona Mirsky and Paul Morgan. We equally acknowledge the kind provision of virus from Keith Leppard and Michel Perricaudet, and the support of Phil Blanning through his excellent technical assistance. Many thanks also go to those authors who provided us with preprints of our work, and who thus helped greatly in allowing this review to be as updated as possible. P.R.L. is a Research Fellow of The Lister Institute of Preventive Medicine.

References

Acsadi, G., Jani, A., Massie, B., Simoneau, M., Holland, P., Blaschuk, K. & Karpeti, G. (1994a) A differential efficiency of adenovirus-mediated *in vivo* gene transfer into skeletal muscle cells of different maturity, *Hum. Mol. Genet.* **3**, 579–584.

Acsadi, G., Jani, A., Huard, J., Blaschuk, K., Massie, B., Holland, P., Lochmuller, H. & Karpati, G. (1994b) Cultured human myoblasts and myotubes show markedly different transducibility by replication-defective adenovirus recombinants. *Gene Ther.* **1**, 338–340.

Akli, S., Caillaud, C., Vigne, E., Stratford-Perricaudet, L.D., Poenaru, L., Perricaudet, M., Kahn, A. & Peschanski, M.R. (1993) Transfer of a foreign gene into the brain using adenovirus vectors. *Nature Genet.* **3**, 224–228.

Allsopp, T.E., Wyatt, S., Paterson, H.F. & Davies, A.M. (1993) The proto-oncogene *bcl*-2 can selectively rescue neurotrophic factor-dependent neurons from apoptosis. *Cell* **73**, 295–307.

Andersen, J.K., Garber, D.A., Meaney, C.A. & Breakefield, X.O. (1992) Gene transfer into mammalian central nervous system using herpes virus vectors: extended expression of bacterial lacZ in neurons using the neuron specific enolase promoter. *Hum. Gene Ther.* **3**, 487–499.

Babiss, L.E., Friedman, J.M. & Darnell Jr., J.E. (1986) Cellular promoters incorporated into the adenovirus genome: effects of viral regulatory elements on transcription rates and cell specificity of albumin and β-globin promoters. *Mol. Cell. Biol.* **6**, 3789–3806.

Badie, B., Hunt, K., Economou, J.S. & Black, K.L. (1994) Stereotactic delivery of a recombinant adenovirus into a C6 glioma cell line in a rat brain tumor model. *Neurosurgery* **35**, 910–916.

Bain, D., Wilkinson, G.W.G., Preston, C.M., Castro, M.G. & Lowenstein, P.R. (1994) Adenovirus vectors to transfer genes into neurons: implications for gene therapy of neurological disorders. *Gene Ther.* **1**, Suppl. 1, S68.

Bajocchi, G., Feldman, S.H., Crystal, R.G. & Mastrangeli, A. (1993) Direct *in vivo* gene transfer to ependymal cells in the central nervous system using recombinant adenovirus vectors. *Nature Genet.* **3**, 229–234.

Bennet, J., Wilson, J., Sun, D., Forbes, B. & Maguire, A. (1994) Adenovirus vector-mediated *in vivo* gene transfer into adult murine retina. *Invest. Ophthal. Vis. Sci.* **35**, 2535–2542.

Berkner, K.L. (1988) Development of adenoviral vectors for the expression of heterologous genes. *Biotechniques* **6**, 616–624.

Besserau, J.L., Stratford-Perricaudet, L.D., Piette, J., Le Poupon, C. & Changeaux, J.C. (1994) *In vivo* and *in vitro* analysis of electrical activity dependent expression of muscle acetylcholine receptor genes using adenovirus. *Proc. Natl. Acad. Sci. USA* **91**, 1304–1308.

Boucher, R.C., Knowles, M.R., Johnson, L.G., Olsen, J.C., Pickles, R., Wilson, J.M. Engelhardt, J. Yang, Y & Grossman, M. (1994) Clinical protocol: gene therapy of cystic fibrosis using E1-deleted adenovirus: a phase I trial in the nasal cavity. *Hum. Gene Ther.* **5**, 615–639.

Bout, A., Imler, H.L., Schultz, H., Perricaudet, M., Zurcher, C., Herbrink, P., Valerio, D. & Pavirani, A. (1994a) *In vivo* adenovirus mediated transfer of human CFTR cDNA to rhesus monkey airway epithelium: efficacy, toxicity and safety. *Gene Ther.* **1**, 385–394.

Bout, A., Perricaudet, M., Baskin, G., Imler, J.L., Scholte, B.J., Pavirani, A. & Valerio, D. (1994b) Lung gene therapy: *in vivo* adenovirus-mediated gene transfer to rhesus monkey airway epithelium. *Hum. Gene Ther.* **5**, 3–10.

Boviatsis, E.J., Chase, M., Wei, M.X., Tamiya, T., Hurford, Jr., R.K., Kowall, N.W., Tepper, R.I., Breakefield, X.O. & Chiocca, E.A. (1994) Gene transfer into experimental brain tumors mediated by adenovirus, herpes simplex virus, and retrovirus vectors. *Hum. Gene Ther.* **5**, 183–191.

Brody, S.L., Jaffe, H.A., Han, S.K., Wersto, R.P. & Crystal, R.G. (1994a) Direct *in vivo* gene transfer and expression in malignant cells using adenovirus vectors. *Hum. Gene Ther.* **5**, 437–447.

Brody, S.L., Metzger, M., Danel, C., Rosenfeld, M.A. & Crystal, R.G. (1994b) Acute responses of non-human primates to airway delivery of an adenovirus vector containing the human cystic fibrosis transmembrane conductance regulator cDNA. *Hum. Gene Ther.* **5**, 821–836.

Byrnes, A.P., Rusby, J.E., Wood, M.J.A. & Charlton, H.M. (1995) Adenovirus gene transfer causes inflammation in the brain. *Neuroscience* **66**, 1015–1024.

Caillaud, C., Akli, S., Vigne, E., Koulakoff, A., Perricaudet, M., Poenaru, L., Kahn, A. & Berwld Netter, Y. (1993) Adenoviral vector as gene delivery system into cultured rat neuronal and glial cells. *Eur. J. Neurosci.* **5**, 1287–1291.

Card, J.P., Whealy, M.E., Robbins, A.K., Moore, R.Y. & Enquist, L.W. (1991) Two a-herpesvirus strains are transported differentially in the rodent visual system. *Neuron* **6**, 957–969.

Card, J.P., Whealy, M.E., Robbins, A.K. & Enquist, L.W. (1992) Pseudorabies virus envelope glycoprotein gI influence both neurotropism and virulence during infection of the rat visual system. *J. Virol.* **66**, 3032–3041.

Chen, S.H., Shine, H.D., Goodman, J.C., Grossman, R.G. & Woo, S.L.C. (1994) Gene therapy for brain tumours: regression of experimental gliomas by adenovirus-mediated gene transfer *in vivo*. *Proc. Natl. Acad. Sci. USA* **91** 3054–3057.

Chou, S.M., Roos, R., Burrell, R., Gutmann, L. & Harley, J.B. (1973) Subacute focal adenovirus encephalitis. *J. Neuropathol. Exp. Neurol.* **32**, 34–50.

Cotten, M., Wagner, E., Zatloukal, K. & Birnstiel, M. (1993) Chicken adenovirus (CELO virus) particles augment receptor-mediated DNA delivery to mammalian cells and yield exceptional levels of stable transformants. *J. Virol.* **67**, 3777–3785.

Davidson, B.L., Allen, E.D., Kozarsky, K.F., Wilson, J.M. & Roessler, B.J. (1993) A model system for *in vivo* gene transfer into the central nervous system using an adenoviral vector. *Nature Genet.* **3**, 219–223.

Davidson, B.L., Doran, S.E., Shewach, D.S., Latta, J.M., Hartman, J.W. & Roessler, B.J. (1994) Expression of *Escherichia coli* b-galactosidase and rat HPRT in the CNS of *Macaca mulatta* following adenoviral mediated gene transfer. *Exp. Neurol.* **125**, 258–267.

Davis, D., Henslee, P.J. & Markesbery, W.R. (1988) Fatal adenovirus meningoencephalitis in a bone marrow transplant patient. *Ann. Neurol.* **23**, 385–389.

Eliot, M., Gilardi-Hebnstreit, P., Toma, B. & Perricaudet, M. (1990) Construction of a defective adenovirus vector expressing the pseudorabies virus glycoprotein gp50 and its use as a live vaccine. *J. Gen. Virol.* **71**, 2425–2431.

Engelhardt, J.F., Simon, R.H., Yang, Y., Zepeda, M., Weber Penleton, S., Doranz, B., Grossman, M. & Wilson, J.M. (1993) Adenovirus-mediated transfer of the CFTR gene to lung of nonhuman primates: biological efficacy study. *Hum. Gene Ther.* **4**, 759–769.

Engelhardt, J.F., Ye, X., Doranz, B. & Wilson, J.M. (1994) Ablation of E2a in recombinant adenoviruses improves transgene persistence and decreases inflammatory response in mouse liver. *Proc. Natl. Acad. Sci. USA* **91**, 6196–6200.

Fischer, L.J. & Gage, F.H. (1993) Grafting in the mammalian central nervous system. *Physiol. Rev.* **73**, 583–616.

Gallichan, W.S., Johnson, D.C., Graham, F.L. & Rosenthal, K.L. (1993) Mucosal immunity and protection after intranasal immunization with recombinant adenovirus expressing herpes simplex glycoprotein B. *J. Infect. Dis.* **168**, 622–629.

Gilardi, P., Courtney, M., Pavirani, A. & Perricaudet, M. (1990) Expression of human alpha 1-antitrypsin using a recombinant adenovirus vectors. *FEBS Lett.* **267**, 60–62.

Ginsberg, H.S., Moldawer, L.L., Sehgal, P.B., Redington, M., Kilian, P.L., Chanock, R.M. & Prince, G.A. (1991) A mouse model for investigating the molecular pathogenesis of adenovirus pneumonia. *Proc. Natl. Acad. Sci. USA* **88**, 1651–1655.

Graham, F.L., Smiley, J., Russell, W.C. & Nairu, R. (1977) Characteristics of a human cell line transformed by DNA from human adenovirus type 5. *J. Gen. Virol.* **36**, 59–72.

Greber, U.F., Willetts, M., Webster, P. & Helenius, A. (1993) Stepwise dismantling of adenovirus 2 during entry into cells. *Cell* **75**, 477–486.

Grubb, B.R., Pickles, R.J., Ye, H., Yankaskas, J.R., Vick, R.N., Engelhardt, J.F., Wilson, J.M., Johnson, L.G. & Boucher, R.C. (1994) Inefficient gene transfer by adenovirus vector to cystic fibrosis airway epithelia of mice and humans. *Nature* **371**, 802–806.

Haddada, H., Lopez, M., Martinache, C., Ragot, T., Abina, M.A. & Perricaudet, M. (1993) Efficient adenovirus mediated gene transfer into human blood monocyte derived macrophages. *Biochem. Biophys. Res. Commun.* **195**, 1174–1183.

Hayashi, Y., DePaoli, A.M., Burant, C.F. & Refetoff, S. (1994) Expression of a thyroid hormone responsive recombinant gene introduced into adult mice livers by replication-defective adenovirus can be regulated by endogenous thyroid hormone receptor. *J. Biol. Chem.* **269**, 23872–23875.

Herz, J. & Gerard, R.D. (1993) Adenovirus-mediated transfer of low density lipoprotein receptor gene acutely accelerates cholesterol clearance in normal mice. *Proc. Natl. Acad. Sci. USA* **90**, 2812–2816.

Holloway, S., Mennie, M., Crosbie, A., Smith, B., Raeburn, S., Dinwoodie, D., Wright, A., May, H., Calder, K., Banon, L. and Brock, D.J.H. (1994) Predictive testing for Huntington's disease: social characteristics and knowledge of applicants, attitudes to the test procedure and decisions made after testing. *Clin. Genet.* **46**, 175–180.

Horellou, P., Vigne, E., Casterl, M.N., Barneoud, P., Colin, P., Perricaudet, M., Delaere, P. & Mallet, J. (1994) Direct intracerebral gene transfer of an adenoviral vector expressing tyrosine hydroxylase in a rat model of Parkinson's disease. *NeuroReport* **6**, 49–53.

Imperiale, M.J., Kao, H.T., Feldman, L.T., Nevins, J.R. & Strickland, S. (1984) Common control of the heat shock gene and early adenovirus gene: evidence for a cellular E1A-like activity. *Mol. Cell. Biol.* **4**, 867–874.

Ishibashi, S.M., Brown, M.S., Goldstein, J.L., Gerard, R.D., Hammer, R.E. & Herz, J. (1993) Hypercholesterolemia in low density lipoprotein receptor knockout mice and its reversal by adenovirus-mediated gene delivery. *J. Clin. Invest.* **92**, 883–893.

Jacobs, S.C., Stephenson, J.R. & Wilkinson, G.W.G. (1992) High level expression of the tick-borne encephalitis virus NS1 protein using an adenovirus-based vector: protection elicited in a murine model. *J. Virol.* **66**, 2086–2095.

Jomary, C., Piper, T.A., Dickson, G., Couture, L.A., Smith, A.E., Neal, M.J. & Jones, S.E. (1994) Adenovirus-mediated gene transfer into murine retinal cells *in vitro* and *in vivo*. *FEBS Lett.* **347**, 117–122.

Kaplitt, M.G., Kwong, A.D., Kleopoulos, S.P., Mobb, C.V., Rabkin, S.D. & Pfaff, D.W. (1994a) Preproenkephalin promoter yields region specific and long term expression in adult brain after direct *in vivo* gene transfer via a defective herpes simplex viral vector. *Proc. Natl. Acad. Sci. USA* **91**, 8979–8983.

Kaplitt, M.G., Leone, P., Samulski, R.J., Xiao, X., Pfaff, D.W., O'Malley, K.L. & During, M.J. (1994b) Long term gene expression and phenotypic correction using adeno-associated virus vectors in the mammalian brain. *Nature Genet.* **8**, 148–154.

Kass-Eisler, A., Falck-Pedersen, E., Alvira, M., Rivera, J., Buttrick, P.M., Wittenberg, B.A., Cipriani, L. & Leinwand, L.A. (1993) Quantitative determination of adenovirus-mediated gene delivery to rat cardiac myocytes *in vitro* and *in vivo*. *Proc. Natl. Acad. Sci. USA* **90**, 11498–11502.

Kass-Eisler, A., Falck-Pedersen, E., Elfenbein, D.H., Alvira, M., Buttrick, P.M. & Leinwand, L.A. (1994) The impact of developmental stage, route of administration and the immune system on adenovirus-mediated gene transfer. *Gene Ther.* **1**, 395–402.

Kelsey, D.S. (1978) Adenovirus meningo-encephalitis. *Pediatrics* **61**, 291–293.

Kozarsky, K.F. & Wilson, J.M. (1993) Gene therapy: adenovirus vectors. *Curr. Opin. Gen. Devel.* **3**, 499–503.

Kozarsky, K.F., McKinley, D.R., Austin, L.L., Raper, S.E., Stratford-Perricaudet, L.D. & Wilson, J.M. (1994) *In vivo* correction of low density lipoprotein receptor deficiency in the Watanabe heritable hyperlipidemic rabbit with recombinant adenoviruses. *J. Biol. Chem.* **269**, 13695–13702.

Le Gal La Salle, G., Robert, J.J., Berrard, S., Ridoux, V., Stratford-Perricaudet, L.D., Perricaudet, M. & Mallet, J. (1993) Adenovirus as a potent vector for gene transfer in neurons and glia in the brain. *Science* **259**, 988–990.

Lemarchand, P., Jaffe, H.A., Danel, C., Cid. M.C., Kleinman, K., Stratford-Perricaudet, L.D., Perricaudet, M., Pavirani, A., Lecocq, J.P. & Crystal, R.G. (1992) Adenovirus mediated transfer of a recombinant human alpha 1-antitrypsin cDNA to human endothelial cells. *Proc. Natl. Acad. Sci. USA* **89**, 6482–6486.

Li, T., Adamina, M., Roof, D.J., Berson, E.L., Dryja, T.P., Roessler, B.J. & Davidson, B.L. (1994) *In vivo* transfer of a reporter gene to the retina mediated by an adenoviral vector. *Invest. Ophthal. Vis. Sci.* **35**, 2543–2549.

Lisovoski, F., Cadusseau, J., Akli, S., Caillaud, C., Vigne, E., Poenaru, L., Stratford-Perricaudet, L., Perricaudet, M., Kahn, A. & Peschanski, M. (1994) *In vivo* transfer of a marker gene to study motoneuronal development. *NeuroReport* **5**, 1069–1072.

Lokensgard, J.R., Bloom, D.C., Dobson, A.T. & Feldman, L.T. (1994) Long term promoter activity during herpes simplex virus latency. *J. Virol*, **68**, 7148–7158.

Lowenstein, P.R. (1994a) Gene transfer into neurons: from basic applications to gene therapy. *Gene Ther. Suppl.* **1**, S1–S98.

Lowenstein, P.R. (1994b) Molecular neurosurgery: mending the broken brain. Gene therapy for neurological disease? *Bio/Technology* **12**, 1075–1079.

Lowenstein, P.R. (1995) Degenerative and inherited neurological disorders. In: *Molecular and Cell Biology of Human Gene Therapeutics* (ed Dickson, G.), pp. 301–349. London, Chapman and Hall.

Lowenstein, P.R. & Enquist, L.W. (eds) (1995) *Protocols for Gene transfer into Neurons: Towards Gene Therapy of Neurological Disorders*. Chichester, John Wiley (in press).

Lowenstein, P.R., Fournel, S., Bain, D., Tomasec, P., Clissold, P.M., Castro, M.G. & Epstein, A.L. (1994) Herpes simplex virus 1 (HSV1) helper co-infection affects the distribution of an amplicon encoded protein in glia. *Neuroreport* **5**, 1625–1630.

Lowenstein, P.R., Fournel, S., Bain, D., Tomasec, P., Clissold, P.M., Castro, M.G. & Epstein, A.L. (1995a) Detection of amplicon and HSV1 helper encoded proteins reveals neurons and astrocytoma cells infected by both viral particles simultaneously: the need to monitor both particles concurrently to identify cells infected exclusively by the amplicon. *Mol. Brain Res.* **30**, 169–175.

Lowenstein, P.R., Bain, D., Shering, A.F., Wilkinson, G.W.G. & Castro, M.G. (1995b) The cytomegalovirus major immediate early promoter within a replication-deficient adenovirus recombinant drives transgene expression in astrocytes and fibroblasts, but not nervous Schwann cells (submitted).

Lowenstein, P.R., Bain, D., Shering, A.F., Wilkinson, G.W.G. and Castro, M.G. (1995c) Cell type specific expression from viral promoters within replication deficient adenovirus recombinants in primary neocortical cultures. *Rest. Neurol. Neurosci.* **8**, 37–39.

MacMillan, J. & Lowenstein, P. (1995) The strategic development of gene therapy approaches to the treatment of human neurological disease. The future of neurological therapies? In *Protocols for Gene transfer into Neurons: Towards Gene Therapy of Neurological Disorders* (eds Lowenstein, P.R. & Enquist, L.W.). Chichester, John Wiley (in press).

Markert, J.M., Malick, A., Coen, D.M. & Martuza, R.L. (1993) Reduction and elimination of encephalitis in an experimental gliomas therapy model with attenuated herpes simplex mutants that retain susceptibility to acyclovir. *Neurosurgery* **32**, 597–603.

Martuza, R.L., Malick, A., Markert, J.M., Ruffner, K.L. & Coen, D.M. (1991) Experimental therapy of human glioma by means of a genetically engineered virus mutant. *Science* **252**, 854–856.

Mastrangeli, A., Danel, C., Rosenfeld, M.A., Stratford-Perricaudet, L., Perricaudet, M., Pavirani, A., Lecocq, J.P. & Crystal, R.G. (1993) Diversity of airway epithelial cell targets for *in vivo* recombinant adenovirus-mediated gene transfer. *J. Clin. Invest.* **91**, 225–234.

Mittereder, N., Yei, S., Bachurski, C., Cuppoletti, J., Whitsett, J.A., Tolstoshev, P. & Trapnell, B.C. (1994) Evaluation of the efficacy and safety of *in vitro*, adenovirus-mediated transfer of the human cystic fibrosis transmembrane conductance regulator cDNA. *Hum. Gene Ther.* **5**, 717–729.

Morin, J.E., Lubeck, M.D., Barton, J.E., Conely, A.J., Davies, A.R. & Hung, P.P. (1987) Recombinant adenovirus induces antibody response to hepatitis B surface antigen in hamsters, *Proc. Natl. Acad. Sci. USA*, **84**, 4626–4630.

Muchardt, C., Li, C., Kornuc, M. & Gaynor, R. (1990) CREB regulation of cellular cyclic AMP responsive and adenovirus early promoters. *J. Virol.* **64**, 4296–4305.

Mulderry, P.K., Chapman, K.E., Lyons, V. & Harmer, A.J. (1993) 5′-Flanking sequences from the rat preprotachykinin gene direct high-level expression of a reporter gene in adult rat sensory neurons transfected in culture by microinjection. *Mol. Cell. Neurosci.* **4**, 164–172.

Nalbantoglu, J., Gilfix, B.M., Bertrand, P., Robitaille, Y., Gauthier, S., Rosenblatt, D.S. & Poirier, J. (1994) Predictive value of apolipoprotein E genotyping in Alzheimer's disease: results of an autopsy series and an analysis of several combined studies. *Ann. Neurol.* **36**, 889–895.

Oldfield, E.H., Culver, K.W., Ram, Z. & Blaese, R.M. (1993) A clinical protocol: gene therapy for the treatment of brain tumours using intra-tumoral transduction with the thymidine kinase gene and intravenous ganciclovir. *Hum. Gene Ther.* **4**, 39–69.

Osamura, T., Mizuta, R., Yoshioka, H. & Fushiki, S. (1993) Isolation of adenovirus type 11 from the brain of a neonate with pneumonia and encephalitis. *Eur. J. Ped.* **152**, 496–499.

Oualikene, W., Gonin, P. & Eloit, M. (1994) Short and long term dissemination of deletion mutants of adenovirus in permissive (cotton rat) and non-permissive (mouse) species. *J. Gen. Virol.* **75**, 2765–2768.

Perez Cruet, M.J., Trask, T.W., Chen, S.H., Goodman, J.C., Woo, S.L.C., Grossman, R.G. & Shine, H.D. (1994) Adenovirus-mediated gene therapy of experimental gliomas. *J. Neurosci. Res.* **39**, 506–511.

Prevec, L., Schneider, M., Rosenthal, K.L., Belbeck, L.W., Derbyshire, J.B. & Graham, F.L. (1989) Use of human adenovirus-based vectors for antigen expression in animals. *J. Gen. Virol.* **70**, 429–434.

Prevec, L., Christie, B.S., Laurie, K.E., Bailey, M.M., Graham, F.L. & Rosenthal, K.L. (1991) Immune response to HIV-1 gag antigens induced by recombinant adenovirus vectors in mice and rhesus macaque monkeys. *J. Acq. Immun. Defic. Synd.* **4**, 568–576.

Quantin, B., Perricaudet, L.D., Tajbakhsh, S. & Mandel, J.L. (1991) Adenovirus as an expression vector in muscle cells *in vivo*. *Proc. Natl. Acad. Sci. USA* **89**, 2581–2584.

Ram, Z., Culver, K.W., Walbridge, S., Blaese, R.M. & Oldfield, E.H. (1993) *In situ* retroviral mediated gene transfer for the treatment of brain tumors in rats. *Cancer Res.* **53**, 83–88.

Ridoux, V., Robert, J.J., Zhang, X., Perricaudet, M., Mallet, J. & Le Gal La Salle, G. (1994a) The use of adenovirus vectors for intracerebral grafting of transfected nervous cells. *NeuroReport* **5**, 801–804.

Ridoux, V., Robert, J.J., Zhang, X., Perricaudet, M., Mallet, J. & Le Gal La Salle, G. (1994b) Adenoviral vectors as functional retrograde neuronal tracers. *Brain Res.* **648**, 171–175.

Roessler, B.J., Allen, E.D., Wilson, J.M., Hartman, J.W. & Davidson, B.L. (1993) Adenoviral mediated gene transfer to rabbit synovium *in vivo*. *J. Clin. Invest.* **92**, 1085–1092.

Rosenfeld, M.A., Siegfried, W., Yoshimura, K., Yoneyama, K., Fukayama, M., Stier, L.E., Paakko, P.K., Gilardi, P., Stratford-Perricaudet, L.D., Perricaudet, M., Jallat, S., Pavirani, A., Lecocq, J.P. & Crystal, R.G. (1991) Adenovirus-mediated transfer of a recombinant α1-antitrypsin gene to the lung epithelium *in vivo*. *Science* **252**, 431–434.

Rosenfeld, M.A., Yoshimura, K., Trapnell, B.C., Yoneyama, K., Rosenthal, E.R., Dalemans, W., Fukayama, M., Bargon, J., Stier, L.E., Stratford-Perricaudet, L., Perricaudet, M., Guggino, W.B., Pavirani, A., Lecocq, J.P. & Crystal, R.G. (1992) *In vivo* transfer of the human cystic fibrosis transmembrane conductance regulator gene to the airway epithelium. *Cell* **68**, 143–155.

Rosenfeld, M.A., Chu, C.S., Seth, P., Danel, C., Banks, T., Yoneyama, K., Yoshimura, K. & Crystal, R.G. (1994) Gene transfer to freshly isolated human respiratory epithelial cells *in vitro* using a replication-deficient adenovirus containing the human cystic fibrosis transmembrane conductance regulator cDNA. *Hum. Gene Ther.* **5**, 331–342.

Ross, D. & Ziff, E. (1992) Defective synthesis of early region 4 mRNAs during abortive adenovirus infections in monkey cells. *J. Virol.* **66**, 3110–3117.

Rubin, B.A. & Rorke, L.B. (1994) Adenovirus vaccines. In *Vaccines* (eds Plotkin, S.A. & Mortimer Jr. E.A.), 2nd edn, pp. 475–501. Philadelphia, W.B. Saunders.

Sambrook, J., Fritsch, E.F. & Maniatis, T. (1989) *Molecular Cloning. A Laboratory Manual*, 2nd edn. Cold Spring Harbor, NY, CSH Laboratory Press.

Saunders, A.M., Strittmatter, W.J., Schmechel, D., St. George-Hyslop, P.H., Pericack-Vance, M.A., Joo, S.H., Rosi, B.L., Gusella, J.F., Crapper MacLachlan, D.R., Alberts, M.J., Hulette, C., Crain, B., Goldgaber, D. & Roses, A.D. (1993) Association of apolipoprotein E allele epsilon 4 with late-onset familial and sporadic Alzheimer's disease. *Neurology* **43**, 1467–1472.

Schwartz, A.R., Togo, Y. & Hornick, R.B. (1974) Clinical evaluation of live, oral types 1, 2, and 5 adenovirus vaccines. *Am. Rev. Respir. Disd.* **109**, 233–238.

Scinto, L.F.M., Daffner, K.R., Dressler, D., Ransil, B.I., Rentz, D., Weintraub, S., Mesulam, M. & Potter, H. (1994) A potential non-invasive neurobiological test for Alzheimer's disease. *Science* **266**, 1051–1054.

Seiger, A., Nordberg, A., Von Holst, H., Backman, L., Ebendal, T., Alafuzoff, I., Amberla, K., Hartvig, P., Herlitz, A., Lilja, A., Lundqvist, H., Langstrom, B., Meyerson, B., Persson, A., Viitanen, M., Winblad, B. & Olson, L. (1993) Intracranial infusion of purified nerve growth factor to an Alzheimer patient: the first attempt of a possible future treatment strategy. *Behav. Brain Sci.* **57**, 255–261.

Shenk, T. (1995) Adenoviridae: the viruses and their replication. In *Virology* (ed. Fields, B.), 3rd edn Raven Press, New York.

Shewach, D.S., Zerve, L.K., Hughes, T.L., Roessler, B.H., Breakefield, X.O. & Davidson, B.L. (1994) Enhanced cytotoxicity of antiviral drugs mediated by adenovirus directed transfer of the herpes simplex virus thymidine kinase gene in rat glioma cells. *Cancer Gene Ther.* **1**, 107–112.

Simon, R.H., Engelhardt, J.F., Yang, Y., Zepeda, M., Weber Pendleton, S., Grossman, M. & Wilson, J.M. (1993) Adenovirus-mediated transfer of the CFTR gene to lung of nonhuman primates: toxicity study. *Hum. Gene Ther.* **4**, 771–780.
Smith, T.A., Mehaffey, M.G., Kayda, D.B., Saunders, J.M., Yei, S., Trapnell, B.C., McClelland, A. & Kaleko, M. (1993) Adenovirus mediated expression of therapeutic plasma levels of human factor IX in mice. *Nature Genet.* **5**, 397–402.
Spergel, J.M. & Chen-Kiang, S. (1991) Interleukin 6 enhances a cellular activity that functionally substitutes for Ela protein in transactivation. *Proc. Natl. Acad. Sci. USA* **88**, 6472–6476.
Stratford-Perricaudet, L.D., Levrero, M., Chasse, J.F., Perricaudet, M. & Briand, P. (1990) Evaluation of the transfer and expression in mice of an enzyme-encoding gene using a human adenovirus vector. *Hum. Gene Ther.* **1**, 241–256.
Stratford-Perricaudet, L.D., Makeh, I., Perricaudet, M. & Briand, P. (1992) Widespread long-term gene transfer to mouse skeletal muscles and heart. *J. Clin. Invest.* **90**, 626–630.
Top, F.H.Jr., Buescher, E.L., Bancroft, W.H. & Russell, P.K. (1971a) Immunization with live types 7 and 4 adenovirus vaccines. II. Antibody response and protective effect against acute respiratory disease due to adenovirus type 7. *J. Infect. Dis.* **124**, 155–160.
Top, F.H.Jr., Grossman, R.A., Bartelloni, P.J. *et al.* (1971b) Immunization with live types 7 and 4 adenovirus vaccines. I. Safety, infectivity, antigenticity, and potency of adenovirus type 7 vaccine in humans. *J. Infect Dis.* **124**, 148–154.
Ugolini, G., Kuypers, H.G.J.M. & Strick, P.L. (1989) Transneuronal transfer of herpes virus from peripheral nerves to cortex and brainstem. *Science* **243**, 89–91.
Welsh, M.J., Smith, A.E., Zabner, J., Rich, D.P., Graham, S.M., Gregory, R.J., Pratt, B.M. & Moscicki, R.A. (1994) Clinical protocol: cystic fibrosis gene therapy using an adenovirus vector: *in vivo* safety and efficacy in nasal epithelium. *Hum. Gene Ther.* **5**, 209–219.
Wickham, T.J., Mathias, P., Cheresh, D.A. & Nemerow, G.R. (1993) Integrins $\alpha v\beta 3$ and $\alpha v\beta 5$ promote adenovirus internalization but not virus attachment. *Cell* **73**, 309–313.
Wickham, T.J., Filardo, E.J., Cheresh, D.A. & Nemerow, G.R. (1994) Integrin $\alpha v\beta 5$ selectively promotes adenovirus mediated cell membrane permeabilization. *J. Cell Biol.* **127**, 257–264.
Wilkinson, G.W.G. (1994) Gene therapy and viral vaccination. *Rev. Med. Microbiol.* **5**, 97–106.
Wilkinson, G.W.G. & Borysiewicz, L.K. (1995) Gene therapy and viral vaccination: the interface. *Br. Med. J.* **51**, 205–216.
Wilkinson, G.W.G., Darley, R. & Lowenstein, P.R. (1994) Viral vectors for gene therapy. In: *From Genetics to Gene Therapy*, (ed.) Latchman, D.S., London, Bios Scientific Publishers. pp. 161–192.
Wilson, J.M., Engelhardt, J.F., Grossman, M. Simon, R.H. & Yang, Y. (1994) Clinical protocol: gene therapy of cystic fibrosis lung disease using E1 deleted adenoviruses: a phase I trial. *Hum. Gene Ther.* **5**, 501–519.
Wold, W.S.M., Hermiston, T.W. & Tollefson, A.E. (1994) Adenovirus proteins that subvert host defenses. *Trends Microbiol.* **2**, 437–443.
Yang, Y., Nunes, F.A., Berencsi, K., Furth, E.E., Gonczol, E. & Wilson, J.M. (1994a) Cellular immunity to viral antigens limit E1-deleted adenoviruses for gene therapy. *Proc. Natl. Acad. Sci. USA* **91**, 4407–4411.
Yang, Y., Nunes, F.A., Berencsi, K., Gonczol, E., Engelhardt, J.F. & Wilson, J.M. (1994b) Inactivation of E2A in recombinant adenoviruses improves the prospect for gene therapy in cystic fibrosis. *Nature Genet.* **7**, 362–369.
Yei, S., Bachurski, C.J., Weaver, T.E., Wert, S.E., Trapnell, B.C. & Whitsett, J.A. (1994a) Adenoviral-mediated gene transfer of human surfactant protein B to respiratory epithelial cells. *Am. J. Respir. Cell Mol. Biol.* **11**, 329–336.
Yei, S., Mittereder, N., Wert, S., Whitsett, J.A., Wilmott, R.W. & Trapnell, B.C. (1994b) *In vivo* evaluation of the safety of adenovirus-mediated transfer of the human cystic fibrosis transmembrane conductance regulator cDNA to the lung. *Hum. Gene Ther.* **5**, 731–744.
Zabner, J., Couture, L.A., Gregory, R.J., Graham, S.M., Smith, A.E. & Welsh, J.M. (1993a)

Adenovirus-mediated gene transfer transiently corrects the chloride transport defect in nasal epithelia of patients with cystic fibrosis. *Cell* **75**, 207–216.

Zabner, J., Petersen, D.M., Puga, A.P., Graham, S.M., Couture, L.A., Keyes, L.D., Lukason, M.J., St George, J.A., Gregory, R.J., Smith, A.E. & Welsh, J.M. (1993b) Safety and efficacy of repetitive adenovirus-mediated transfer of CFTR cDNA to airway epithelia of primates and cotton rats. *Nature Genet.* **6**, 75–83.

Zabner, J., Couture, L.A., Smith, A.E. & Welsh, J.M. (1994) Correction of cAMP-stimulated fluid secretion in cystic fibrosis airway epithelia: efficiency of adenovirus-mediated gene transfer *in vitro*. *Hum. Gene Ther.* **5**, 585–593.

Zemanick, M.C., Strick, P.L. & Dix, R.D. (1991) Direction of transneuronal transport of herpes simplex virus 1 in the primate motor system is strain-dependent. *Proc. Natl. Acad. Sci. USA* **88**, 8048–8051.

CHAPTER 3

ADENOVIRUS: A NEW TOOL TO TRANSFER GENES INTO THE CENTRAL NERVOUS SYSTEM FOR TREATMENT OF NEURODEGENERATIVE DISORDERS

Philippe Horellou, Frederic Revah, Olivier Sabaté,
Marie-Hélène Buc-Caron, Jean-Jacques Robert and Jacques Mallet

C 9923 CNRS, Laboratoire de Génétique Moléculaire de la Neurotransmission et des Processus Dégénératifs, 1 Avenue de la Terrasse, 91198 Gif-sur-Yvette, France

Table of Contents

3.1 Introduction

Over the last few years gene therapy has emerged as a potentially powerful approach to the treatment of neurological diseases. The discovery of an increasing number of neurotrophic factors which can either inhibit neurodegenerative processes or stimulate regeneration, and the improved understanding of the neurotransmitter mechanisms underlying the major disease symptoms provide the basis for current gene therapy strategies toward new treatments for patients with neurological disease.

Neurodegenerative disorders originate from the progressive death of certain categories of neurons resulting in impairment of major neurophysiological functions. For

Genetic Manipulation of the Nervous System
ISBN 0-12-437165-5

instance, up to 70–90% of dopaminergic nigrostriatal neurons degenerate in Parkinson's disease. The GABAergic striatal neurons are those most affected in Huntington's disease and cholinergic neurons are implicated in Alzheimer's diseases. The mechanisms of this specific cell death are yet to be found. However, the impairments that characterize these diseases have been shown to result from the degeneration of the specific categories of neurons affected. There are today no efficient cures for these diseases. There are no fundamental treatments for Huntington's and Alzheimer's disease. In Parkinson's disease, the current symptomatic treatment of the motor impairments present in this condition involves oral administration of L-dopa. The molecule is transported across the blood–brain barrier and converted to dopamine by residual dopaminergic neurons as well as other types expressing dopa-decarboxylase, leading to a substantial improvement of motor function. However, after a few years the degeneration of dopaminergic neurons still progresses, the effects of L-dopa are reduced and side effects appear (Marsden and Parkes, 1976). At that stage, new therapeutical approaches for Parkinson's disease appear necessary.

In this context, neural grafting represents a possible tool for the substitution of damaged neurons. Transplantation studies in an animal model of Parkinson's disease were started in the late 1970s (Björklund and Stenevi, 1979; Perlow *et al.*, 1979). Clinical trials to evaluate transplantation of human fetal dopaminergic cells as a therapy for Parkinson's disease have given promising results and offer the hope of a valuable treatment for patients, in particular when they suffer from severe 'on–off' phenomena under optimal therapy (Lindvall *et al.*, 1990). Indeed, 3 years after intracerebral grafting of cells originating from the ventral mesencephalon of four fetuses, patients with Parkinson's disease still display clear clinical improvement, manifested by a reduction of the severity of symptoms, as well as of the time spent in the 'off' phase, and by a prolongation of the effect of a single dose of L-dopa (Lindvall *et al.*, 1994). However, neuronal grafts require access to fetal material of a specific age and even when successful, only a small proportion of the grafted neurons survive transplantation.

New therapeutic approaches can easily be tested in animal models which have been developed using rodents and non-human primates. The rat model of Parkinson's disease is particularly convenient as it generates statistical results relatively soon and at a low cost (Ungerstedt and Arbuthnott, 1970). These animals receive an injection of 6-hydroxydopamine (6-OHDA) unilaterally into the ascending mesostriatal pathway originating from the dopaminergic neurons of the substantia nigra (A9) and the left mesostriatal pathway originating from the dopaminergic neurons located in the left side of the ventral tegmental area (A10). This causes denervation of the ipsilateral striatum from its dopaminergic afferents. In this model, the induced depletion of dopamine in the denervated striatum correlates with a quantitative sensorimotor asymmetry. The denervated rats respond by a turning behaviour to low doses of apomorphine, an agonist of the dopamine receptors that stimulates the denervation-induced supersensitive dopamine receptors. The rate of rotation can be reduced by treatment that compensates for the dopamine deficiency of the denervated striatum.

3.2 The adenovirus as a tool to express genes in the brain

The recombinant adenoviruses commonly used are derived from type 5 adenovirus which has never been found to be associated with tumoral pathology in humans most likely because the probability of integration of the adenoviral DNA into the genome of the host cell is low. The effects of wild-type adenoviral infection are widely documented, and more than 10 million patients have received live adenovirus tablets as oral vaccine with no major side effects (Rubin and Rorke, 1988). Finally the natural tropism of adenovirus is the respiratory tract, and wild-type viruses are found only rarely in the human brain (Kelsey, 1978). As a consequence, when a recombinant adenovirus defective for replication is administered into the brain there is very little risk that replication of this recombinant virus occurs by trans-complementation by a resident wild-type virus. This is not the case when using herpes simplex virus-1 (HSV1)-based vectors, since a large proportion of the population has encountered herpes viruses and this virus is neurotropic.

The biology of the adenovirus is well understood. The human adenoviral DNA is large (36 kb). It contains many genes which are classified into early (from E1 to E4) and late (L1 to L5) genes, according to their period of expression prior to and after viral DNA replication, respectively. The virus life cycle includes attachment to cell-specific receptors, internalization into endosomes, release in the cytoplasm by endosmolysis, and translocation to the nucleus, where activation of the early regions occurs followed by DNA replication and activation of the late regions. Thereafter, the viral DNA is packaged into capsids, and the infected cells die releasing the viruses. Unlike HSV, no latent state has been described for adenoviruses. Compared to HSV1-derived expression vectors for instance, the transcription machinery and life cycle of adenoviruses is relatively simple, allowing straightforward strategies for obtaining purified replication-defective recombinant viruses.

Since the wild-type adenovirus causes a lytic infection, recombinant adenovirus for gene transfer experiments must be defective for some essential genes. The E1 region contains genes that can be deleted and E1 defective viruses can be complemented in trans by 293 cells (Graham *et al.*, 1977). The 293 cells contain the E1 genes integrated into the genome so that upon infection of this cell line, the E1-deleted viruses can replicate and be amplified. The E3 region, which is not necessary for viral propagation in cell culture, can also be deleted. The viruses described here can accommodate transgenes up to 7.5 kb in length, and can be obtained at titres as high as 10^{11}–10^{12} pfu ml^{-1}.

In spite of the advantages of adenovirus, and although the first results obtained using adenoviruses encoding β-galactosidase from *Escherichia coli* (Ad-RSVβGal) for both an *in vivo* and an *ex vivo* gene therapy approach are promising, one must be aware that several problems will have to be solved before human application can be envisioned. These problems include the fine tuning of the cell specificity, time course and levels of transgene expression. One will also have to understand and circumvent the cytopathic effects of adenoviruses that can sometimes be observed.

3.3 *Ex vivo* gene therapy

There are several important issues for *ex vivo* gene transfer to the brain, including: (1) extent of transgene expression; (2) stability of transgene expression; (3) supply of the cells; and (4) safety of the procedure. These issues differ according to the transgene, the vector and the cell type used. Initially, as discussed in other chapters, retroviruses as well as the transfection of cells were largely used to transfer the gene of interest.

The initial step in the study of gene transfer to the brain consisted of modification of cell lines to secrete dopa or dopamine. These cell lines of neuronal, endocrine or fibroblast origins have generated tools to study the feasibility of neurotransmitter delivery in animal models of Parkinson's disease (Horellou *et al.*, 1989, 1990a,b; Wolff *et al.*, 1989; Uchida *et al.*, 1990). They also have allowed analysis of the relative importance of dopa versus dopamine in the promotion of functional recovery after grafting (Horellou *et al.*, 1990).

Some primary cell types originating from the periphery have been used. Primary fibroblasts (Fisher *et al.*, 1991) and myoblasts (Jiao *et al.*, 1993) expressing the tyrosine hydroxylase (TH) gene have been shown to partially restore dopamine levels in the rodent Parkinson's disease model. The main advantage of this approach is the possible grafting of cells from the patient to be grafted himself, thereby solving the supply problem and reducing the need for immunosuppression to prevent rejection.

Central nervous system-derived cells may have interesting potential with respect to intracerebral transplantation. The use of immortalized neural progenitors illustrate this aspect. Recently, a temperature-sensitive SV40-T antigen-transformed neural progenitor cell line, HiB5, has been isolated and the cells transplanted into neonatal hippocampus and cerebellum where they have acquired morphologic characteristics of the neurons and glial cells (Renfranz *et al.*, 1991). Other immortalized neural progenitor cell lines have been isolated with similar properties (Snyder *et al.*, 1992; Onifer *et al.*, 1993). In the perspective of human application, it seems important to avoid the transformation of cells by conditional oncogene, and primary cells stimulated to grow with mitogens that would retain properties of neural progenitors may represent the ideal cell type for intracerebral transplantation.

As they may be amplified *in vitro*, primary glial cells represent a possible candidate for gene transfer in the perspective of intracerebral transplantation. These cells have been modified to express both TH and brain-derived neurotrophic factor (BDNF) using recombinant retroviruses (Lundberg *et al.*, 1991; Lin *et al.*, 1994). These modified cells have been found to integrate into the rat brain following their transplantation in the rat model of Parkinson's disease.

We have recently used human progenitor cells as a vehicle for adenovirus (Buc-Caron, 1995). First, we established conditions allowing the majority (>65%) of the cells to express β-galactosidase *in vitro* without toxicity. Survival of neural cells after grafting is a major concern of investigators in the field of intracerebral transplantation. In our experiments, large numbers of human neuroblasts expressing β-galactosidase were observed in three out of four rats grafted with at least 10^6 neural progenitors infected *in vitro*. Injection of such high numbers of cultivated progenitors is probably necessary because of large-scale cell death immediately after grafting. Our grafts continued to express the transgene 2 or 3 weeks post-grafting, the longest time tested (Sabaté *et al.*, 1995). It is likely that long-term expression can be obtained since human neurons *in vitro* can express the reporter gene for up to 3 months (Levallois *et al.*, 1994), and grafted rodent astrocytes have been found to express β-galactosidase up to 5 months after transplantation into the rat brain (Ridoux *et al.*, 1994).

Nevertheless, following severe loss of nigral cell bodies and synaptic contacts at late stages of the disease, it is unlikely that recombinant viruses alone would restore the appropriate release of L-dopa or dopamine into the striatum. Thus, in such cases *ex vivo* gene therapy may be the best approach to providing therapeutic factors.

However, a more complete correction of the phenotypical defect might require restoration of regulated release at the synaptic level. This may be best approached by mesencephalic dopaminergic cells, whose progenitors may be expanded *in vitro* allowing the production of a large quantity of cells suitable for transplantation. Their capacity to function fully in the striatal circuitry will have to be determined. It is likely that neural progenitors would differentiate according to their final location, as previously established with a neural progenitor cell line (Renfranz *et al.*, 1991).

At any rate, the combined use of human neural progenitors amplified *in vitro* and adenoviruses may potentially allow the treatment of neurodegenerative disorders. Enough cells expressing a gene of interest, encoding a trophic factor or a neurotransmitter synthesizing enzyme, could be obtained from a single fetus to graft several patients. In addition, the *in vitro* step to amplify the cells allows testing for the absence in the fetal tissue of contaminating agents such as viruses and thereby results in improved safety.

3.4 *In vivo* gene therapy

As discussed extensively in other chapters of this book, three viruses genetically modified, recombinant HSV, adeno-associated virus and adenovirus, have been recently shown to mediate gene expression after intracerebral injection.

A defective HSV was developed to deliver heterologous genes into the rat brain after stereotaxic injection (Palella *et al.*, 1989; Fink *et al.*, 1992; Wolfe *et al.*, 1992; Lokensgard *et al.*, 1994; During *et al.*, 1994). In these *in vivo* experiments, the transgene was shown to be expressed in neurons of different brain areas surrounding the injection site as well as in neurons whose axons projected to the injection site. However, only a restricted number of cells were expressing the transgene (Wolfe *et al.*, 1992; During *et al.*, 1994). The adeno-associated virus has also been recently shown to allow an expression in neurons (Kaplitt *et al.*, 1994). However, the risk of contamination of recombinant virus by competent viruses is still a major problem with both recombinant viruses.

Recently, direct gene transfer to the brain has been developed with a recombinant human adenovirus carrying the *E. coli lacZ* reporter gene (Le Gal La Salle *et al.*, 1993;

Davidson *et al.*, 1993; Akli *et al.*, 1993; Bajocchi *et al.*, 1993). Stereotaxic intracerebral injection of this virus resulted in an evident and robust labelling of a large number of cells around the inoculation site. The labelled cells (i.e. infected with the recombinant human adenovirus 5 and expressing β-galactosidase) were found to correspond to neurons, astrocytes, oligodendrocytes, microglial and ependymal cells. The efficiency of the *in vivo* infection is relatively high: about one cell appeared to express β-galactosidase for 100 injected recombinant adenovirus (Akli *et al.*, 1993). The neuropathological consequences of low inocula were minimal in rats, and a neuronal death with gliosis and vascular inflammatory response was observed only with high titres of virus injection (Le Gal La Salle *et al.*, 1993; Akli *et al.*, 1993).

To test the functional efficacy of adenovirus-mediated gene transfer, we expressed the TH gene in the rat model of Parkinson's disease. A defective human adenovirus recombinant for the human TH-1 cDNA was generated for direct injection of the TH gene. The entire coding sequence of TH was placed under the transcriptional control of the long terminal repeat of the Rous sarcoma virus promoter (in place of the E1 gene of the adenovirus). A high titre of recombinant adenovirus encoding the human TH (Ad-RSVhTH) ([$1 \pm 0.16\ 10^{12}$ plaque forming units (pfu) ml^{-1}] was obtained by caesium chloride ultracentrifugation followed by dialysis to eliminate caesium.

We have improved both the expression level and the dispersion of the transgene in the striatum, by optimizing the conditions of intracerebral injection. The dose of 15×10^7 pfu was found to be optimum for the level of expression of the transgene and to induce a limited tissue necrosis as assessed by microscopic analysis of striatum stained with neutral red. We improved the dispersion of transgene expression by increasing the number of sites from three to nine and by decreasing the rate of injection from 1 μl min^{-1} to 0.5 μl/min^{-1}. The increase of the number of injection sites was achieved by releasing the virus at three dorsoventral positions (equidistant at 0.4 mm) while removing the cannula from each of the three anteroposterolateral coordinates. This modification resulted in a distinctly more dispersed pattern of expression through the dorsoventral axis. The lower rate of injection (0.5 μl min^{-1}) increased the diffusion of the virus as indicated by the greater dispersion of the cells expressing β-galactosidase.

To test the capacity of the Ad-RSVhTH to allow functional expression of the TH enzyme *in vivo*, the purified Ad-RSVhTH was injected into the striatum of 6-OHDA denervated rats. Young adult female Sprague–Dawley rats were used in all experiments, and surgery was performed under equithesin anaesthesia. At least 10 days after the 6-OHDA lesion and 1–2 weeks before intracerebral injection, rats were tested for apomorphine-induced turning. A total of $1.5 \pm 0.25 \times 10^8$ pfu of Ad-RSVhTH was then injected into nine sites of the striatum. Control animals received $1.5 \pm 0.25 \times 10^8$ pfu of the control purified Ad-RSVβGal virus.

The injected animals were tested again for apomorphine-induced turning 7 and 14 days after intracerebral injection. One week post-injection the mean decrease in the TH-treated animals, about 30%, was statistically significant ($P = 0.011$). Two weeks post-injection, there was again a significant mean decrease of 22% ($P = 0.048$). One week post-injection, the control group of rats injected with the Ad-RSVβGal virus showed no mean modification of apomorphine-induced turning behaviour

($P=0.752$). Two weeks post-injection none of the control rats showed a significant decrease (>30%) and, in fact, the group showed a mean increase of 19% ($P=0.056$). The difference between the Ad-RSVhTH and Ad-RSVβGal groups was significant ($P=0.05$) as determined, using one-way analysis of variance followed by Scheff's F-test and Fischer's PLSD, and the non-parametric Mann–Whitney U-test (Horellou *et al.*, 1994).

Two weeks after the intrastriatal injections of adenovirus, the efficiency of gene transfer to the striatum was determined in the Ad-RSVhTH group by immunohistological staining with an anti-TH antibody, and in the Ad-RSVβGal group by reaction with 5-bromo-4-chloro-3-indolyl β-galactopyranoside (X-gal). In most animals, a large number of cells expressing the transgene was detected within the denervated striatum. The labelled cells were dispersed throughout the entire striatum. Expression was confined to the striatum, from its rostral to caudal regions, and to the anterior part of the external capsule. This wide distribution was presumably the result of the optimized injection protocol described above. Double glial fibrillary acidic protein (GFAP)/TH immunostaining, and the morphological aspect, indicated that most infected cells were reactive astrocytes. Some cells (a minority) had a clear neuronal morphology. To evaluate the extent of neural tissue destruction, inflammatory response and gliosis, adjacent sections were analysed by acetylcholinesterase (AChE) histochemistry and GFAP immunohistochemistry. At the site of injection, discrete zones of tissue destruction were seen with no difference between the Ad-RSVhTH- and the Ad-RSVβGal-injected rats. A similar pattern of GFAP-reactive astrocytes was found around the injection site in both groups of animals.

3.5 Immunological aspects and improvement of the recombinant adenoviral tool

Indeed, adenovirus-mediated gene transfer for restoration of functions has already been described in models of diseases affecting organs other than the brain. In a mouse model of ornithine transcarbamylase (OTC) deficiency, systemic administration of a recombinant adenovirus carrying the OTC gene led to expression of the enzyme in the liver and induced a phenotypic correction of the liver disease in newborn mice (Stratford-Perricaudet *et al.*, 1990). In cystic fibrosis, local administration of a recombinant adenovirus for cystic fibrosis transmembrane conductance regulator (CFTR) in the lung gave an improvement in human clinical trials (Zabner *et al.*, 1993). In newborn mouse mutants lacking dystrophin, intramuscular injection of a recombinant adenovirus for a dystrophin minigene protected the muscle fibres against the degeneration process that affects dystrophin-deficient myofibres (Vincent *et al.*, 1993). However, these experiments either allowed only transient expression, in the case of the CFTR study (Zabner *et al.*, 1993), or were performed in newborn animals, a condition allowing long-term expression (Stratford-Perricaudet *et al.*, 1990; Vincent *et al.*, 1993).

Transient transgene expression has recently been demonstrated for E1-deleted adenoviruses. The transient nature of expression is linked to the synthesis of adenoviral proteins by the infected cells, as shown using adenoviral-mediated gene transfer to the liver. Despite the deletion of the E1 gene in the recombinant adenovirus, cellular E1-like factors can induce a basal expression of the adenoviral promoter leading to specific cellular immune responses and destruction of the genetically modified cells (Yang *et al.*, 1994). In the brain, gene expression from adenovirus vectors has previously been found to be restricted or to fade with time (Le Gal La Salle *et al.*, 1993; Akli *et al.*, 1993). The immunologically isolated nature of the brain does not fully protect the infected cells from the immune response, and clearance of viruses from infected brains is due to either infiltration of mononuclear cells across the blood–brain barrier or to a mechanism involving cytotoxic T lymphocytes (Townsen and Baringer, 1979; Tishon *et al.*, 1993).

In some viral infections, mononuclear cells have been shown to be recruited into the CNS and to induce an inflammation. This inflammation has been shown to be immunologically specific and dependent on a population of T lymphocytes sensitized to the inoculated virus (McFarland *et al.*, 1972). The T lymphocytes release lymphokines that attract macrophages that constitute the major component of the inflammatory response. There are several examples of virus-induced, immune-mediated reactions leading to cell injury. For instance, in the biphasic disease of mice caused by Theiler's virus, a cell-mediated immune response appears as a component of the late demyelination since immunosuppression with anti-thymocyte serum, albeit potentiating the initial grey matter infection, prevents the late demyelination (Lipton and Dal Canto, 1976).

Improved recombinant adenoviruses expected to allow prolonged transgene expression are being generated by inactivation of other adenoviral genes, such as E2a, that induce the expression of immunodominant viral proteins (Engelhardt *et al.*, 1994; Yang *et al.*, 1994). It may become possible to use such improved vectors to obtain long-lasting expression in the brain. Our work with an E1/E3-deleted adenovirus shows that adenovirus-mediated TH gene transfer compensates for the apomorphine-induced turning behaviour in 6-OHDA-lesioned rats, suggesting that dopamine depletion in the striatum of denervated rats can be compensated. It would be valuable to study this compensation after longer periods with an E1/E2a-inactivated adenovirus and thus envisage the development of gene therapy in Parkinson's disease.

3.6 Conclusions

The use of recombinant adenoviruses as vehicles to transfer genes encoding neurotransmitter-synthesizing enzymes or trophic factors into somatic cells is of particular interest because first, they allow efficient infection of brain cells after intracerebral injection of purified recombinant viruses, and second, it should be possible to develop a safe and efficient system. Such tools may offer the possibility to transfer genes *in situ* by direct intracerebral injection with biologically active genes. These tools allow the modification of cells that do not divide, therefore they can be used to direct the expression of genes into neurons. The possibility of using tissue-specific promoters is also of considerable interest so that the expression can be cell specific.

References

Akli, S., Caillaud, C., Vigne, E., Stratford-Perricaudet, L.D., Poenaru, L., Perricaudet, M., Kahn, A. & Peschanski, M. (1993) Transfer of a foreign gene into the brain using adenovirus vectors. *Nature Genet.* **3**, 224–228.

Bajocchi, G., Feldman, S.H., Crystal, R.G. & Mastrangeli, A. (1993) Direct *in vivo* gene transfer to ependymal cells in the central nervous system using recombinant adenovirus vectors. *Nature Genet.* **3**, 229–234.

Björklund, A. & Stenevi, U. (1979) Reconstruction of the nigrostriatal dopamine pathway by intracerebral nigral transplants. *Brain Res.* **177**, 555–560.

Buc-Caron, M.-H. (1995) Neuroepithelial progenitor cells explanted from human fetal brain proliferate and differentiate *in vitro*. *Neurobiol. Dis.* (in press).

Davidson, B.L., Allen, E.D., Kozarsky, K.F., Wilson, J.M. & Roessler, B.J. (1993) A model system for *in vivo* gene transfer into the central nervous system using an adenoviral vector. *Nature Genet.* **3**, 219–223.

During, M.J., Naegele, J.R., O'Malley, K.L. & Geller, A.I. (1994) Long-term behavioural recovery in parkinsonian rats by an HSV vector expressing tyrosine hydroxylase. *Science* **266**, 1399–1403.

Engelhardt, J.F., Ye, X.H., Doranz, B., Wilson, J.M. (1994) *Proc. Natl. Acad. Sci. USA* **91**, 6196.

Fink, D.J., Sternberg, L.R., Weber, P.C., Mata, M., Goins, W.F. & Glorioso, J.C. (1992) *In vivo* expression of β-galactosidase in hippocampual neurons by HSV-mediated gene transfer. *Hum. Gene Ther.* **3**, 12–19.

Fisher, L.J., Jinnah, H.A., Kale, L.C., Higgins, G.A. & Gage, F.H. (1991) Survival and function of intrastriatally grafted primary fibroblasts genetically modified to produce L-dopa. *Neuron* **6**, 371–380.

Graham, F.L., Smiley, J., Russell, W.C. & Nairn, R. (1977) Characteristics of a human cell line transformed by DNA from human adenovirus 5. *J. Gen. Virol.* **36**, 59–72.

Horellou, P., Guibert, B., Leviel, V. & Mallet, J. (1989) Retroviral transfer of a human tyrosine hydroxylase cDNA in various cell lines: regulated release of dopamine in mouse anterior pituitary AtT-20 cells. *Proc. Natl. Acad, Sci. USA* **86**, 7233–7237.

Horellou, P., Brundin, P., Kalén, P., Mallet, J. & Björklund, A. (1990a) *In vivo* release of DOPA and dopamine from genetically engineered cells grafted to the denervated rat striatum. *Neuron* **5**, 393–402.

Horellou, P., Marlier, L., Privat, A. & Mallet, J. (1990b) Behavioural effect of engineered cells that synthesize L-DOPA or dopamine after grafting into the rat neostriatum, *Eur. J. Neurosci.* **2**, 116–119.

Horellou, P., Vigne, E., Castel, M.N., Barnéoud, P., Colin, P., Perricaudet, M., Delaère, P. & Mallet, J. (1994) Direct intracerebral gene transfer of an adenoviral vector expressing tyrosine hydroxylase in a rat model of Parkinson's disease. *NeuroReport* **6**, 49–53.

Jiao, S., Guerevich, V. & Wolff, J.A. (1993) Long-term correction of rat model of Parkinson's disease by gene therapy. *Nature* **362**, 450–453.

Kaplitt, M.G. *et al.* (1994) Long-term gene expression and phenotypic correction using adeno-associated virus vectors in the mammalian brain. *Nature Genet.* **8**, 148–153.

Le Gal La Salle, G., Robert J.J., Berrard, S., Ridoux V., Stratford-Perricaudet, L.D., Perricaudet M. & Mallet J. (1993) An adenovirus vector for gene transfer into neurons and glia in the brain. *Science* **259**, 988–990.

Levallois, C., Privat, A. & Mallet, J. (1994) Adenovirus insertion encoding the Lac Z gene in human nervous cells in primary dissociated cultures. *C.R. Acad. Sci. Paris, Life Sci.* **317**, 495–498.

Lin, O., Yoshimoto, Y., Collier, T.J., Frim, D.M., Breakfield, X.O. & Bohn, M.C. (1994) BDNF secreting astrocytes grafted into the striatum of a partially lesioned rat model of Parkinson's disease ameliorate amphetamine-induced rotational behavior. *Soc. Neurosci. Abstr.* 454.3

Lindvall, O., Brundin, P., Widner, H., Rehncrona, S., Gustavii, B., Frackowiak, R., Leenders, K.L., Sawle, G., Rothwell, J.C., Marsden, C.D. and Björklund, A. (1990) Grafts of fetal dopamine neurons survive and improve motor function in Parkinson's disease, *Science* **247**, 574–577.

Lindvall, O. *et al.* (1994) Evidence for long-term survival and function of dopaminergic grafts in progressive Parkinson's disease. *Ann. Neurol.* **35**, 172–180.

Lipton, H.L. & Dal Canto, M.C. (1976) *Science* **192**, 62–64.

Lokensgard, J.R., Bloom, D.C., Dobson, A.T., Feldman, L.T. (1994) Long-term promoter activity during herpes simplex virus latency. *J. Virol.* **68**, 7148–7158.

Lundberg, C., Horellou, P., Brundin, P., Wictorin, K., Kalén, P., Colin, P., Julien, J.F., Björklund, A. & Mallet, J. (1991) Transplantation of primary glial cells that produce DOPA after retroviral TH gene transfer in the rat model of Parkinson's disease. *Soc. Neurosci. Abstr.* 229.6.

Marsden, C.D. & Parkes, J.D. (1976) 'On–off' effects in patients with Parkinson's disease on chronic levodopa therapy. *Lancet* **1**, 292–296.

McFarland, H.F., Griffin, D.E. & Johnson, R.T. (1972) *J. Exp. Med.* **136**, 216–226.

Onifer, S.M., Whittemore, S.R. & Holets, V.R. (1993) Variable morphological differentiation of a raphe-derived neuronal cell line following transplantation into the adult rat CNS. *Exp. Neurol.* **122**, 130–142.

Palella, T.D., Hidaka, Y., Silverman, L.J., Levine, M., Glorioso, J. & Kelley, W.N. (1989) Expression of human HPRT mRNA in brains of mice infected with a recombinant herpes simplex virus. *Gene* **80**, 137–144.

Perlow, M.J., Freed, W.J., Hoffer, B.J., Seiger, Å., Olson, L. & Wyatt, R.J. (1979) Brain grafts reduce motor abnormalities produced by destruction of nigrostriatal dopamine system, *Science* **204**, 643–646.

Renfranz, P.J., Cunningham, M.G. & McKay, R.D.G. (1991) Region-specific differentiation of the hippocampal stem cell line HiB5 upon implantation into the developing mammalian brain. *Cell* **66**, 713–729.

Ridoux, V. *et al.* (1994) The use of adenovirus vectors for intracerebral grafting of transfected nervous cells. *NeuroReport* **5**, 801–804.

Sabaté, O., Horellou, P., Vigne, E., Colin, P., Perricaudet, M., Buc-Caron, M.H. & Mallet, J. (1995) Transplantation to the rat brain of human neural progenitors which were genetically modified using adenoviruses. *Nature Genet.* **9**, 256–260.

Snyder, E.Y. *et al.* (1992) Multipotent neural cell lines can engraft and participate in development of mouse cerebellum. *Cell* **68**, 33–51.

Stratford-Perricaudet, L.D., Levrero, M., Chase, J.F., Perricaudet, M. & Briand, P. (1990) Evaluation of the transfer and expression in mice of an enzyme encoding gene using a human adenovirus vector. *Hum. Gene Ther.* **1**, 241–256.

Tishon, A., Eddleston, M., de la Torre, J.C., Oldstone, M.B. (1993) *Virology* **197**, 463.

Townsen, J.J. & Baringer, J.R. (1979) *Lab. Invest.* **40**, 178.

Uchida, K., Ishii, A., Kaneda, N., Toya, S., Nagatsu, T. & Kohsaka, S. (1990) Tetrahydrobiopterin-dependent production of L-DOPA in NRK fibroblasts transfected with tyrosine hydroxylase cDNA: future use for intracerebral grafting. *Neurosci. Lett.* **109**, 282–286.

Ungerstedt, U. & Arbuthnott, G.W. (1970) Quantitative recording of rotational behavior in rats after 6-hydroxydopamine lesions of the nigrostriatal dopamine system, *Brain Res.* **24**, 485–493.

Vincent, N. *et al.* (1993) Long-term correction of mouse dystrophic degeneration by adenovirus-mediated transfer of a minidystrophin gene. *Nature Genet.* **5**,130–134.

Wolfe, J.D., Deshmane, S.L. & Fraser, N.W. (1992) Herpes virus vector gene transfer and expression of β-glucuronidase in the central nervous system of MPS VII mice. *Nature Genet.* **1**, 379–384.

Wolff, J.A., Fisher, L.J., Xu, L., Jinnah, H.A., Langlais, P.J., Iuvone, P.M., O'Malley, K.L., Rosenberg, M.B., Shimohama, S., Friedmann, T. & Gage, F.H. (1989) Grafting fibroblasts genetically modified to produce L-dopa in a rat model of Parkinson disease. *Proc. Natl. Acad. Sci. USA* **86**, 9011–9014.

Yang, Y., Nunes, F.A., Berensci, K., Göncczöl, E., Englehardt, J.F. & Wilson, J.M. (1994) Inactivation of E2a in recombinant adenoviruses improves the prospect for gene therapy in cystic fibrosis. *Nature Genet.* **7**, 362–369.

Zabner, J., Couture, L.A., Gregory, R.J., Graham, S.M., Smith, A.E. & Welsh, M.J. (1993) Adenovirus-mediated gene transfer transiently corrects the chloride transport defect in nasal epithelia of patients with cystic fibrosis. *Cell* **75**, 207–216.

CHAPTER 4

ADENOVIRUS-MEDIATED GENE THERAPY OF TUMORS IN THE CENTRAL NERVOUS SYSTEM

H. David Shine*† and Savio L. C. Woo†‡

Departments of Neurosurgery and Cell Biology,† and the Howard Hughes Medical Institute‡, Baylor College of Medicine, Houston, TX 77030, USA*

Table of Contents

4.1 Introduction

4.1.1 Brain and spinal cord tumors

Each year approximately 15 000 brain tumors and 4000 spinal cord tumors are diagnosed (Nelson *et al.*, 1993), and approximately 11 000 patients die from these tumors. Next to leukemia, brain tumors are the most common cause of death from cancer in children. In adults, 50% of primary brain tumors are gliomas, and 50% of all gliomas are glioblastoma multiforme (GBM; Zimmerman, 1969; Levin *et al.*, 1989). Patients with a GBM have an especially poor prognosis. Even with the best current therapy, including resection, irradiation and chemotherapy, the average life span following diagnosis of a GBM is less than 1 year (Salcman, 1980).

Brain and spinal cord tumors have properties that make them difficult to treat by conventional therapies. Unlike the many peripheral organs, destruction by tumor growth or resection of even a small portion of the brain or spinal cord can cause devastating consequences and, unlike muscle, skin or liver, central nervous system (CNS) tissue lacks the ability to regenerate following injury. Evidence of both the importance

Genetic Manipulation of the Nervous System
ISBN 0-12-437165-5

and fragility of the CNS is seen in the early evolution of the bony cranium and vertebral canal for protection. However, this defense also hinders surgical access and limits the extent that a tumor can grow before it causes significant symptoms or death. Diffuse infiltration of tumor cells into the surrounding tissue, limited penetration of chemotherapeutic agents through the blood–brain barrier, and the sensitivity of neural tissue to ionizing radiation further reduce the efficacy of conventional therapies.

4.1.2 Potential of gene therapy for treating CNS tumors

In addition to the difficulty of successfully treating many CNS tumors using conventional therapies, solid CNS tumors have particular attributes that make them potentially good candidates for gene therapy.

(1) CNS tumors rarely metastasize. Hence, only a single site would need to be treated.
(2) CNS tumors are often difficult and sometimes impossible to resect or irradiate without causing functional deficits. Gene therapy can be accurately administered using modern stereotactic neurosurgical techniques, potentially allowing treatment of hitherto untreatable tumors while at the same time preserving more function.
(3) The more malignant CNS tumors result in predictably short survival times, often measured in weeks. Success of gene therapy can be measured in days, making CNS tumors especially well suited to test the effects of gene therapy.
(4) CNS tumors can be accurately monitored with modern imaging techniques, making surgical exposure and exploration less important and allowing accurate appraisal of the efficacy of gene therapy.
(5) In the adult brain and spinal cord, CNS tumor cells are generally the only rapidly dividing cells. Some gene therapy strategies target dividing cells, making CNS tumors particularly good targets.
(6) CNS tumor cells are often highly malignant and aggressive which makes total surgical removal impossible. Immunocompetence may be critical for destroying any remaining incidental tumor cells after resection. Gene therapy is more likely to safeguard or bolster immunocompetence than traditional forms of cancer therapy.
(7) The bleak prognosis, despite optimal therapy, of the more malignant CNS tumors compels investigation of alternative therapeutic directions.

4.2 Viral-mediated gene therapy for CNS tumors

4.2.1 Overview

There are a number of gene therapy approaches using a variety of genes to treat cancer (reviewed by Culver and Blaese, 1994). These strategies include introducing genes that change the immunogenicity of the tumor cells, directing genes to immune cells to increase their antitumoral activity, blocking expression of oncogenes, inserting tumor suppressor genes, introducing protective genes so that the patient can withstand harsher chemotherapeutic agents, blocking mechanisms of metastasis, inserting toxic genes into tumor cells, and introducing genes that render tumor cells sensitive to toxic agents. The last of these strategies may be particularly effective for CNS tumors. Using the herpes simplex virus thymidine kinase gene (HSV-tk) to render the compound ganciclovir (GCV) cytotoxic to tumor cells, as proposed by Moolten (Moolten, 1986; Moolten and Wells, 1990), is particularly promising. Cells expressing HSV-tk phosphorylate the guanosine analog GCV (9-[1,3-dihydroxy-2-propoxymethyl)]guanine; Martin *et al.*, 1983) to the mono-phosphorylated form. Endogenous cellular kinases further phosphorylate the mono-phosphorylated GCV to the tri-phosphorylated form which inhibits DNA synthesis, thereby killing dividing cells. In addition, neighboring non-transduced dividing cells are often killed as well. This as-yet poorly understood phenomenon, called the 'bystander effect', makes HSV-tk particularly suited for the treatment of solid CNS tumors (Kolberg, 1992) since the tumoricidal activity of the treatment is not limited to only the transduced cells.

An important component of any gene therapy strategy is the method of introducing the gene into the cells. Several methods have been developed to transfer foreign genes into mammalian cells. These methods include direct transfer such as microinjection, particle bombardment and electroporation; chemical methods such as calcium phosphate transfection; fusion using liposomes; receptor-mediated endocytosis using DNA–protein complexes; and virally mediated transfer. Although all of these methods could potentially be used to genetically treat solid tumors, the approach that currently shows the most promise, primarily due to the superior transduction efficiency, is the viral transduction of tumor cells.

Retroviruses were the first viral vectors effectively used to perform genetic manipulations and are currently used in a variety of gene therapy applications. They are particularly advantageous for the treatment of genetic diseases where long-term expression is needed, and where not all target cells must be transduced. The fact that retroviruses target only dividing cells (Miller *et al.*, 1990) makes them especially applicable to the treatment of CNS tumors since mature CNS cells are not usually dividing. The capacity for retroviruses to transduce tumor cells in the rodent brain was demonstrated by Short *et al.* (1990) using a retroviral vector that contained the *lacZ* gene under control of the Moloney murine leukemia virus (MMLV) promoter (Cepko, 1988). However, efficient transduction required grafting retrovirus-producing mouse fibroblasts to the tumor site. Nevertheless, a retrovirus was the first virus vector used to demonstrate the efficacy of the HSV-tk/GCV system in treating experimental CNS tumors. In *ex vivo* experiments, cells of a glioma cell line were transduced in culture with a retrovirus vector carrying the HSV-tk gene and then implanted into the flanks of nude mice (Ezzeddine *et al.*, 1991). After several days when the implanted cells had formed tumors, GCV was administered. Subsequent inspection of these animals revealed that the tumors had regressed. In an *in vivo* experiment, tumors were established in the brains of rats and then cells producing a retrovirus vector carrying the HSV-tk gene were injected into tumors (Culver *et al.*, 1992; Ram *et al.*, 1993a). When the animals were treated with GCV, the tumors regressed in 23 of 30 animals. This effect was also demonstrated in experimental liver tumors (Caruso *et al.*, 1993). The success of *ex vivo* and *in vivo* experiments using a retrovirus vector to treat experimental tumors has provided justification for clinical trials to assess the efficacy and toxicity of retroviral-mediated treatment of brain tumors in humans (Oldfield *et al.*, 1993; Raffel *et al.*, 1994; Culver *et al.*, 1994).

While notable progress has been made using the retrovirus as a vector, it has some limitations. The relatively low titer (10^6 pfu ml^{-1}) that can be produced using current technology requires that mouse fibroblasts that produce the vector be grafted to the tumor bed. Presumably, in the HSV-tk/GCV system these producer cells will be killed in the same manner as the cells transduced with the virus but the long term consequences of the xenograft have not been determined.

Other potential viral vectors for gene therapy of CNS tumors include the herpes simplex virus, adenovirus and adenovirus-associated virus (AAV). Of these, the adenovirus shows particular promise and may most closely suit the requirements for gene therapy of CNS tumors.

4.2.2 Adenovirus vectors

Adenoviruses have relatively low toxicity in humans and they do not integrate with the host genome but rather reside in an episomal position. This is a benefit in acute treatments such as treatment of tumors. Wild-type adenoviruses do not have neurotoxic effects and are therefore good vectors for gene therapy in the nervous system. In addition, adenovirus vectors can be produced in high titers (approaching 10^{12} pfu ml^{-1}). They also transduce cells with a high efficiency and infect a broad range of cell types.

Viruses from the Adenoviridae family were first isolated from tonsils and adenoidal tissue of children in 1953 (Rowe *et al.*, 1953). There are six subgroups of human adenoviruses that encompass 41 serotypes (Horwitz, 1990). Serotypes 2 (Ad2) and 5 (Ad5) of the C subgroup are currently being developed for gene therapy (Kozarsky and Wilson, 1993). The adenovirus causes acute infections of the mucous membranes of the upper respiratory tract, eyes and lymph glands to produce symptoms similar to the common cold. Most individuals have been exposed to adenoviruses and have antibodies to them. Adenoviruses are non-enveloped, 65–80 nm wide icosahedrons. Each icosahedron is composed of subunits called capsomeres and has fibers projecting from each of its 12 vertices. The adenovirus genome is linear double-stranded DNA approximately 36 kb in size. Adenoviruses bind to and are internalized by target cells via as-yet unidentified receptors. They enter clathrin-coated pits that form endocytic vesicles called receptosomes. Once in the receptosome, the pH drops which alters the virion which, in turn, causes the vesicle to rupture. The virus is transported across the cytoplasm to the nuclear pores where the DNA is released into the nucleus where it resides separate from the cell's genome. Once positioned within the nucleus, adenoviral expression occurs roughly in two phases that generally reflect the expression of regulatory (early) and structural (late) proteins. Deleting the early genes makes the adenovirus replication defective.

Adenoviral vectors have been used in a variety of *in vitro* and *in vivo* studies and clinical applications of gene therapy. For example, adenoviral vectors have been used to transfer the cystic fibrosis transmembrane conductance receptor gene to lung epithelium (Rosenfeld *et al.*, 1992; Simon *et al.*, 1993), the dystrophin gene to muscle cells (Ragot *et al.*, 1993), the ornithine transcarbamylase gene to hepatocytes (Stratford-Perricaudet *et al.*, 1990), the α_1-antitrypsin gene to lung epithelial cells (Rosenfeld *et al.*, 1991) and to endothelial cells (Lemarchand *et al.*, 1992), and the *lacZ* gene to hepatocytes (Jaffe *et al.*, 1992; Li *et al.*, 1993) and to neural cells (Akli *et al.*, 1993; Bajocchi *et al.*, 1993; Davidson *et al.*, 1993; Le Gal La Salle *et al.*, 1993). Using adenoviral vectors carrying the *lacZ* gene under control of either the human cytomegalovirus promoter (AdLacZ) or the Rous sarcoma virus promoter (Av1LacZ4; Genetic Therapy, Inc., Gaithersburg, MD), two laboratories demonstrated that experimental CNS tumors could be efficiently transduced *in vivo*. In the first study, tumors generated by injecting C_6 astrocytoma cells into rat brains were efficiently transduced by the AdLacZ. Approximately 25–30% of the cells were positive for β-galactosidase activity 4 days after virus injection (Badie *et al.*, 1994). In the second study, experimental brain and leptomeningeal tumors were generated by injecting 9L gliosarcoma cells into the cerebral hemisphere or into the subarachnoid space, respectively. Injection of Av1LacZ4 into the cerebral tumors or the subarachnoid space efficiently transduced the tumor cells (Viola *et al.*, 1995).

The success of experiments using retroviral vectors carrying HSV-tk to render tumors vulnerable to GCV, coupled with the higher transduction efficiency and low neurotoxicity observed in experiments using an adenovirus vector, suggested to us that an adenovirus could be used effectively to transduce experimental CNS tumors with HSV-tk. Additionally, the high transduction efficiency characteristic of adenovirus vectors suggested that they may have efficacy superior to the retrovirus-based vectors. However, unlike retrovirus vectors that only infect dividing cells, adenovirus vectors have the capacity to infect quiescent cells of the CNS. This raised the concern that adenoviral therapy might not be specific for tumor cells. Nevertheless, because adenoviral vectors have shown little or no tendency to migrate from the injection site and because phosphorylated GCV is toxic only to rapidly dividing cells, we surmised that the lack of specificity for dividing tumor cells would not undermine the efficacy of the technique. To test this idea, an adenovirus lacking the E1 region and carrying the HSV-tk gene driven by the Rous sarcoma virus (RSV) long terminal repeat (LTR) was constructed (ADV/RSV-tk; Figure 1; Chen *et al.*, 1994). The resulting adenovirus is replication defective and expresses the HSV-tk gene (Figure 2a). Transduction with ADV/RSV-tk rendered cultured gliomas susceptible to the cytotoxic effect of GCV (Figure 2b; Chen *et al.*, 1994). These results suggested that the treatment would be effective *in vivo* as well.

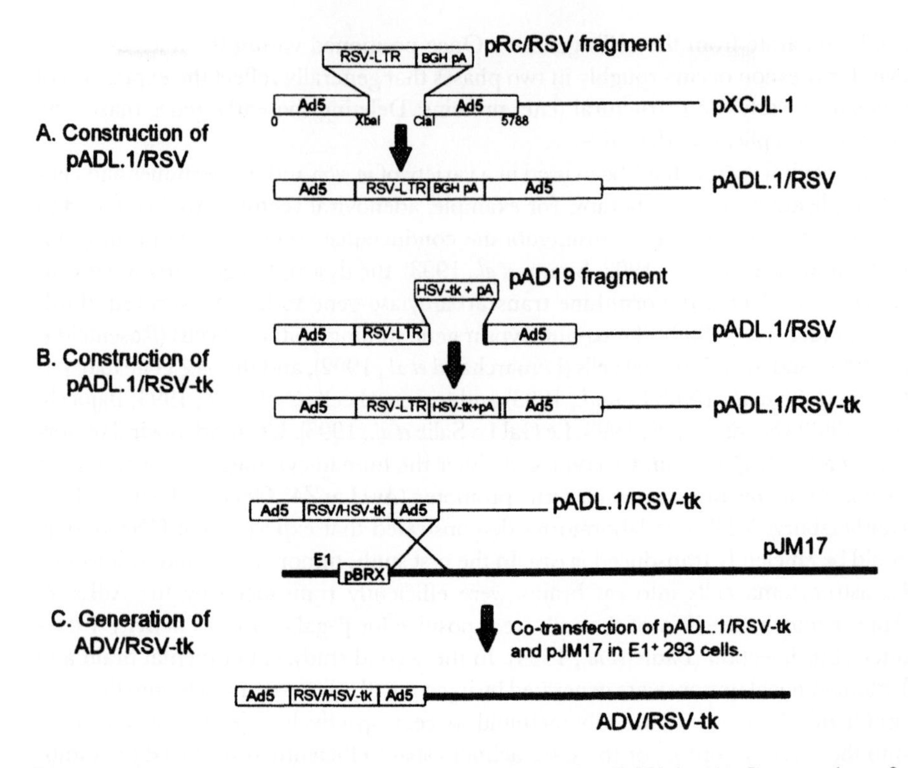

Figure 1 Schematic diagram of the construction of ADV/RSV-tk. (A) Construction of pADL.1/RSV. The 783 bp Hha I fragment containing the RSV-LTR and the bovine growth hormone 3′ polyadenylation signal sequence (BGH pA) was excised from pRc/RSV (Invitrogen) and ligated into the Xba I and Cla I sites of the pBR322-based vector pXCJL.1 (Frank Graham, McMaster University) to produce pADL.1/RSV (Fang *et al.*, 1994). (B) Construction of pADL.1/RSV-tk. The 2.8 kb Bgl II/Bam HI fragment containing the HSV-tk gene and its endogenous poly (A) tail was excised from pAD19 (Allen Bradley, Baylor College of Medicine) and inserted into pADL.1/RSV, destroying the BGH pA site. (C) Generation of ADV/RSV-tk. pADL.1/RSV-tk was co-transfected with pJM17, a plasmid containing a replication-defective copy of the adenovirus genome (McGory *et al.*, 1988), into the 293 transformed human kidney cell line by calcium phosphate precipitation. The 293 cell line contains and expresses the left end of the Ad5 genome so that the E1-deleted virus can replicate and form virions (Graham *et al.*, 1977; Graham and Prevec, 1991).

4.3 Experimental models of ADV/RSV-tk and ganciclovir therapy for CNS tumors

4.3.1 Overview

We tested the efficacy of ADV/RSV-tk in rendering tumor cells susceptible to GCV toxicity in four rodent models of CNS tumor using three tumor cell lines. In each model, cultured tumor cells were injected into the CNS using stereotactic methods.

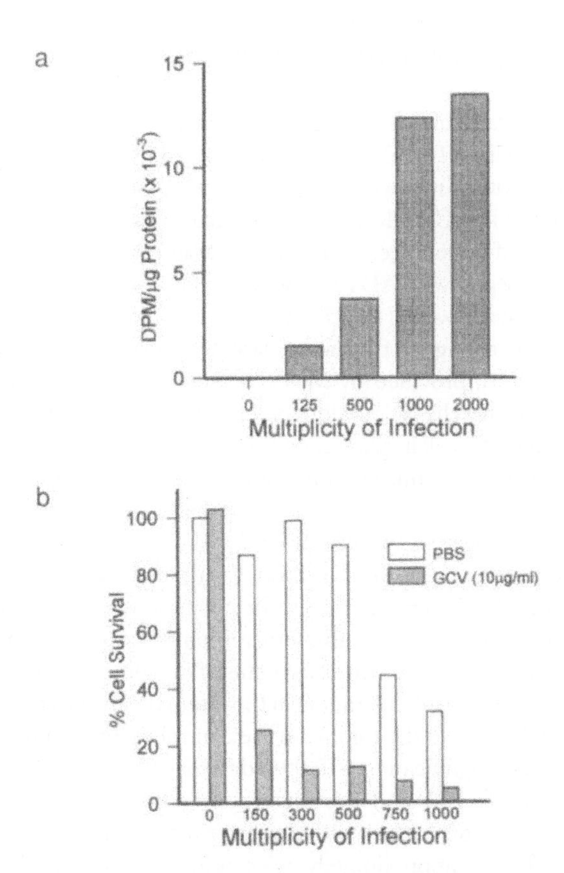

Figure 2 Transduction of C_6 glioma cells with ADV/RSV-tk *in vitro.* (a) Expression of HSV-tk in C_6 cells after ADV/RSV-tk transduction. Cells (5×10^6) were plated on 1.5 cm plates transduced with ADV/RSV-tk at the multiplicity of infection (MOI) indicated. Two days later a protein extract was prepared from the cells by freeze-thawing and thymidine kinase activity was determined by phosphorylation of [^{3}H] acyclovir (Fyfe *et al.*, 1978). (b) GCV susceptibility of C_6 cells transduced with ADV/RSV-tk. Two plates of C_6 cells were incubated for 6 h with ADV/RSV-tk at six MOI values varying from 0 to 1000. After transduction, the cells were treated with either phosphate buffered saline (PBS) or GCV (10 µg ml^{-1}). The surviving cells were counted 68 h later and the percentages of cell survival were calculated by comparing cell counts with those of cultures of non-transduced cells treated with PBS.

After 7 or 8 days, when tumors had formed at the sites of implantation ADV/RSV-tk was injected into the tumors and on the following day the animals began 6 days of GCV treatment. The effects of the gene therapy were determined by measuring the size of the tumor at 16–20 days after tumor cell implantation, or by noting the length of time that the treated animals survived compared to controls. The results of each experiment demonstrated that treatment with ADV/RSV-tk and GCV has a powerful tumoricidal effect.

4.3.2 C_6 brain tumors in athymic nude mice

In our first experiments, brain tumors were generated in athymic nude mice (Chen *et al.*, 1994) using the C_6 cell line. The C_6 cell line was derived from a Sprague–Dawley rat that had been treated with *N*-methylnitrosourea and resembles astrocytoma tumor cells (Benda *et al.*, 1968). When C_6 cells were implanted using a stereotactic technique into the caudate nucleus of nude mice, tumors grew rapidly. Eight days after C_6 cell injection, treatment was begun with ADV/RSV-tk and GCV or one of three control treatments. Control treatments were: (1) ADV/RSV-tk and phosphate-buffered saline (PBS); (2) a control virus carrying the β-galactosidase (βgal) gene under control of the RSV-LTR (ADV/RSV-βgal; provided by M. Perricaudet, Institut Gustave Roussy, Centre National de la Recherche Scientifique); and (3) GCV, and ADV/RSV-βgal and PBS. Twenty days after C_6 cell implantation the control animals had large tumors that filled the area of the caudate nucleus, and had invaded the cerebral cortex, and spread through the bur hole in the skull into the subgaleal space (Figure 3a). In contrast, the animals that received ADV/RSV-tk and GCV had either no detectable tumors or small tumors (Figure 3b). Brain sections from mice that had been treated with ADV/RSV-tk and GCV were examined for deleterious effects. No necrosis, demyelination, loss of neurons, or inflammatory response was evident beyond the original tumor site. Computer-assisted measurements of the tumors showed a 23-fold difference between the mean size of the tumors in the group that received ADV/RSV-tk and GCV and the mean size of the tumors in the control groups (Figure 4). There was a slight but statistically significant ($P < 0.001$) difference in the tumor size of the control animals that received GCV and those that received PBS, suggesting that GCV alone may have an inhibitory effect on tumor growth. Animals treated with ADV/RSV-tk and GCV lived approximately twice as long as the control animals or untreated animals. Ultimately, however, the animals succumbed to cortical and scalp tumors secondary to the primary tumors.

4.3.3 Primary 9L brain tumors in Fischer rats

Although the ADV/RSV-tk and GCV treatment had a substantial tumoricidal effect on the C_6 tumors, athymic nude mice lack a competent immune system, a factor that may play a role in complete tumor ablation following HSV-tk and GCV treatment. To address the question of the involvement of the host's immune system, the efficacy of ADV/RSV-tk and GCV treatment was tested in a syngeneic rat tumor model using the 9L tumor cell line (Perez-Cruet *et al.*, 1994). The 9L cell line was cloned from a tumor induced in a CD Fischer rat by repetitive injections of *N*-methylnitrosourea (Jänisch and Schreiber, 1976; Weizsaecker *et al.*, 1981). Tumors generated with 9L cells are described as mixed glioblastoma multiforme and sarcomas or gliosarcomas (Barker *et al.*, 1973), and have been used extensively to test various treatments, including gene therapy, for CNS tumors (Culver *et al.*, 1992; Barba *et al.*, 1993, 1994; Ram *et al.*, 1993a, b, 1994; Boviatsis *et al.*, 1994; Tapscott *et al.*, 1994; Viola *et al.*, 1995). We stereotactically injected 9L cells into the caudate nucleus of Fischer rats. Eight days

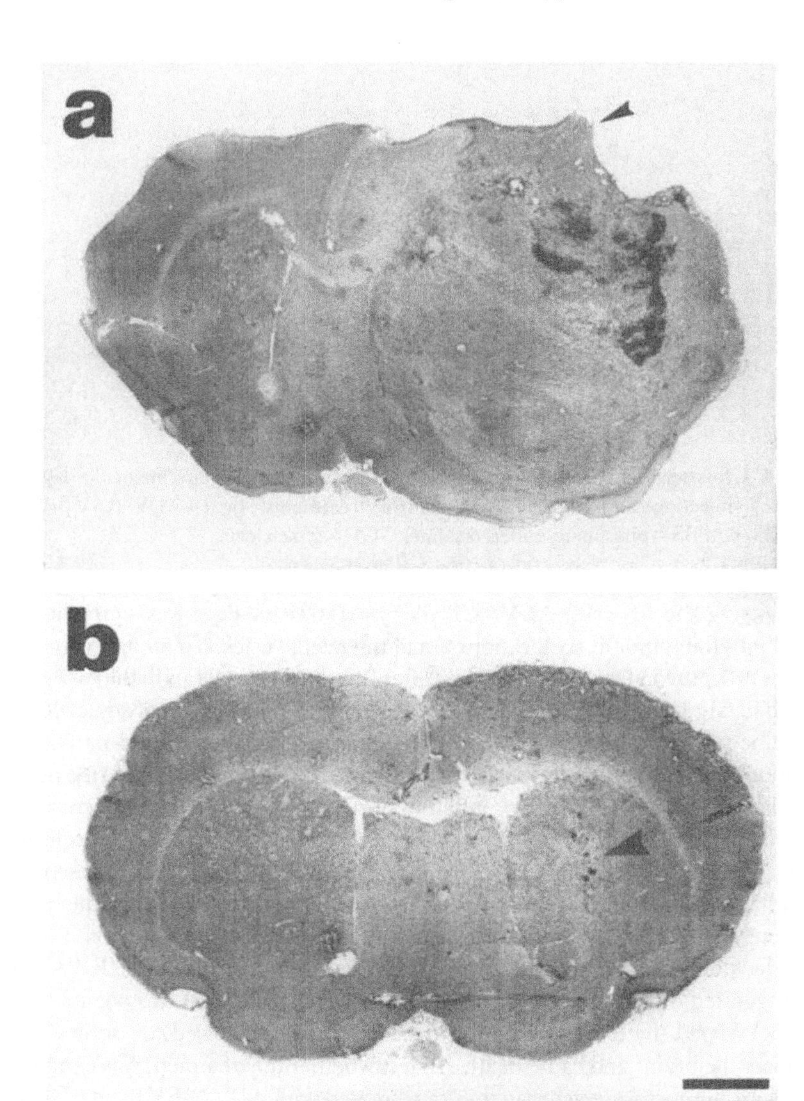

Figure 3 Coronal sections of brains from nude mice 20 days after C_6 tumor cell injection and experimental or control treatments. (a) Control animal treated with ADV/RSV-tk and saline. A large tumor is visible that occupies nearly the whole right hemisphere, obliterates the lateral ventricle and forces the midline to the left. The black substance is hemorrhage. An arrow marks where the tumor extended through the bur hole and invaded the subgaleal space. (b) Experimental animal with a C_6 brain tumor treated with ADV/RSV-tk and GCV. An arrow-head marks the former tumor site in which carbon grains are visible. Hematoxylin and eosin. Bar = 1 mm.

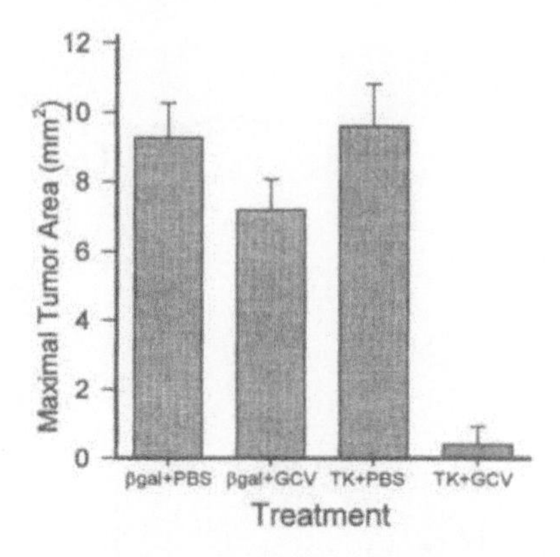

Figure 4 Cross-sectional areas of tumors in the C_6 /nude mouse experiment 20 days after tumor cell injection and experimental control treatments. βgal = ADV/RSV-βgal; tk-ADV/RSV-tk; PBS = phosphate-buffered saline; GCV = ganciclovir.

after injecting the 9L cells, ADV/RSV-tk or ADV/RSV-βgal was stereotactically injected into the tumor at six locations along the needle track. Treatment with GCV or PBS was begun on Day 9. Twenty days after 9L cell injection, rats in the two control groups had large tumors (Figures 5a and 6) that caused compression of adjacent brain tissue. The tumors were generally well circumscribed with some focal perivascular infiltration into adjacent tissue, and were characterized by hypercellurarity, nuclear pleomorphism and mitoses. No inflammatory cell infiltration or necrosis was observed. In the animals treated with ADV/RSV-tk and GCV, no tumor cells were visible (Figures 5b and 6). Microscopic inspection revealed the presence of macrophages, lymphocytes, neutrophils, necrosis and hemorrhage at the former tumor sites, consistent with an inflammatory reaction to tumor destruction.

The lack of tumor cells in the brains of the rats that received ADV/RSV-tk and GCV suggested that the tumors were totally ablated. However, it was possible that a few cells escaped the treatment and, if the animals were allowed to survive, would repopulate the brains and cause death. To test whether the treatment was completely effective, 9L tumors were generated in rats, treated with ADV/RSV-tk and GCV or ADV/RSV-βgal and GCV and allowed to survive. The animals that received the control treatment developed massive tumors and died within 22 days (Figure 7). Some animals treated with ADV/RSV-tk died within 150 days after tumor cell injection. Upon inspection of their brains, no tumors were observed and the cause of death was attributed to peritonitis as a consequence of the repetitive intraperitoneal injections of GCV. Those animals that survived the GCV treatment now have lived over 400 days. Because of the rapid growth of 9L cells it is highly unlikely that any residual 9L cells exist in these survivors or if they do, their biological behavior is modified so that they cannot grow into new tumors.

To determine whether this treatment could be extended to metastatic tumors of the CNS, we developed a rat model of mammary tumors in the brain using a

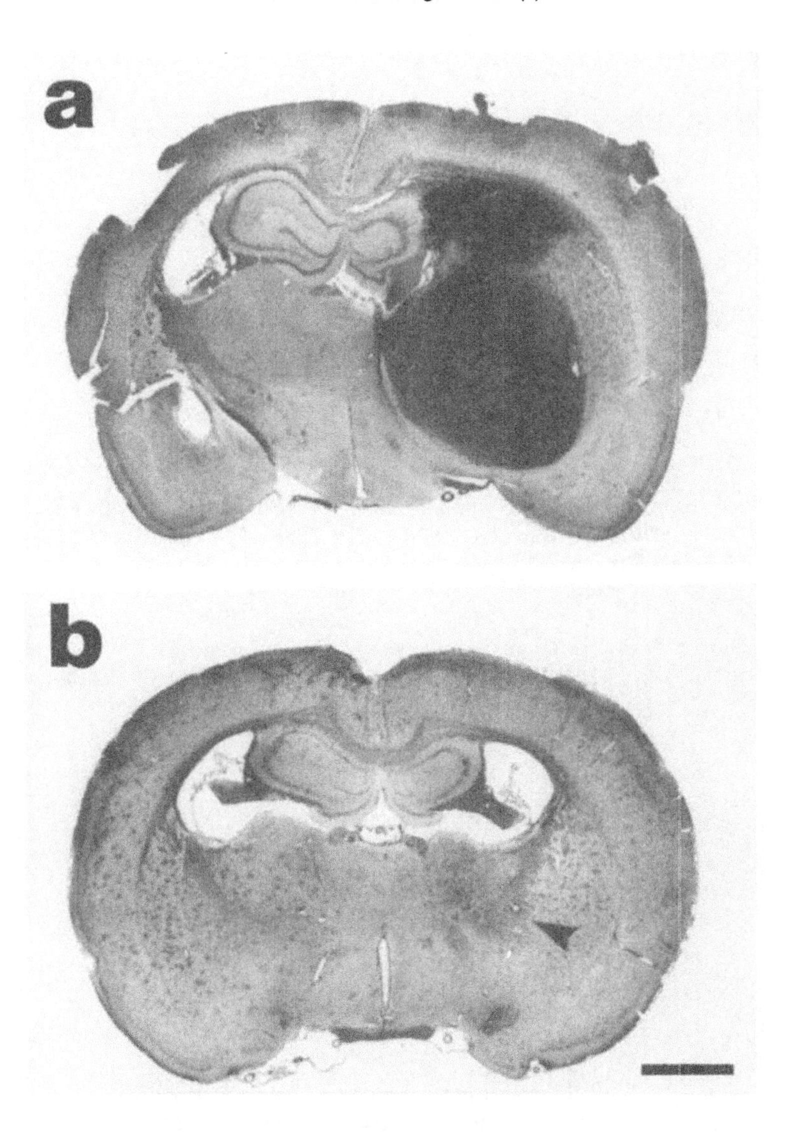

Figure 5 Coronal sections of brains from Fischer rats that had received 9L tumor cell injections and experimental or control treatments. Animals were sacrificed 20 days after tumor cell injection. (a) Control animal treated with ADV/RSV-βgal and GCV. Note the large tumor that compresses the adjacent brain and deflects the midline. (b) Experimental animal treated with ADV/RSV-tk and GCV. An arrowhead marks the former tumor site. Hematoxylin and eosin. Bar = 2 mm.

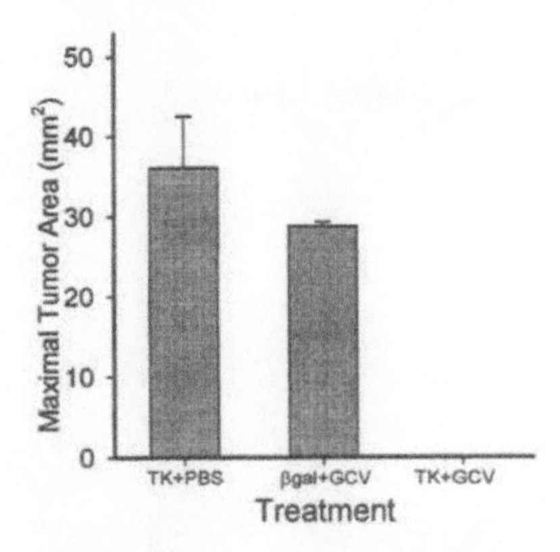

Figure 6 Cross-sectional areas of brain tumors in the 9L/Fischer rat experiments treated with one of three treatments: ADV/RSV-tk and saline (TK+PBS), ADV/RSV-βgal and GCV (βgal+GCV), or ADV/RSV-tk and GCV (TK+GCV). The rats were sacrificed 20 days after tumor cell injection.

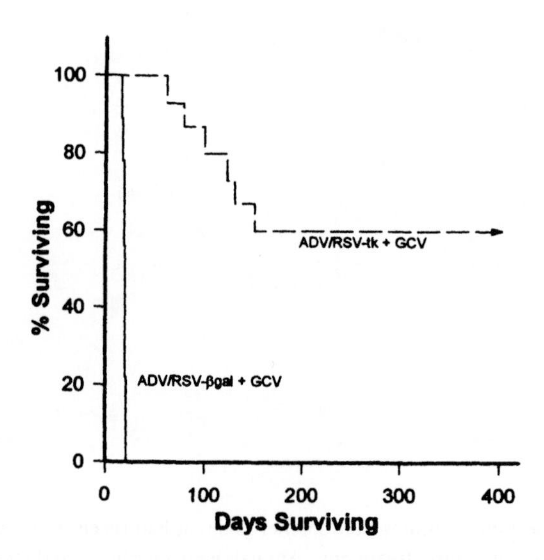

Figure 7 Long-term survival of rats with 9L tumors and treated with either ADV/RSV-βgal and GCV or ADV/RSV-tk and GCV.

mammary adenocarcinoma cell line (13762 MAT B III; Çolak *et al.*, 1995a). ADV/RSV-tk and GCV treatment ablated tumors and increased survival in these animals similar to that observed in the 9L primary brain tumor model.

4.3.4 Spinal cord

Our experiments and those of others usually test tumors generated in the brains of host animals. However, intramedullary glial tumors of the human spinal cord may also be amenable to gene therapy. Patients with spinal gliomas generally have a poor prognosis for functional recovery even when intensive surgical and radiation therapy effectively controls the tumor because of the relatively small size and functional complexity of the spinal cord (Boring *et al.*, 1994; Cristante *et al.*, 1994). Spinal cord gliomas also have a much smaller and better circumscribed volume than do cerebral gliomas, characteristics potentially advantageous for gene therapy. To investigate the applicability of adenovirus-mediated gene therapy for spinal cord tumors, we developed a rat model using the 9L tumor cell line and tested ADV/RSV-tk and GCV treatment in this model (Çolak *et al.*, 1995b). In this study, therapeutic efficacy was assessed by measuring neurological function as well as tumor size and length of survival. As in the three models described earlier, the ADV/RSV-tk and GCV treatment inhibited tumor growth, prevented neurologic deterioration and substantially increased survival.

4.4 Discussion

4.4.1 Efficacy of ADV/RSV-tk and GCV for CNS tumors

The high degree of tumoricidal efficacy that we observed with HSV-tk and GCV treatment in each of these four models of CNS tumors and that others have reported using retroviral vectors is noteworthy when the degree of transduction is taken into account. To date, all reports of viral-mediated transduction of tumor cells *in vivo* describe transduction efficiencies between 5% and 50% whether retroviral or adenoviral vectors are used. Nevertheless, results in these experiments have regularly indicated that non-transduced tumor cells also succumb to the treatment. This poorly understood phenomenon has been termed the 'bystander effect' (Kolberg, 1992, 1994). *In vitro* mixing experiments using a variety of cell lines have shown that as little as 10% of the cells need to be transformed with HSV-tk in order to achieve killing of all of the cells. Several mechanisms have been proposed for the bystander effect. Freeman *et al.* (1993) demonstrated that culturing Kirsten virus-transformed sarcoma tumor cells (KBALB) with KBALB cells that carried the HSV-tk gene (KBALB-STK) at ratios of 9:1, respectively, resulted in 100% of the cells killed when they were treated with GCV. If the KBALB and KBALB-STK cells were separated by a 0.4 μm filter the effect was not observed. The authors surmised that apoptotic vesicles were generated by the dying cells and then phagocytized by the neighboring untransduced cells. Bi *et al.* (1993) demonstrated that when human fibrosarcoma cells that expressed HSV-tk were co-cultured with the same cell line that did not express HSV-tk both populations succumbed to GCV treatment when the two cell lines were cultured at densities that permitted cell contact. If the cells were cultured at low densities so that the majority of the cells were not in contact, the cells that did not express HSV-tk were spared. They interpreted these data to suggest that low molecular weight molecules were transported from the HSV-tk-expressing cells to the non-expressing cells through gap junctions. Further support of this hypothesis was demonstrated by co-culturing transduced and untransduced cells together and showing that [^{3}H]ganciclovir was transferred between, and localized in, the nucleus of untransduced cells that touched transduced cells but not in those that did not touch transduced cells. *In vivo*, additional mechanisms may play a role in tumor ablation. Using Doppler ultrasound imaging to measure the degree of tumor vascularity and necrosis, Ram *et al.* (1994) showed that transduction of subcutaneous 9L tumors in Fischer rats with HSV-tk and treatment with GCV resulted in a significant reduction in tumor vasculature and in increased necrosis that was not seen in untransduced tumors. They suggest that vascular endothelial cells in the tumors are transduced with HSV-tk and destroyed by GCV treatment. The resultant reduction in blood supply, they surmise, produces tumor ischemia and tumor cell death. Finally, several laboratories have presented data that suggest that an immune response is raised to the dying tumor cells that is then targeted to the non-transduced tumor cells (Caruso *et al.*, 1993; Barba *et al.*, 1993, 1994; Perez-Cruet *et al.*, 1994; Vile *et al.*, 1994). Further investigations into the mechanisms underlying the bystander effect will be instrumental in developing more effective strategies to treat tumors using HSV-tk and GCV.

4.4.2 Toxicity

The efficacy of adenoviral HSV-tk transduction followed by GCV administration demonstrated in several rodent models of CNS tumors suggests that this treatment may be effective for human CNS tumors as well. However, before clinical trials can be considered, the effects of adenoviral vectors on the normal CNS must be determined. To address this, we have tested the toxicity of ADV/RSV-tk with and without GCV in the baboon (*Papio cynocephalus*) and the cotton rat (*Sigmodon hispidus*). We tested the toxic effects of ADV/RSV-tk and GCV in normal baboon brain because the relative proportion of astrocytes in the primate CNS is greater than in rodents. Astroctyes have the capacity to re-enter the cell cycle and divide, which would make them vulnerable to the cytotoxic effects of GCV. Widespread destruction of normal astrocytes could result in unacceptable adverse effects of this treatment in humans. Preliminary results of these tests indicate that the ADV/RSV-tk vector does have cytotoxic effects on normal cells in the CNS, the cytopathic effect is virus dose-dependent, and GCV accentuates the toxicity (Goodman *et al.*, 1995). However, the toxic effects appear to be localized to the site of injection which in an actual clinical case would contain tumor tissue and are of the sort that could be clinically managed.

The cotton rat is permissible to infections with wild-type adenovirus (Ginsberg *et al.*, 1991). We have tested ADV/RSV-tk toxicity in the brains of cotton rats and preliminary results indicate that intracerebral injections do not raise excessive cytopathic reactions at the site of injection although the ependyma and choroid plexus become inflamed. These reactions also could be managed clinically.

4.4.3 Human trials

The results of our efficacy and toxicity experiments indicate that the use of ADV/RSV-tk and GCV could be effective against CNS tumors in humans with acceptable toxicity. Unfortunately, there are no other useful animal models to test further this treatment strategy. Hence, the next rational step is a limited trial with patients with highly aggressive CNS tumors that have not been controlled by conventional therapies. Recognizing this we have begun the approval process that is required before clinical trials can begin. In the US this process includes approval by the Recombinant DNA Advisory Committee of the National Institutes of Health and the US Food and Drug Administration. Even though human trials are expected to begin soon, it must be emphasized that these are only the first forays into this area. Even if the first trials show promise, many refinements will undoubtedly be made before this therapy becomes a standard therapeutic method.

4.4.4 Future directions

One major hurdle in carrying this therapy to clinical application is modifying it to treat large tumors. Human tumors typically have a mass that is several orders of magnitude larger than the experimental rodent tumors that we and others have used in experimental trials. Simply increasing the viral dose to match the larger tumor size would be difficult using today's vectors. For instance, based upon our experience in baboon toxicity experiments, injecting a dose of ADV/RSV-tk into a human brain tumor at a dose equivalent to that injected into a rodent tumor based on the relative tumor size could have adverse consequences to the patient. A single site intratumoral injection of virus at doses that are not unacceptably toxic may not have sufficient tumoricidal effect since we and others have observed that adenovirus does not diffuse very far from the injection site. It is unlikely that the bystander effect will be sufficient in these instances to eradicate the tumor. Thus a means to disperse more widely the virus throughout the tumor bed must be devised.

Another area of treatment development must be in the area of vector improvement. Today's first-generation vectors, while effective, may have cytopathic effects that will make them less desirable for use in humans at the higher doses. It is probable that second-generation vectors will be designed with less toxicity by further engineering the vector's backbone sequence and increasing the expression of the recombinant gene so that fewer virions are needed for equivalent gene expression. Specificity can also be improved by designing vectors that employ tissue-specific promoters that target only the tumor cells. Finally, the success of the HSV-tk and

GCV system should stimulate the development of alternative systems with greater tumoricidal effects and less toxicity to normal tissues.

Acknowledgments

We thank Dr Winifred J. Hamilton for editorial assistance. S.L.C.W. is an investigator of the Howard Hughes Medical Institute. A portion of this research was supported by a grant from the Texas Higher Education Coordinating Board Advanced Technology Program.

References

Akli, S., Caillaud, C., Vigne, E., Stratford-Perricaudet, L.D., Poenaru, L., Perricaudet, M., Kahn, A. & Peschanski, M.R. (1993) Transfer of a foreign gene into the brain using adenovirus vectors. *Nature Genet.* **3**, 224–228.

Badie, B., Hunt, K., Economou, J.S. & Black, K.L. (1994) Stereotactic delivery of a recombinant adenovirus into a C6 glioma cell line in a rat brain tumor model. *Neurosurgery* **35**, 910–916.

Bajocchi, G., Feldman, S.H., Crystal, R.G. & Matrangeli, A. (1993) Direct *in vivo* gene transfer to ependymal cells in the central nervous system using recombinant adenovirus vectors. *Nature Genet.* **3**, 229–234.

Barba, D., Hardin, J., Jasodrhara, R. & Gage, F.H. (1993) Thymidine kinase-mediated killing of rat brain tumors. *J. Neurosurg.* **79**, 729–735.

Barba, D., Hardin, J., Sadelain, M. & Gage, F.H. (1994) Development of anti-tumor immunity following thymidine kinase-mediated killing of experimental brain tumors. *Proc. Natl. Acad. Sci. USA* **91**, 4348–4352.

Barker, M., Hoshino, T., Gurcay, O., Wilson, C.B., Nielsen, S.L., Downie, R. & Eliason, J. (1973) Development of an animal brain tumor model and its response to therapy with 1,3-bis (2-chloroethyl)-1-nitrosourea. *Cancer Res.* **33**, 976–986.

Benda, P., Lightbody, J., Sato, G., Levine, L., Sweet, W. (1968) Differentiated rat glial cell strain in culture. *Science* **161**, 370–371.

Bi, W.L., Parysek, L.M., Warnick, R. & Stambrook, P.J. (1993) *In vitro* evidence that metabolic cooperation is responsible for the bystander effect observed with HSV-tk retroviral gene therapy. *Hum. Gene Ther.* **4**, 725–731.

Boring, C.C., Squires, T.S., Tong, T. & Montgomery, S. (1994) Cancer Statistics, 1994. *Cancer J. Clin.* **44**, 7–26.

Boviatsis, E., Chase, M., Wei, M.X., Tamiya, T., Hurford, Jr., R.K., Kowall, N.W., Tepper, R.I., Breakefield, X.O. & Chiocca, E.A. (1994) Gene transfer into experimental brain tumors mediated by adenovirus, herpes simplex virus, and retrovirus vectors. *Hum. Gene Ther.* **5**, 183–191.

Caruso, M., Panis, Y., Gagandeep, S., Houssin, D., Salzmann, J.-L. & Klatzmann, D. (1993) Regression of established macroscopic liver metastases after in situ transduction of a suicide gene. *Proc. Natl. Acad. Sci. USA* **90**, 7024–7028.

Cepko, C. (1988) Retrovirus vectors and their applications in neurobiology. *Neuron* **1**, 345–353.

Chen, S.-H., Shine, H.D., Goodman, J.C., Grossman, R.G. & Woo, S.L.C. (1994) Gene therapy for brain tumors: Regression of experimental gliomas by adenovirus-mediated gene transfer *in vivo*. *Proc. Natl. Acad. Sci. USA* **91**, 3054–3057.

Çolak, A., Goodman, J.C., Chen, S.-H., Woo, S.L.C., Grossman, R.G. & Shine, H.D. (1995a) Adenovirus-mediated gene therapy in an expermental model of breast cancer metastatic to the brain. *Hum. Gene Ther.* **6**, 1317–1322.

Çolak, A., Goodman, J.C., Chen, S.-H., Woo, S.L.C., Grossman, R.G. & Shine, H.D. (1995b) Adenovirus-mediated gene therapy for experimental spinal cord tumors: Tumoricidal efficacy and functional outcome. *Brain Res.* **691**, 76–82.

Cristante, L. & Herrmann, H.-D. (1994) Surgical management of intramedullary spinal cord tumors: functional outcome and sources of morbidity. *Neurosurgery* **35**, 69–74.

Culver, K.W. & Blaese, R.M. (1994) Gene therapy for cancer. *Trends Genet.* **10**, 174–178.

Culver, K.W., Ram, Z., Walbridge, S., Ishii, H., Oldfield, E.H. & Blaese, R.M. (1992) *In vivo* gene transfer with retroviral vector-producer cells for treatment of experimental brain tumors. *Science* **256**, 1550–1552.

Culver, K.W., Van Gilder, J., Link, C.J., Carlstrom, T., Buroker, T., Yuh, W., Koch, K., Schabold, K., Doornbas, S., Wetjen, B. & Blaese, R.M. (1994) Gene therapy for the treatment of malignant brain tumors with *in vivo* tumor transduction with the herpes simplex thymidine kinase gene/ganciclovir system. *Hum. Gene Ther.* **5**, 343–379.

Davidson, B.L., Allen, E.D., Kozarsky, K.F., Wilson, J.M. & Roessler, B.J. (1993) A model system for *in vivo* gene transfer into the central nervous system using an adenoviral vector. *Nature Genet.* **3**, 219–222.

Ezzeddine, Z.D., Martuza, R.L., Platika, D., Short, M.P., Malick, A., Choi, B. & Breakefield, X.O. (1991) Selective killing of glioma cells in culture and *in vivo* by retrovirus transfer of the herpes simplex virus thymidine kinase gene. *New Biol.* **3**, 608–614.

Fang, B., Eisensmith, R.C., Li, X.H.C., Finegold, M.J., Shedlovsky, A., Dove, W. & Woo, S.L.C. (1994) Gene therapy for phenylketonuria: phenotypic correction in a genetically deficient mouse model by adenovirus-mediated hepatic gene transfer. *Gene Ther.* **1**, 247–254.

Freeman, S.M., Abboud, C.N., Whartenby, K.A., Packman, C.H., Koeplin, D.S., Moolten, F.L. & Abraham, G.N. (1993) The 'bystander effect': tumor regression when a fraction of the tumor mass is genetically modified. *Cancer Res.* **53**, 5274–5283.

Fyfe, J.A., Keller, P.M., Furman, P.A., Miller, R.L. & Elion, G.B. (1978) Thymidine kinase from herpes simplex virus phosphorylates the new antiviral compound 9-(2-hydroxyethoxymethyl) guanine. *J. Biol. Chem.* **253**, 8721–8727.

Ginsberg, H.S., Moldawer, L.L., Sehgal, P.B., Redington, M., Kilian, P.L., Chanock, R.M. & Prince, G.A. (1991) A mouse model for investigating the molecular pathogenesis of adenovirus pneumonia. *Proc. Natl. Acad. Sci. USA* **88**, 1651–1655.

Goodman, J.C., Trask, T.W., Chen, S.-H., Woo, S.L.C., Grossman, R.G., Carey, K.D., Hubbard, G., Carrier, D.A., Rajagopalan, S., Aguilar-Cordova, C.E. & Shine, H.D. (1995) Adenoviral-mediated thymidine kinase gene transfer into the primate brain followed by systemic ganciclovir: Pathologic, radiologic and molecular studies. *Hum. Gene Ther.* (in press).

Graham, F. & Prevec, L. (1991) Manipulation of adenovirus vectors. In: *Methods in Molecular Biology: Gene transfer and Expression Protocols* (ed. Murray, E.J.), Vol. 7, pp. 109–128. Clifton, NJ, Humana Press.

Graham, F.L., Smiley, J., Russell, W.C. & Nairn, R. (1977) Characteristics of a human cell line transformed by DNA from human adenovirus type 5. *J. Gen. Virol.* **36**, 59–74.

Horwitz, M.S. (1990) Adenoviridae and their replication. In: *Virology* (eds Fields, B.N. & Knipe, D.M.), pp. 1679–1721. New York, Raven Press.

Jaffe, H.A., Daniel, C., Longenecker, G., Metzger, M., Setoguchi, Y., Rosenfeld, M.A., Gant, T.W., Thorgeirsson, S.S., Stratford-Perricaudet, L.D., Perricaudet, M., Pavirani, A., Lecocq, J.-P. & Crystal, R.G. (1992) Adenovirus-mediated *in vivo* gene transfer and expression in normal rat liver. *Nature Genet.* **1**, 372–378.

Jänisch, W. & Schreiber, D. (1976) *Experimental Tumors of the Central Nervous System* (English version eds Bigner, D.D. & Swenberg, D.). Kalamazoo, Upjohn.
Kolberg, R. (1992) Gene therapists test puzzling 'bystander effect.' *J. NIH Res.* **4**, 68–74.
Kolberg, R. (1994) The bystander effect in gene therapy: great, but how does it work? *J. NIH Res.* **6**, 62–64.
Kozarsky, K.F. & Wilson, J.M. (1993) Gene therapy: adenovirus vectors. *Curr. Opin. Gen. Devel.* **3**, 499–503.
Le Gal La Salle, G., Robert, J.J., Berrard, S., Ridoux, V., Stratford-Perricaudet, L.D., Perricaudet, M., Mallet, J. (1993) An adenovirus vector for gene transfer into neurons and glia in the brain. *Science* **259**, 988–990.
Lemarchand, P., Jaffe, H.A., Danel, C., Cid, M.C., Kleinman, H.K., Stratford-Perricaudet, L.D., Perricaudet, M., Pavirani, A., Lecocq, J.-P., and Crystal, R.G. (1992) Adenovirus-mediated transfer of a recombinant human α_1-antitrypsin cDNA to human endothelial cells. *Proc. Natl. Acad. Sci. USA* **89**, 6482–6486.
Levin, V.A., Sheline, G.E. & Gutin, P.H. (1989) Neoplasms of the central nervous system. In: *Cancer: Principles and Practice of Oncology* (eds V.T. DeVita, Jr., S. Hellman, and S.A. Rosenberg) pp. 1557–1611. J.B. Lippincott Co., Philadelphia.
Li, Q., Kay, M.A., Finegold, M., Stratford-Perricaudet, L.D. & Woo, S.L.C. (1993) Assessment of recombinant adenoviral vectors for hepatic gene therapy. *Hum. Gene Ther.* **4**, 403–409.
Martin, J.C., Dvorak, C.A., Smee, D.F., Matthews, T.R., Verheyden, J.P.H. (1983) 9-[(1,3-dihydroxy-2-propoxymethyl)]guanine: A new potent and selective antiherpes agent. *J. Med. Chem.* **26**, 759–761.
McGrory, J., Bautista, D. & Graham, F.L. (1988) A simple technique for the rescue of early region I mutations into infectious human adenovirus type 5. *Virology* **163**, 614–617.
Miller, D.G., Adam, A.M. & Miller, A.D. (1990) Gene transfer by retrovirus vectors occurs only in cells that are actively replicating at the time of infection. *Mol. Cell Biol.* **10**, 4239–4242.
Moolten, F.L. (1986) Tumor chemosensitivity conferred by inserted herpes thymidine kinase genes: Paradigm for a prospective cancer control strategy. *Cancer Res.* **46**, 5276–5281.
Moolten, F.L. & Wells, J.M. (1990) Curability of tumors bearing herpes thymidine kinase genes transferred by retroviral vectors. *J. Natl. Cancer Inst.* **82**, 297–300.
Nelson, D.F., McDonald, J.V., Lapham, L.W., Qazi, R. & Rubin, P. (1993) Central nervous system tumors. In *Clinical Oncology* (ed. P. Rubin) pp. 617–644. Saunders, Philadelphia.
Oldfield, E.H., Ram, Z., Culver, K.W., Blaese, R.M., DeVroom, H.L. & Anderson, W.F. (1993) Gene therapy for the treatment of brain tumors using intra-tumoral transduction with the thymidine kinase gene and intravenous ganciclovir. *Hum. Gene Ther.* **4**, 39–69.
Perez-Cruet, M.J., Trask, T.W., Chen, S.-H., Goodman, J.C., Woo, S.L.C., Grossman, R.G. & Shine, H.D. (1994) Adenovirus-mediated gene therapy for experimental gliomas. *J. Neurosci. Res.* **39**, 506–511.
Raffel, C., Culver, K.W., Kohn, D., Nelson, N., Siegel, S., Gillis, F., Link, C.J., Villablanca, J.G. & Anderson, W.F. (1994) Gene therapy for the treatment of recurrent pediatric malignant astrocytomas with *in vivo* tumor transduction with the herpes simplex thymidine kinase gene/ganciclovir system. *Hum. Gene Ther.* **5**, 863–890.
Ragot, T., Vincent N., Chafey, P., Vigne, E., Gilgenkrantz, H., Couton, D., Cartaud, J., Briand, P., Kaplan, J.C., Perricaudet, M. & Kahn, A. (1993) Efficient adenovirus-mediated transfer of a human minidystrophin gene to skeletal muscle of mdx mice. *Nature* **361**, 647–650.
Ram, Z., Culver, K.W., Walbridge, S., Blaese, R.M. & Oldfield, E.H. (1993a) *In situ* retroviral-mediated gene transfer for the treatment of brain tumors in rats. *Cancer Res.* **53**, 83–88.
Ram, Z., Culver, K.W., Walbridge, S., Frank, J.A., Blaese, R.M. & Oldfield, E.H. (1993b) Toxicity studies of retroviral-mediated gene transfer for the treatment of brain tumors. *J. Neurosurg.* **79**, 400–407.
Ram, Z., Walbridge, S., Shawker, T., Culver, K.W., Blaese, R.M. & Oldfield, E.H. (1994) The

effect of thymidine kinase transduction and ganciclovir therapy on tumor vasculature and growth of 9L gliomas in rats. *J. Neurosurg.* **81**, 256–260.

Rosenfeld, M.A., Siegfried, W., Yoshimura, K., Yoneyama, K., Fukayama, M., Stier, L.E., Paakko, P., Gilardi, P., Stratford-Perricaudet, L.D., Perricaudet, M., Pavirani, A., Lecocq, J.-P. & Crystal, R.G. (1991) Adenovirus-mediated transfer of a recombinant α_1-antitrypsin gene to the lung epithelium *in vivo*. *Science* **252**, 431–434.

Rosenfeld, M.A., Yoshimura, K., Trapnell, B.C., Yoneyama, K., Rosenthal, E.R., Dalemans, W., Fukayama, M., Bargon, J., Stier, L.E., Stratford-Perricaudet, L., Perricaudet, M., Guggino, W.B., Pavirani, A., Lecocq, J.-P. & Crystal, R.G. (1992) *In vivo* transfer of the human cystic fibrosis transmembrane conductance regulator gene to the airway epithelium. *Cell* **68**, 143–155.

Rowe, W.P., Huebner, R.J., Gilmore, L.K., Parrot, R.H. & Ward, T.G. (1953) Isolation of a cytopathogenic agent from human adenoids undergoing spontaneous degeneration in tissue culture. *Proc. Soc. Exp. Med.* **84**, 570–573.

Salcman, M. (1980) Survival in glioblastoma: historical perspective. *Neurosurgery* **7**, 435–439.

Short, M.P., Choi, B.C., Lee, J.K., Malick, A., Breakefield, X.O. & Martuza, R.L. (1990) Gene delivery to glioma cells in rat brain by grafting of a retrovirus packaging cell line. *J. Neurosci. Res.* **27**, 427–433.

Simon, R.H., Engelhardt, J.F., Yang, Y., Zepeda, M., Weber-Pendleton, S., Grossman, M. & Wilson, J.M. (1993) Adenovirus-mediated transfer of the CFTR gene to lung of nonhuman primates: toxicity study. *Hum. Gene Ther.* **4**, 771–780.

Stratford-Perricaudet, L.D., Levrero, M., Chasse, J.-F., Perricaudet, M. & Braind, P. (1990) Evaluation of the transfer and expression in mice of an enzyme-encoding gene using a human adenovirus vector. *Hum. Gene Ther.* **1**, 241–256.

Tapscott, S.J., Miller, A.D., Olson, J.M., Berger, M.S., Groudine, M. & Spence, A.M. (1994) Gene therapy of rat 9L gliosarcoma tumors by transduction with selectable genes does not require drug selection. *Proc. Natl. Acad. Sci. USA* **91**, 8185–8189.

Vile, R.G. & Hart, I.R. (1993) Use of tissue-specific expression of the herpes simplex virus thymidine kinase gene to inhibit growth of established murine melanomas following direct intratumoral injection of DNA. *Cancer Res.* **53**, 3860–3864.

Vile, R.G., Nelson, J.A., Castleden, S., Chong, H. & Hart, I.R. (1994) Systemic gene therapy of murine melanoma using tissue specific expression of the *HSVtk* gene involves an immune component. *Cancer Res.* **54**, 6228–6234.

Viola, J.J., Ram, Z., Walbridge, S., Oshiro, E.M., Trapnell, B., Tao-Cheng, J.-H. & Oldfield, E.H. (1995) Adenovirally mediated gene transfer into experimental solid brain tumors and leptomeningeal cancer cells. *J. Neurosurg.* **82**, 70–76.

Weizsaecker, M., Deen, D.F., Rosenblum, M.L., Hoshino, T., Gutin, P.H. & Barker, M. (1981) The 9L rat brain tumor: Description and application of an animal model. *J. Neurol.* **224**, 183–192.

Zimmerman, H.M. (1969) Brain tumors: Their incidence and classification in man and their experimental production. *Ann. NY Acad. Sci.* **159**, 337–359.

CHAPTER 5

GENE DELIVERY USING ADENO-ASSOCIATED VIRUS

L. Tenenbaum and E. L. Hooghe-Peters

Department of Pharmacology, Neuroendocrinology Unit, Faculty of Medicine and Pharmacy, Free University of Brussels, Laarbeeklaan, 103, B1090 Brussels, Belgium

Table of Contents

5.1 Introduction

Adeno-associated virus is a member of the Parvoviridae family which are among the smallest and structurally simplest of the DNA animal viruses. The family Parvoviridae has been divided into three genera (Siegl *et al.*, 1985): parvoviruses which can grow in dividing cells of vertebrates, densoviruses which multiply in insects and dependoviruses which infect a broad range of vertebrates but usually require a co-infection with an adenovirus (Hoggan *et al.*, 1966) for a productive infection in cell culture. Because of the frequent association with adenovirus stocks, the dependoviruses have been called adeno-associated viruses (AAV).

Different AAV serotypes from birds to humans have been isolated. Human AAV serotypes can replicate in cell cultures derived from several different species as long as the cells are co-infected with a helper virus for which the cells are permissive (reviewed in Berns and Bohenzky, 1987).

Viruses of different families unrelated to the adenovirus family, such as herpesvirus (Buller *et al.*, 1981) and vaccinia virus (Schlehöfer *et al.*, 1986), can also serve as helpers to support AAV replication. Therefore, it has been suggested that cellular functions

Genetic Manipulation of the Nervous System
ISBN 0-12-437165-5

induced by the helper rather than viral functions are responsible for the helper effect. In support of this hypothesis, exposure of several cell lines to agents which interfere with cellular DNA synthesis, such as chemical carcinogens, UV irradiation or cycloheximide (Schlehöfer *et al.*, 1983, 1986; Yalkinoglu *et al.*, 1988; Yakobson *et al.*, 1989), as well as cell-synchronizing treatments (Yakobson *et al.*, 1987), render the cells permissive for viral production or at least for AAV macromolecular synthesis.

In the absence of a helper virus the dependovirion penetrates into the cell nucleus where the DNA is uncoated and is then integrated into the cell genome (Berns *et al.*, 1975; Handa *et al.*, 1977). The genome can be rescued from the integrated state after superinfection of the cell with a helper virus (Cheung *et al.*, 1980).

At an epidemiologic level, although human infections are common – most adults (85–90%) are seropositive for antibodies to AAV – AAV has never been implicated as an aetiological agent. Seroconversion occurs during childhood and is usually concomitant with an adenovirus infection (Blacklow and Rowe, 1968; Blacklow *et al.*, 1968, 1971). The infection appears to be benign or even beneficial. Indeed, several studies have found that women with cervical carcinomas are seronegative for antibodies to AAV (Sprecher-Goldberger *et al.*, 1971; Mayor *et al.*, 1976; Georg-Fries *et al.*, 1984). These data raise the possibility that AAV might actually have a protective effect.

Cryptic infections are also common. Up to 20% of the lots of primary African green monkey kidney cells and 1–2% of primary lots of human embryonic kidney cells were found to be naturally latently infected with AAV which can be rescued by adenoviral infection (reviewed in Berns and Bohenzky, 1987).

5.2 The virion

The non-enveloped AAV virion has a diameter of 20–22 nm and contains a linear single-stranded DNA genome of 4.7 kb encapsidated within a simple icosahedral capsid composed of three proteins. AAV DNA strands of both polarities are packaged into separate virions and with equal frequency so that half the particles contain plus strands and the other half minus strands. The AAV particles are stable in a wide pH range between 3 and 9 and are resistant to heating at 56°C for at least 1 h. The particle can be dissociated by exposure to either a combination of papain and ionic detergents or an alkaline pH (reviewed in Berns and Bohenzky, 1987).

In a typical productive infection, i.e. in the presence of adenovirus, an AAV virus titre as high as 10^{11}–10^{12} infectious particles ml^{-1} can be obtained. The contaminating adenovirus can be inactivated by heating at 56°C for 1 h and the AAV virus can be further purified and concentrated by CsCl gradient centrifugation (Yakobson *et al.*, 1987; Ruffing *et al.*, 1992).

5.3 Genome organization

The concomitant isolation of two different molecular clones of AAV type 2 by ligation of the double-stranded form of the virus in pBR322 plasmid has opened the door to the genetic analysis of the virus (Samulski *et al.*, 1982; Laughlin *et al.*, 1983). The cloned genome is infectious when transfected into permissive cells, i.e. upon transfection of a cloned viral genome into a permissive cell, the viral genes are expressed and the viral genome is rescued from the vector, replicated and then encapsidated to form infectious virus. The entire sequence of AAV is known (Srivastava *et al.*, 1983) and many mutants have been created, propagated in *Escherichia coli* and their phenotype assessed (Hermonat *et al.*, 1984; Chejanovsky and Carter, 1989, 1990).

The genome has been arbitrarily divided into 100 map units (see Figure 1).

AAV DNA has an inverted terminal repeat (ITR) of 145 bases. The terminal 125 bases are palindromic. The overall palindrome is interrupted by two shorter palindromes symmetrically disposed on either side of the centre of the larger palindrome. As a consequence, when the terminal 125 bases are folded, a T-shaped structure is formed which is thought to serve as a primer for DNA replication. The ITR is also the only *cis*-acting element necessary for efficient encapsidation of the single-stranded viral DNA as well as for the integration into the cellular genome and the rescue from plasmidic and cellular sequences (McLaughlin *et al.*, 1988; Samulski *et al.*, 1989).

There are two large non-overlapping open reading frames, each occupying about one-half of the genome. The left side codes for regulatory proteins and the right side for structural proteins.

Three promoters have been identified and named according to their approximate map position: p5, p19 and p40. At least one transcript has been found to start at each of the promoters. All transcripts are copied from the minus strand and can be spliced at an intron between map units 41 and 46. An alternatively spliced p40 mRNA which uses the same donor splice site as the major intron but a different acceptor site has also been identified. Messenger RNAs initiated at the three different promoters all terminate at the same polyadenylation site at map position 96. Promoters p5 and p19 direct the expression of regulatory proteins (the Rep proteins) and p40 controls structural proteins (the Cap proteins).

Four regulatory proteins have been identified. Rep78 and Rep68 are derived from promoter p5, Rep52 and Rep40 from promoter p19; Rep68 and Rep40 are the products from the spliced mRNAs. All four Rep proteins are translated using the same reading frame.

The AAV virion is composed of three structural proteins, VP-1, VP-2 and VP-3. VP-3 represents about 80% of the total mass. All three are translated from a spliced p40 transcript. VP-2 and VP-3 are synthesized from the major spliced p40 mRNA, VP2 using an alternate initiation codon (ACG); VP1 is synthesized from the alternatively spliced p40 mRNA (reviewed in Berns, 1990; Muzyczka, 1992).

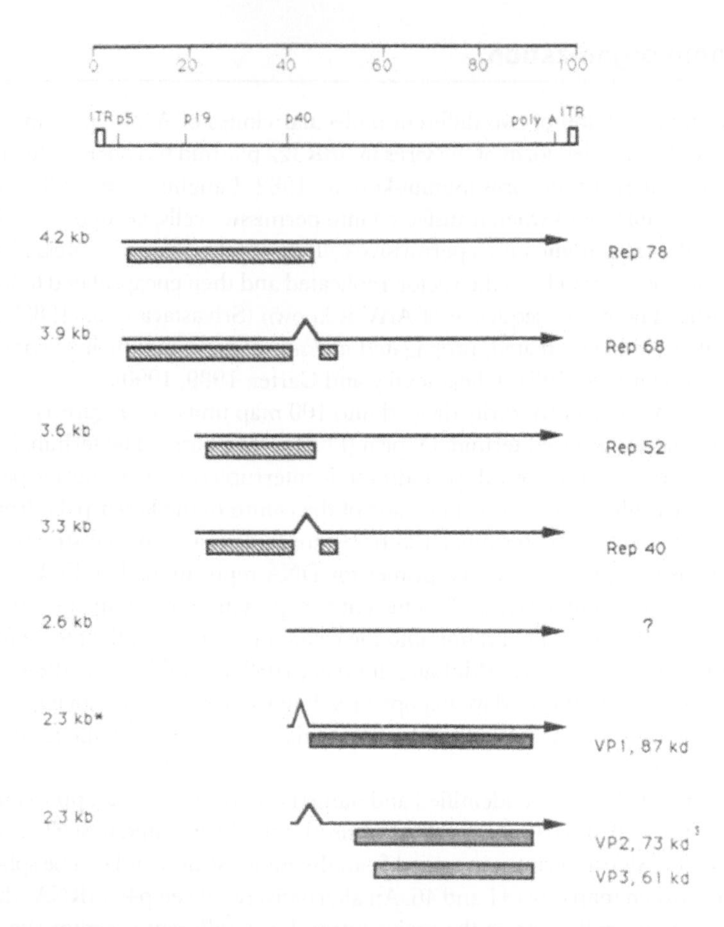

Figure 1 Adeno-associated virus genome. The position of the inverted terminal repeats (ITR), the three AAV promoters (p5, p19 and p40) and the polyadenylation signal (polyA) is shown on the top. The size of each mRNA is shown on the left. The major intron is from nucleotide 2107 to nucleotide 2227; the alternatively spliced p40 mRNA (*) uses the same donor splice site as the major intron (nucleotide 2107) but a different acceptor site (nucleotide 2201). The proteins synthesized from each mRNA are shown on the right. The reading frames used in each mRNA are also shown (■, reading frame 2; ▨, reading frame 3). The question mark indicates that no known protein is translated from this message. [$]VP2 is synthesized from the same mRNA as VP3 but an alternative ACG start codon is used.

5.4 Adeno-associated virus life cycle

AAV can be propagated either in a latent state as an integrated provirus that can be rescued by superinfection with adenovirus or by lytic infection in the presence of a co-infecting adenovirus.

5.4.1 The lytic cycle

5.4.1.1 DNA replication

The current model for AAV replication is described in detail in Berns (1990) and Berns and Bohenzky (1987) (see Figure 2).

As a linear single-stranded DNA, the AAV genome represents an unusual template for DNA replication. It is believed that the terminal palindrome of AAV can serve as a primer for AAV DNA replication.

Upon entering the cell, the AAV virion particles de-coat in the nucleus and release their single-stranded DNA genome. The initial conversion of single-stranded DNA into duplex DNA occurs by using the 3′-end hairpin as a primer for the fill-in replication, to form a large T-shaped hairpin intermediate (RF1 'turn-around' form). As a result, the entire coding region as well as the ITR on the 5′-end is completely replicated. To replicate the 3′-end ITR, a nick is made by the Rep proteins on the parental strand at the terminal resolution site (TRS) in the 3′-ITR region to generate a new internal 3′-hydroxyl group. This nick is then used as a primer to finish the replication of the 3′-ITR. At this point the entire genome is replicated. This duplex replication form DNA (RF1 'extended' form) serves as a template for strand displacement synthesis generating a new RF1 turn-around replicative form which can be subsequently amplified and a single-stranded AAV genome which is packaged to produce mature virions. Dimeric duplex DNA (RF2) is also observed and presumably results from a second round of replication on monomeric duplexes (RF1) which have not been resolved.

Rep78 and Rep68 can bind to the AAV replication origin (ITR; Snyder *et al.*, 1990; McCarty *et al.*, 1994; Chiorini *et al.*, 1994). They display a site-specific and strand-specific nickase activity, a helicase and ATPase activity which are critical for AAV DNA replication (Im and Muzyczka, 1990). Rep52 and Rep40 seem to play a role in the viral single-stranded DNA accumulation (Chejanovski and Carter, 1989).

The cellular functions required for AAV DNA replication are poorly defined. The recent development of *in vitro* AAV DNA replication assays will help better defining of the cellular functions involved as well as the respective role of the four Rep proteins (Hong *et al.*, 1994; Ni *et al.*, 1994).

5.4.1.2 Transcription control

When cells are infected with AAV and adenovirus, the initial step is the synthesis of the adenovirus E1A gene product which induces transcription from the AAV p5 and p19 promoters and leads to the synthesis of a small amount of the Rep proteins (Chang *et al.*, 1989). Beside their function in DNA replication, the Rep proteins also regulate gene expression from both AAV and heterologous promoters. In the presence of helper virus, the Rep proteins induce the synthesis of mRNA from all three AAV promoters to a higher level (Labow *et al.*, 1986, Tratschin *et al.*, 1986, Redeman *et al.*, 1989).

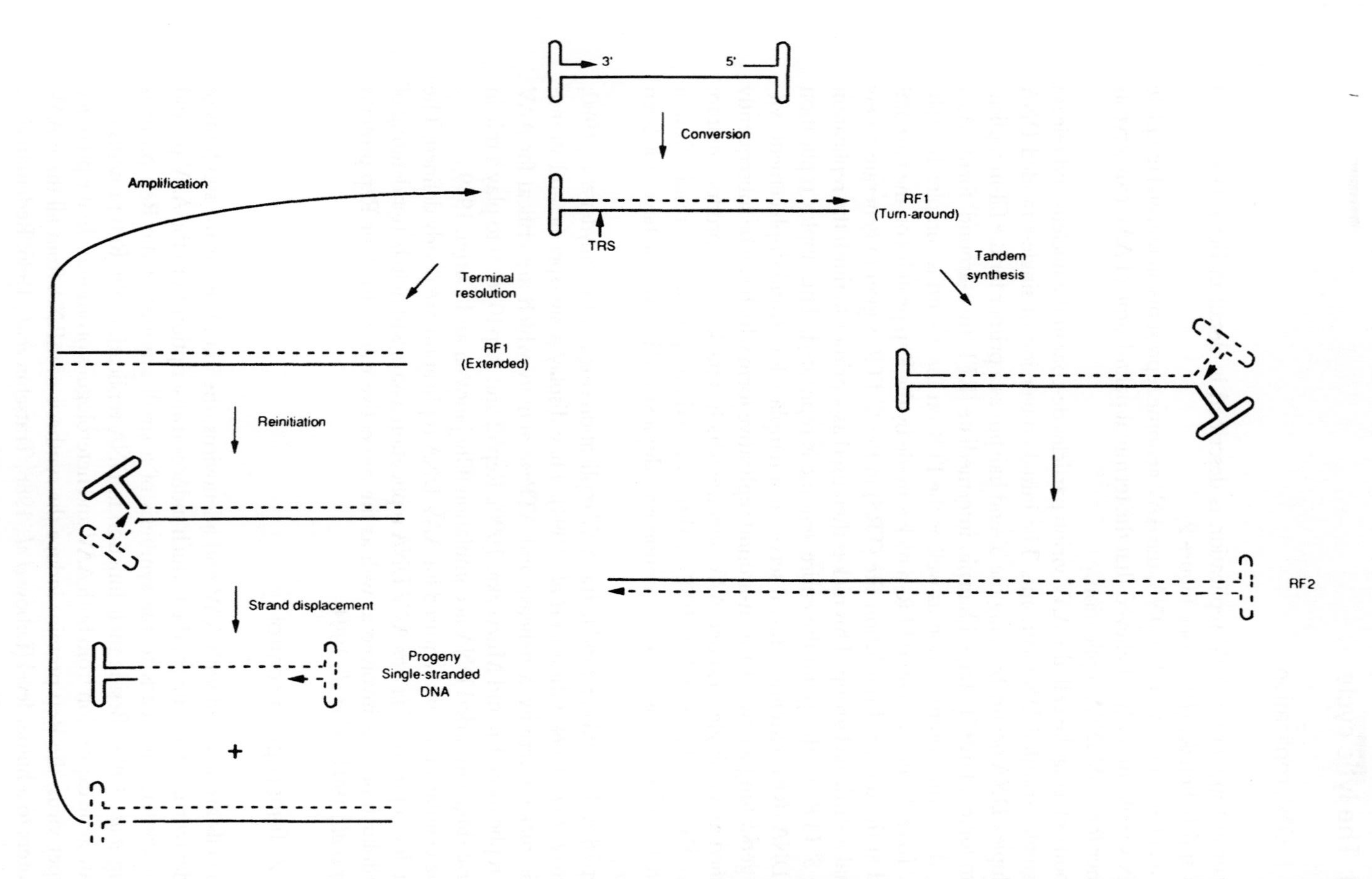

Figure 2 Adeno-associated virus DNA replication. Arrows indicate the 3′OH ends. Solid lines represent parental DNA molecules; dotted lines represent newly synthesized DNA molecules. TRS is the terminal resolution site. RF1 represents the monomeric (one genome) double-stranded replicative form; RF2 represents the dimeric (tandem genomes) double-stranded replicative form.

In the absence of adenovirus, the *rep* gene has been found to inhibit expression from both AAV and heterologous promoters. Rep proteins inhibit gene expression of several heterologous viruses, such as adenovirus (Laughlin *et al.*, 1978), human papillomavirus type 16 (Hermonat, 1994a) and human immunodeficiency virus-1 (HIV-1; Antoni *et al.*, 1991; Oelze *et al.*, 1994). Furthermore, markers under the control of a variety of promoters are repressed when co-transfected with plasmids containing the *rep* gene. These include the SV40 promoter, the mouse metallothionein promoter and the herpes thymidine kinase promoter (Labow *et al.*, 1987), the HIV-1 long terminal repeat (LTR; Antoni *et al.*, 1991) and the human c-*fos* and c-*myc* proto-oncogene promoters (Hermonat, 1994b). AAV promoters were also shown to be inhibited by Rep proteins (Tratschin *et al.*, 1986; Mendelson *et al.*, 1988a; Beaton *et al.*, 1989).

5.4.2 Latency

Berns and collaborators demonstrated that a latent infection could be achieved in cell culture by simply infecting human cell lines with AAV in the absence of adenovirus (Berns *et al.*, 1975; Cheung *et al.*, 1980). Such cultures remained latently infected for more than 100 passages. At high multiplicities of infection (250 infectious units per cell), AAV has been observed to cause latent infection in 10–30% of the infected cells. The viral DNA was found to be integrated into cell DNA as a tandem repeat of several copies, with the termini of the viral genome close to the junction with cellular sequences (Kotin and Berns, 1989).

5.4.2.1 Site-specific integration

Digestion of genomic DNA from latently infected cells with several restriction endonucleases and hybridization with AAV DNA-specific probes revealed that the sizes of the putative viral–cellular junction fragments were different in every clone. It was therefore concluded that the AAV DNA integrated into random sites in the cellular genome (Laughlin *et al.*, 1986). Later on, the cellular sequences flanking AAV provirus were isolated from one of these clones. When these junction DNA fragments were used as probes to characterize the AAV integration site in other latently infected cell lines, it was discovered that the proviruses in the majority (68%) of the cells examined were linked to the same cellular flanking sequences, suggesting that AAV integrated into host DNA in a site-specific manner (Kotin *et al.*, 1990; Samulski *et al.*, 1991). The cellular sequence was called AAVS1 and was localized on human chromosome 19 using rodent–human hybrid cell lines containing various human chromosomes. Moreover, the AAV proviruses were also directly visualized by *in situ* hybridization on q13.4-ter of chromosome 19 (Kotin *et al.*, 1991; Samulski *et al.*, 1991). Evidence for transcriptional activity within AAVS1 was obtained by reverse transcription and polymerase chain reaction (PCR) amplification of total mRNA from embryonic human foreskin fibroblasts. The obtained PCR fragment hybridized to a cDNA clone from a library of human foreskin fibroblasts that contained a fragment of the AAVS1 sequence (Kotin *et al.*, 1992).

Sites of integration have been mapped to multiple points within a 100 bp sequence in AAVS1. Frequent rearrangements both in viral and cellular sequences have been observed explaining the variety of cellular–viral junction fragments observed in the early studies.

There is no homology between the AAVS1 site and the AAV genome. Thus the integration event must presumably involve non-homologous recombination. Recently, it was shown that AAV site-specific integration is directed by the primary sequence of the chromosome 19 pre-integration locus and the recombinogenic signals were localized to a 500 nucleotide region at this locus (Giraud *et al.*, 1994).

Rep proteins probably play a role in site-specific integration. In a recent report, AAV vectors which contain an intact *rep* gene and the ITR integrate into AAVS1 (Shelling and Smith, 1994). On the contrary, AAV vectors containing only the terminal repeats are highly proficient for integration but in several cases they were shown not to target AAVS1 (Kumer and Leffak, 1991; Walsh *et al.*, 1992, Muro-Cacho *et al.*, 1992, Flotte *et al.*, 1993a). Further data suggesting that Rep proteins are involved in the targeting to AAVS1 come from *in vitro* assays showing that Rep78 and Rep68 specifically bind to a 109 bp DNA fragment from human chromosome 19 near sites of viral integration in AAVS1. A Rep recognition sequence similar to the Rep binding site on AAV ITR was identified in AAVS1. Furthermore, Rep proteins mediate the formation of a complex between the DNA of human chromosome 19 and AAV hairpin DNA (Weitzman *et al.*, 1994).

AAV site-specific integration has been reviewed by Samulski (1993).

5.4.2.2 Transcription from integrated AAV

Early studies showed that during latent infection with wild-type AAV, very little expression of AAV genes is detected. This low level of AAV gene expression might be explained by the ability of AAV to negatively regulate its own gene expression in the absence of helper virus. However, AAV recombinant genomes in which the *cap* gene was replaced by foreign reporter genes under the control of the p40 promoter express these genes at a high level in both transient and stable expression assays (Tratschin *et al.*, 1984, 1985; Mendelson *et al.*, 1988a,b).

5.4.2.3 Perturbation of the cellular phenotype by AAV

In several instances, AAV infection or the presence of integrated AAV has been shown to alter the phenotype of cells in culture.

Inhibition of growth of transformed cells (Bantel-Schaal, 1991) as well as increased sensitivity of transformed cells to genotoxic treatments (Walz *et al.*, 1992) after AAV infection have been reported. Clones of transformed cells harbouring integrated AAV also show a reduced growth rate and a reduced ability to form clones in agar (Walz and Schlehöfer, 1992) as well as an increased sensitivity to UV irradiation (Winocour *et al.*, 1992). The comparison of the behaviour of cell clones containing integrated *rep*$^+$ and *rep*$^-$ vectors suggested that the expression of the Rep proteins is required to observe these effects (Winocour *et al.*, 1992).

Normal cell growth and progression through the cell cycle was also shown to be affected by AAV infection (Winocour *et al.*, 1988). However, very high multiplicities were necessary and no AAV gene expression was required, since heavily irradiated virus had the same effect as unirradiated virus. In one instance, transient alteration of the adherence properties of cells has been observed (Bantel-Schaal and Stohr, 1992).

5.4.2.4 Rescue of integrated AAV sequences

Infection with helper virus can result in rescue and replication of the AAV genome to produce infectious progeny virus (Cheung *et al.*, 1980).

Little is known about the mechanism of excision of AAV genome from the cellular genome beside the fact that the 145 bp ITRs are the only sequences necessary *in cis* for this to occur. Gottlieb and Muzyczka (1988) have isolated a cellular enzyme which they called endo R, that is capable of excising AAV sequences from prokaryotic plasmids *in vitro* at low frequencies. This enzyme recognizes GC-rich, at least nine nucleotides long polypurine–polypyrimidine sequences. Nevertheless, whether endo R is involved in the rescue of AAV DNA from plasmid DNA or from cellular chromosomes *in vivo* is not known.

5.5 Host range

5.5.1 Human cells

All the human cells tried so far could be successfully infected by AAV or transduced with AAV-based vectors. Most of the studies have been done with established cell lines such as HeLa, D6, KB and 293. Several leukaemia cell lines (HL60, KG1a, U937 and K562) as well as a normal lymphoblastoid line (NC37) have also been successfully transduced (Lebkowski *et al.*, 1988).

However, the behaviour of AAV vectors designed for gene transfer *in vivo* or gene therapy should be tested in primary cultures as cell lines are often aneuploid and are thus questionable models. Primary cultures tested so far include human fibroblasts (Podsakoff *et al.*, 1994; Russel *et al.*, 1994), haematopoietic progenitor cells (Zhou *et al.*, 1994 and lymphocytes (Mendelson *et al.*, 1992; Muro-Cacho *et al.*, 1992). Human cells that have in addition been examined for their ability to acquire an AAV provirus include established transformed cell lines (HeLa, KB, D6, 293); Epstein–Barr virus (EBV)-transformed B lymphocytes, peripheral blood mononuclear cells, $CD4^+$ T-cell clones and $CD34^+$ cells (Muro-Cacho *et al.*, 1992); an erythroleukaemia line (K562; Walsh *et al.*, 1992); a bronchial epithelial cell line (IB3; Flotte *et al.*, 1993b); and primary fibroblasts (Russel *et al.*, 1994; Podsakoff *et al.*, 1994).

5.5.2 Life cycle of AAV in cells of the human CNS

We have evaluated AAV infectivity towards and integration ability into human brain tumour cells. A preliminary report of these data has already been published (Tenenbaum *et al.*, 1994).

As a model system we used the U373-MG glial cell line derived from a human astrocytic tumour. In the presence of adenovirus, AAV replicates in U373-MG cells, yielding replicative forms of expected sizes: monomeric duplexes (RF1), dimeric duplexes (RF2) as well as single-stranded DNA. Replicated DNA was produced in higher amounts in U373-MG than in HeLa cells (Figure 3A). We next asked if this result reflected a more efficient replication of AAV DNA in U373–MG cells or a higher infectivity of AAV virions towards these cells. An AAV stock with a titre of 10^{11} infectious units ml^{-1} produced on HeLa cells was titrated on U373-MG cells, using the method of *in situ* focus hybridization assay (Yakobson *et al.*, 1989). The number of replicative foci was found to be 10- to 20-fold higher on U373-MG than on HeLa cells (Figure 3B). It can be concluded that the AAV stock contained more replication-proficient virus than detected by the *in situ* focus hybridization assay performed on HeLa cells. Whether U373-MG cells reveal all the replication-proficient virus particles was not determined. However, consistently with our results, Winocour and collaborators (1988) found that for a particular AAV stock, one AAV multiplicity unit as detected by the *in situ* focus hybridization assay performed on HeLa cells was equivalent to 20 virus particles as evaluated by electron microscopy. It could be that the receptor for AAV virus particles (as yet unidentified) is more efficiently expressed in U373-MG cells than in HeLa cells.

To determine if AAV can integrate in human glial cells, U373-MG cells were infected with wild-type AAV at a multiplicity of 10 virus per cell in the absence of adenovirus. Thirty-three isolated clones were expanded and the genomic DNA of 10 of them analysed. Figure 3C shows a Southern blot for one of these clones (VUB-C6) containing integrated AAV sequences. As a positive control, we show D528, a subclone (isolated and kindly provided by S. Etkin and E. Winocour) of the D5 cell line which is a derivative of the D6 cell line containing integrated AAV. The integrated AAV sequences in the VUB-C6 cell line were rescuable by infection with adenovirus as shown in Figure 3D. The positive control, D528, yields the typical monomeric and dimeric replicative forms. The higher molecular weight DNA yielded by VUB-C6 probably consists of AAV multimers containing imperfect (unexcisable) genomes bordered by intact palindromic ends. As AAV concatemers are usually in a head-to-tail orientation in the integrated state, it was expected that digestion of the replicated DNA with a restriction enzyme that cuts only once in the AAV genome would result in the disappearance of the high molecular weight replicative forms and in the appearance of a lower band of approximately one genome length. Consistently, digestion of the replicated AAV DNA obtained by adenovirus infection of the VUB-C6 clone produced a AAV DNA band of approximately 4.5 kb (data not shown). The lower intensity of the band resulting from rescue of integrated AAV sequences in VUB-C6 as compared to D528 probably reflects the lower efficiency of replication of larger parvoviral genome as already observed with molecular clones of the mouse MVM parvovirus containing inserts of increasing length (Dupont, 1993).

It remains to be determined if AAV integration occurs at the AAVS1 site on chromosome 19 in these glial cells. Whether the cellular environment is important for the specificity of integration is not known.

Wild-type AAV also replicated in adenovirus-infected primary cultures of human gliomas (see Figure 4). The yields of replicated DNA were comparable in these cultures and in HeLa cells.

5.5.3 Other mammalian cells

AAV infects simian (COS), Chinese hamster (Yakobson *et al.*, 1987, 1989) and mouse cells (LaFace *et al.*, 1988; Chou *et al.*, 1993). The AAVS1 sequence is conserved in human and monkey but it was not detected in rat, mouse, canine, bovine and rabbit (Samulski *et al.*, 1991). However, AAV was shown to integrate in Chinese hamster cells (Winocour *et al.*, 1992). Long-term expression of a transgene was shown to occur in rabbit epithelial cells (Flotte *et al.*, 1993a) and in rat brain cells (Kaplitt *et al.*, 1994) after *in vivo* transduction. Transduction efficiencies were high in both systems: 30% and 10%, respectively.

5.6 Adeno-associated virus as a transducing vector

Genetic analysis of AAV mutants indicated that the only *cis*-acting sequences required for AAV rescue from the plasmid, DNA replication and encapsidation are the ITRs. Consequently, most of the AAV genome can be deleted and substituted by foreign DNA (reviews: Carter, 1992; Kotin, 1994).

5.6.1 Design of the vectors

The first recombinant AAV vectors were constructed by introducing prokaryotic reporter genes under the control of the AAV p19 or p40 promoters or of SV40 early promoter. They either retained or not a Rep^+ phenotype but all of them were deficient for the expression of capsid proteins.

Transient expression from AAV vectors was evaluated using constructs in which the *cat* gene (coding for chloramphenicol acetyltransferase) was cloned downstream to AAV p19 (rep^- cap^- vector) or p40 promoters (rep^+ cap^- vector). Both promoters provided constitutive high level expression of *cat* even in the absence of adenovirus. The p19 promoter but not the p40 could be further stimulated by adenovirus (Tratschin *et al.*, 1984; Mendelson *et al.*, 1988a). Stable expression from AAV vectors was studied with constructs in which the *neo* gene (coding for neomycin phosphotransferase conferring resistance to aminoglycoside antibiotics) was set under the control of AAV p40 (rep^+ cap^- vector) or of the SV40 early promoter (rep^+ or rep^- cap^- vectors). Selection for

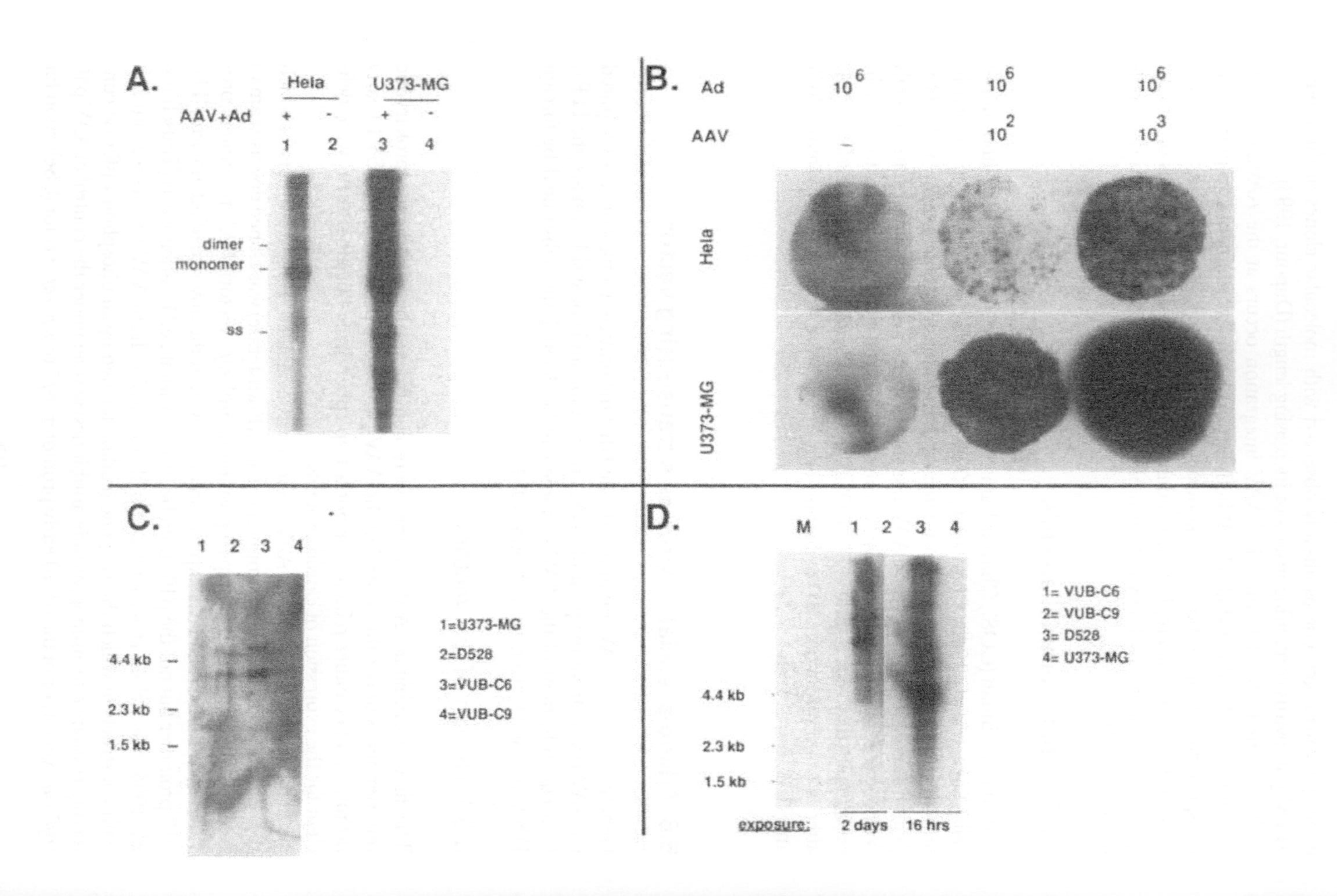

A.
Hela
U373-MG
AAV+Ad
+
-
+
-
1
2
3
4
dimer
monomer
ss
B.
Ad
AAV
10^6
10^6
10^6
10^2
10^3
Hela
U373-MG
C.
1
2
3
4
4.4 kb
2.3 kb
1.5 kb
1=U373-MG
2=D528
3=VUB-C6
4=VUB-C9
D.
M
1
2
3
4
4.4 kb
2.3 kb
1.5 kb
exposure:
2 days
16 hrs
1= VUB-C6
2= VUB-C9
3= D528
4= U373-MG

Figure 3 Life cycle of wild-type adeno-associated virus in human astrocytoma cells. (A,B) AAV DNA replication in the U373-MG astrocytoma cell line in the presence of adenovirus. (A) Replicative forms. HeLa (lanes 1 and 2) or U373-MG (lanes 3 and 4) cells were infected (lanes 1 and 3) or not (lanes 2 and 4) with AAV type 2 and adenovirus type 2 at a multiplicity of 10 infectious units per cell for both viruses. Two days post-infection, low molecular weight DNA was isolated by the extraction method of Hirt (1967). Samples corresponding to 5×10^4 cells were fractionated by agarose gel electrophoresis and transferred to a nylon membrane. The blot was hybridized at 65°C with random-primed ^{32}P-labelled AAV DNA and exposed for 16 h to an X-ray film in the presence of an intensifying screen. The approximate position of double-stranded monomers (RF1) and dimers (RF2) as well as of single-stranded AAV DNA (ss) are shown. This figure is a partial reproduction from Tenenbaum *et al.* (1994). (B) Infectious particles. Subconfluent HeLa or U373-MG cells were infected with adenovirus 2 at a multiplicity of 10 infectious units per cell, and various dilutions of AAV-2. Twenty-eight hours post-infection, the cells were transferred to nitrocellulose filters. After DNA denaturation by NaOH, the filters were hybridized at 65°C with random-primed ^{32}P-labelled AAV DNA fragment and exposed for 16 h to an X-ray film in the presence of an intensifying screen. (C) Integration of wild-type AAV in the U373–MG astrocytoma cell line in the absence of adenovirus. U373-MG cells were infected with AAV-2 at a multiplicity of 10 infectious units per cell. Isolated clones were expanded and their genomic DNA extracted. The DNA of the parental U373-MG cell line (lane 1) and of the D528 cell line (lane 2) as well as of two clones of AAV-infected U373-MG cells (VUB-C6, lane 3 and VUB-C9, lane 4) was analysed by Southern blot. Ten micrograms of genomic DNA were digested with 20 units of *Bam*HI enzyme for 2 h and electrophoresed in a 1% agarose gel. After blotting, the nylon membrane was hybridized with an AAV probe. The membrane was autoradiographed for 3 weeks in the presence of an intensifying screen. (D) Rescue of AAV DNA from an astrocytoma clone containing integrated AAV. VUB-C6 (lane 1) and VUB-C9 (lane 2) clones as well as D528 (lane 3) and U373-MG (lane 4) cells were infected with adenovirus 2 at a multiplicity of 100 infectious units per cell. Two days post-infection, low molecular weight DNA was isolated and Southern blotting performed as described in (A). The membrane was exposed for 16 h or 2 days. M, size marker consisting of pSM620 plasmid (an AAV molecular clone; Samulski *et al.*, 1982) DNA digested with *Pst*I restriction enzyme.

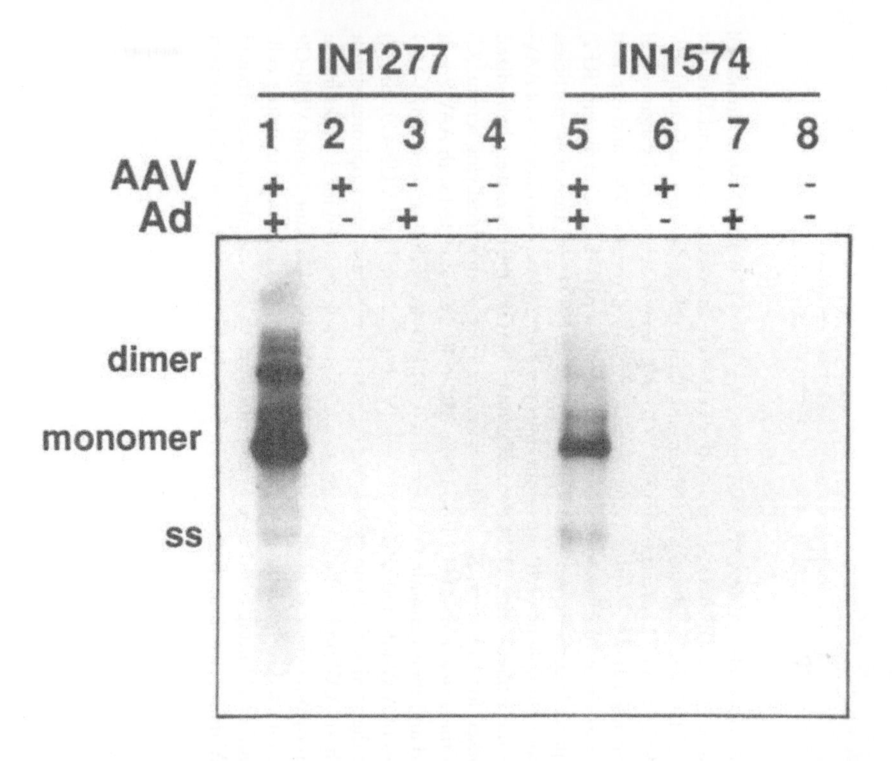

Figure 4 Adeno-associated virus replication in primary cultures of human oligodendroglioma. Short-term cultures of human oligodendroglioma derived from two different patients, IN1277 and IN1574, were obtained from J.L. Darling and infected at passage 4. Two $\times 10^5$ cells were infected with AAV type 2 and adenovirus type 2 at a multiplicity of 10 infectious units per cell for both viruses. Two days post-infection, low molecular weight DNA was isolated by the extraction method of Hirt (1967). Samples corresponding to 5×10^4 cells were fractionated by agarose gel electrophoresis and transferred to a nylon membrane. The blot was hybridized at 65°C with random-primed ^{32}P-labelled AAV DNA fragment and exposed for 16 h to an X-ray film in the presence of an intensifying screen. The approximate positions of double-stranded monomers (RF1) and dimers (RF2) as well as of single-stranded AAV DNA (ss) are shown.

resistance to geneticin (G418) using the Rep$^+$ vectors showed a 2–3% transduction frequency in HeLa cells when the expression of the *neo* gene was driven by p40 (Tratschin *et al.*, 1985) and 1–10% when the expression was driven by SV40 promoter (Hermonat and Muzyczka, 1984; McLaughlin *et al.*, 1988). At first sight, surprisingly, the Rep$^-$ vector could transduce cells to geneticin resistance at a much higher frequency, up to 70%. This transduction inhibition by the Rep proteins could be explained either by inhibition of transcription from the SV40 promoter or by inhibition of integration by the Rep proteins. The former explanation is consistent with the previously observed negative regulation of heterologous promoters by Rep proteins in other systems. The level of expression of neomycin phosphotransferase might not be high enough to confer resistance to geneticin at the high dosage used to avoid the appearance of spontaneously resistant clones. As an example, *rep*$^+$ and *rep*$^-$ AAV vectors carrying the *neo* gene under the control of the SV40 early promoter are depicted in Figure 5C and D.

It must be noticed, however, that *rep*$^-$ vectors seem not to target the AAVS1 site. Whether they integrate randomly or at one or few specific sites is not known yet. Probably due to the rigidity of the parvoviral capsid, recombinant genomes larger than the wild-type AAV genome cannot be encapsidated. Viral production from a recombinant AAV genome 3% larger than the wild type was drastically reduced (Tratschin *et al.*, 1984). As a consequence the size of the inserts that can be cloned in AAV vectors is limited to about 4 kb for the *rep*$^-$ vectors and 2 kb for the *rep*$^+$ vectors.

The *rep*$^-$ vectors thus accept foreign inserts of larger size and seem to transduce cells with a higher efficiency. Therefore they were further studied and used to introduce genes of therapeutic interest.

Due to the exquisite recombinogenic properties of the ITR, wild-type and recombinant AAV molecular clones must be propagated into totally recombination-deficient (*recA, recBC, sbcB, recF*) host bacteria to avoid sequence rearrangements (Boissy and Astell, 1985).

5.6.2 Encapsidation of the AAV recombinant genome

In order to excise, replicate and encapsidate vector sequences from recombinant plasmids, three elements are necessary: AAV Rep proteins, AAV Cap proteins and adenovirus. Up to now the AAV proteins were supplied by a helper plasmid co-transfected with the recombinant vector plasmid. The adenovirus infection can be performed either before or after the co-transfection.

The first encapsidation trials used a plasmid containing the entire sequence of wild-type AAV. This method yielded an enormous excess of wild type toward recombinant virus probably reflecting preferential encapsidation of the wild-type DNA (Tratschin *et al.*, 1984).

An AAV molecular clone containing a foreign inert DNA insert of 1.1 kb into a region not important for AAV gene expression and resulting in a DNA too large to be encapsidated was then used to supply AAV Rep and Cap proteins (Hermonat and Muzyczka, 1984; McLaughlin *et al.*, 1988; see Figure 5E). However, in this case also a contamination of recombinant stocks with about 10–30% of wild-type AAV was observed. Wild-type AAV genomes are thought to appear in this system as the result of homologous recombination between the recombinant AAV and helper plasmids. Therefore, systems in which the helper plasmid harbours no homology with the vector plasmid were developed. The most widely used system, developed by Samulski and collaborators (1987, 1989), consists of:

(1) A cloning vector which retains only the last 191 bp of the AAV genome duplicated in order to bracket the transgene between two copies of the ITR. The extra 45 bp inside the ITRs are part of a non-coding region in AAV and were not reported to play any role in the regulation of gene expression (Figure 5F).

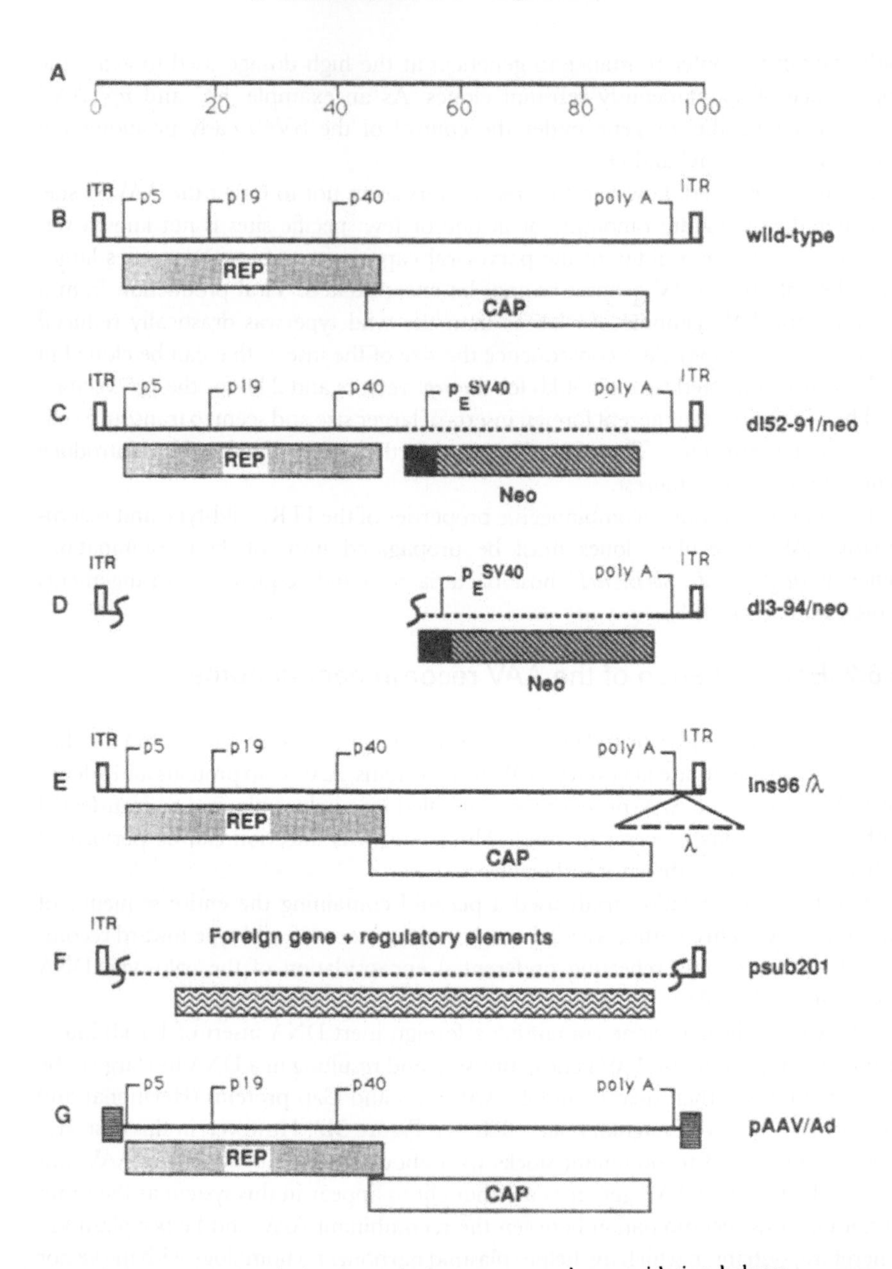

Figure 5 Adeno-associated virus transducing vectors and encapsidation helpers. ———, AAV DNA; ------, foreign DNA of interest; — — —, bacteriophage λ DNA. ▩ REP open reading frame; □, CAP open reading frame; ■, neomycin phosphotransferase; ▧, foreign protein; ▧, SV40 cassette (nt 5171–270); ▤, adenovirus terminal sequences. (A) AAV genome in map units. (B) Wild-type AAV. The inverted terminal repeats (ITR), the three AAV promoters (p5, p19 and p40) and the polyadenylation signal (polyA) are shown. (C) dl52–91/neo contains the SV40 origin of replication and regulatory elements as well as the neomycin resistance

(2) A helper plasmid containing the whole AAV sequence from nucleotide 191 to 4489, containing the entire AAV coding sequence including the polyA signal. However, this plasmid did not provide an optimal expression from p5 promoter as the 45 bp immediately adjacent to the left ITR play a role in p5 regulation (McCarty *et al.*, 1991). Therefore, in order to enhance the expression of AAV gene in this helper plasmid the terminal 107 bp of adenovirus 5 DNA was introduced at both ends of the plasmid (Figure 5G).

Helper-free recombinant AAV stocks with a titre of approximately 10^6 viral particles per transfected (10 cm) plate were obtained. A transduction efficiency as high as 65% was obtained using an AAV/*neo* vector.

The development of encapsidation cell lines comparable to those constructed to prepare retroviral recombinant stocks could possibly allow to obtain higher virus titres. A cell line containing the *rep* and *cap* transcription units and no ITR was isolated by Vincent and collaborators (1990). However, when this cell line was used to provide Rep proteins in trans, only low titre stocks were obtained. The authors suggested that the level of expression of AAV genes in this cell line was too low to efficiently complement the defective recombinant genomes.

Cell lines expressing constitutively the Rep proteins at high level have so far not been isolated probably due to the antiproliferative activity of these proteins. A cell line expressing the Rep proteins under the control of a conditional promoter has been recently described and might be the basis of an efficient encapsidation cell line (Yang *et al.*, 1994).

An interesting recent report describes the use of liposomes to encapsulate AAV recombinant genomes (Philip *et al.*, 1994). This approach takes advantage of the integrative properties of AAV vectors and thus of a stable transgene expression but in addition it allows to transfer a much higher amount of recombinant DNA, and furthermore there is virtually no size limitation to the insert.

5.6.3 Titration of AAV recombinant virus stocks

Wild-type as well as recombinant AAV viruses cannot be titrated by conventional plaque assays since they are defective. Titration of viral stocks has been performed by

gene inserted in a deletion from map unit 52 to 91 in the AAV genome. The polyA signal used is from AAV. AAV REP open reading frame is intact. No capsid protein can be expressed from this vector. (D) dl3-94/neo contains the SV40 origin of replication and regulatory elements as well as the neomycin resistance gene inserted in a deletion from map unit 3 to 94 in the AAV genome. The polyA signal used is from AAV. No AAV protein can be expressed from this vector. (E) The ins96/λ helper plasmid contains the entire AAV genome. A 1.1 kb fragment of bacteriophage λ DNA is inserted at map unit 96. (F) psub201 differs from wild-type AAV in that two XbaI cleavage sites have been added at sequence positions 190 and 4484 and the right-end 191 bp have been substituted for the left-end 190 bp. Foreign DNA can be inserted between the two XbaI restrictions sites. (G) pAAV/Ad is the internal XbaI fragment of psub201 containing the whole AAV coding region bracketed by the terminal 107 bp of adenovirus type 5 DNA several methods, some of them based upon the quantitation of viral DNA, others upon the measure of the expression of the transgene.

Dot blot analysis of recombinant DNA in comparison with serial dilutions of AAV DNA of known concentrations gives a rough estimation of the number of viral particles (Samulski *et al.*, 1989). The '*in situ* focus hybridization assay' (Yakobson *et al.*, 1989) measures precisely the number of replicative foci in the presence of adenovirus and wild-type AAV. A limitation of this method is that the titre depends upon the cell line used (see our data in Section 5.5.2). Methods based upon the expression of the insert give titres which depend upon the transduction efficiency since they detect only the recombinant particles which gave an efficient transduction. When transducing a gene coding for resistance to antibiotics, the number of resistant clones can be evaluated (Samulski *et al.*, 1989; Russel *et al.*, 1994). This method yields titres that can be biased by the selection procedure. When a histochemical or an immunohistochemical measure of the expression of the insert is possible, direct visualization of transduced clones can be performed (Flotte *et al.*, 1993a; Russel *et al.*, 1994).

5.6.4 Transduction of quiescent cells?

Whether AAV can stably transduce non-dividing cells is of particular interest for gene transfer applications to the nervous system. First of all, it is not known whether integration is required for gene expression from AAV vectors but integration should ensure a higher stability of the vector sequences and hence a more stable expression.

Recently, Kaplitt and collaborators (1994) obtained long-term expression of a reporter *lacZ* gene driven by the CMV promoter in the adult rat brain after *in vivo* transduction using an AAV vector. The vector was injected into various regions of the brain and numerous cells were labelled within each region. Vector sequences were also detected *in situ* PCR. The transduction efficiency was approximately 10% as evaluated by the number of labelled cells relative to the number of recombinant virus particles injected. Whether the vector integrated or stayed episomal was not looked for in this study. The same authors also constructed an equivalent vector expressing the human tyrosine hydroxylase cDNA. When this vector was injected into the caudate nucleus, the transgene was expressed up to 4 months near the site of injection. However, the number of positive cells diminished. This could be explained by the loss of possibly non-integrated vector sequences. Alternatively, other factors reducing the stability of gene expression could be involved including selective loss of transduced cells or loss of expression of the transgene. The latter effect has been observed in several other systems when a viral promoter was used to drive the expression of the transgene.

In three other studies, transduction of quiescent or slowly dividing cells was examined in more details in culture. Rapidly cycling and contact-inhibited human primary fibroblast cultures were compared for the transduction efficiency by AAV vectors (Russel *et al.*, 1994). The expression of the alkaline phosphatase gene under the control of the retroviral MLV promoter was 200-fold higher in cycling cultures. In contact-inhibited cultures few positive cells were observed. However these cells were not totally quiescent as 4% of the cells enter the S phase in 24 h. Most of the positive cells

(90%) were dividing as shown by [^{3}H]thymidine incorporation. Nevertheless, non-dividing cells were also transduced (10% of the positive cells) but at a much lower frequency. The same authors showed that transduction of non-dividing cells becomes as efficient as in dividing cells if the cells are stimulated to proliferate several days after infection. The transduction efficiency does not decrease when stimulation is performed up to 12 days after transduction. The authors concluded that AAV vectors can persist in stationary cultures and be recruited for transduction when the cells are stimulated to divide. Also using primary human normal fibroblasts, Podsakoff and collaborators (1994) observed a low transduction efficiency (less than 0.1%) of contact-inhibited cells. However, in this study proliferative fibroblasts were also poorly transduced. The same authors examined the transducibility of 293 cells arrested in the S phase by treatments which inhibit DNA synthesis. In this case the transduction efficiency raised 25–35% and was similar in S phase-arrested cells and in cycling cells. This study raises the possibility that efficient transduction might require a function expressed during the S phase rather than DNA synthesis itself.

Zhou and collaborators (1994), using slow cycling primary hematopoietic CD34$^+$ progenitor cells from the human umbilical cord, showed a high transduction frequency (50–75%) of the *neo* marker. Transduction was not enhanced by pre-stimulation of these cells with cytokines. However, the outcome of visible neomycin-resistant clones required cell division and CD34$^+$ progenitors are slow cycling but eventually divide. Zhou and collaborators do not mention the division time of these cells in their cultures. Their results could be explained in the same way as those of Russel and collaborators, if one assumes that the recombinant AAV genome stayed latent after infection and was expressed only when the cells entered S phase.

Even if AAV vectors require the cells to pass through S phase for transduction, they still offer an enormous advantage towards retroviruses which require cell division at the infection (Miller *et al.*, 1990; Roe *et al.*, 1993).

With regard to the study by Russel *et al.*, the 10% transduction efficiency obtained in the rat brain seems particularly high as neurons and glia are post-mitotic in adult brain. Contact-inhibited fibroblasts might not be an optimal model in that they might not reflect the cellular environment of quiescent cells *in vivo*. Transduction of primary cultures of quiescent cells of the CNS just after dissociation could possibly better reflect the *in vivo* situation.

5.6.5 Correction of CNS disorders using AAV vectors

An AAV vector expressing human tyrosine hydroxylase was injected into the denervated striatum of unilateral 6-hydroxydopamine-lesioned rats, a model of Parkinson's disease. Tyrosine hydroxylase immunoreactivity was detectable in striatal neurons and glia for up to 4 months and significant behavioural recovery in lesioned rats treated with the AAV*th* vector versus the control AAV*lacZ* vector was found (see Figure 6). As few as 10^3 transduced cells were sufficient to observe a corrective effect (Kaplitt *et al.*, 1994). These results open the route to the use of AAV vectors for the efficient direct *in vivo* transfer of genes of therapeutic interest in the CNS.

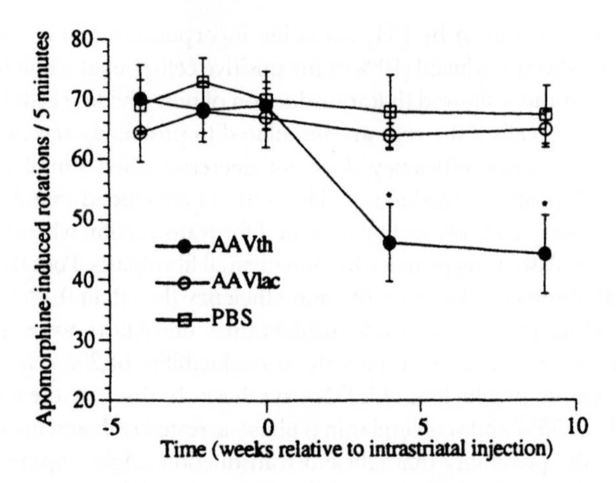

Figure 6 Effect of an AAV transducing vector expressing tyrosine hydroxylase in a rat model of Parkinson's disease. Rotational behaviour was induced in 6-hydroxydopamine-lesioned rats by administration of apomorphine. Animals injected with AAVth demonstrated a significant decrease in rotational rate compared with AAVlacZ or PBS-injected animals. [Reproduced with permission from Kaplitt *et al.* (1994).]

5.6.6 Potential use of AAV vectors to isolate CNS cell lines

CNS cell lines are lacking and normal neurons and oligodendrocytes do not grow in culture. One of the current techniques used to establish cell lines from primary cultures consists of the transduction of oncogenes via a retrovirus vector. In particular, murine O-2A (oligodendrocyte precursors) cells were immortalized using a retroviral vector expressing a thermosensitive mutant of SV40 T antigen (Almazan and McKay, 1992). Such a cell line provides high amounts of dividing cells at 33°C (temperature at which the *tsA58* T antigen is active) which can be switched back to their primary state at 37°C (temperature at which *tsA58* T antigen is inactive). However, retroviruses transduce only actively dividing cells and such an approach is thus not feasible with human neurons and oligodendrocytes which do not grow in culture.

The integration into slowly dividing and possibly, even at low frequency, into non-dividing cells, raises the possibility of using AAV vectors carrying immortalizing genes rather than retroviral vectors to immortalize slowly cycling and possibly quiescent cells of the CNS.

Acknowledgement

We thank Professor Ernest Winocour for critically reading this manuscript.

References

Almazan, G. & McKay, R. (1992) An oligodendrocyte precursor cell line from rat optic nerve. *Brain Res.* **579**, 234–245.

Antoni, B.A., Rabson, A.B., Miller, I.L., Tremple, J.P., Chejanovski, N. & Carter, B.J. (1991) Adeno-associated virus Rep protein inhibits human immunodeficiency virus type 1 production in human cells. *J. Virol.* **65**, 396–404.

Bantel-Schaal, U. (1991) Infection with adeno-associated parvovirus leads to increased sensitivity of mammalian cells to stress. *Virology* **182**, 260–268.

Bantel-Schaal, U. & Stohr. M. (1992) Influence of adeno-associated virus on adherence and growth properties of normal cells. *J. Virol.* **66**, 773–779.

Beaton, A., Palumbo, P. & Berns, K.I. (1989) Expression from the adeno-associated virus p5 and p19 promoters is negatively regulated *in trans* by the Rep protein. *J. Virol.* **63**, 4450–4454.

Berns, K.I. (1990) Parvovirus replication. *Microbiol. Rev.* **54**, 316–329.

Berns, K.I. & Bohenzky, R.A. (1987). Adeno-associated viruses: an update. *Adv. Virus Res.* **32**, 243–306.

Berns, K.I., Pinkerton, T.C., Thomas, G.F. & Hoggan, M.D. (1975) Detection of adeno-associated virus (AAV)-specific nucleotide sequences in DNA isolated from latently infected Detroit 6 cells. *Virology* **68**, 556–560.

Blacklow, N.R. & Rowe, W.P. (1968). Serological evidence for human infection with adeno-associated viruses. *J. Natl. Cancer Inst.* **40**, 314–317.

Blacklow, N.R., Hoggan, M.D., Kapikian, A.Z., Austin, J.B. & Rowe, W.P. (1968) Epidemiology of adeno-associated virus infection in a nursery population. *Am. J. Epidemiol.* **8**, 368–378.

Blacklow, N.R., Hoggan, M.D., Sereno, M.S., Brandt, C.D., Kim, H.W., Parrott, R.H. & Chanock, R.H. (1971) A seroepidemiologic study of adeno-associated virus infections in infants and children. *Am. J. Epidemiol.* **94**, 359–366.

Boissy, R. & Astell, C.R. (1985) *Escherichia coli recBC sbcB recF* hosts permits the deletion-resistant propagation of plasmid clones containing the 5′-terminal palindrome of minute virus of mice. *Gene* **35**, 179–185.

Buller, R.M.L., Janik, J.E., Sebring, E.D. & Rose, J.A. (1981) Herpes simplex types 1 and 2 completely help adenovirus-associated virus replication. *J. Virol.* **40**, 241–247.

Carter, B.J. (1992) Adeno-associated virus vectors. *Curr. Opin. Biotechnol.* **3**, 533–539.

Chang, L.S., Shi, Y. & Shenk, T. (1989) Adeno-associated virus p5 promoter contains an adenovirus E1A inducible element and a binding site for the major late transcription factor. *J. Virol.* **63**, 3479–3488.

Chejanovsky, N. & Carter, B.J. (1989) Mutagenesis of an AUG codon in the adeno-associated virus *rep* gene: effects on viral DNA replication. *Virology* **173**, 120–128.

Chejanovsky, N. & Carter, B.J. (1990) Mutation of a consensus purine nucleotide binding site in the adeno-associated virus *rep* gene generates a dominant negative phenotype for DNA replication. *J. Virol.* **64**, 1764–1770.

Cheung, A., Hoggan, M.D., Hauswirth, W.W. & Berns, K.I. (1980) Integration of the adeno-associated virus genome into cellular DNA in latently infected human Detroit 6 cells. *J. Virol.* **33**, 739–748.

Chiorini, J.A., Weitzman, M.D., Owens, R.A., Urcelay, E., Safer, B. & Kotin, R.M. (1994) Biologically active Rep proteins of adeno-associated virus type 2 produced as fusion proteins in *Escherichia coli*. *J. Virol.* **68**, 797–804.

Dupont, F. (1993) Analyse et exploitation de l'activité antinéoplasique des parvovirus autonomes. PhD thesis, Free University of Brussels, Belgium.

Flotte, T.R., Afione, S.A., Conrad, C., McGrath, S.A., Solow, R., Oka, H., Zeitlin, P.L., Guggino, W.B. & Carter, B.J. (1993a) Stable *in vivo* expression of the cystic fibrosis transmembrane conductance regulator with an adeno-associated virus vector. *Proc. Natl. Acad. Sci. USA* **90**, 10613–10617.

Flotte, T.R., Afione, S.A., Solow, R., Drumm, M.L., Markakis, D., Guggino, W.B., Zeitlin, P.L. & Carter, B.J. (1993b) Expression of the cystic fibrosis transmembrane conductance regulator from a novel adeno-associated virus promoter. *J. Biol. Chem.* **268**, 3781–3790.
Georg-Fries, B., Bederlack, S., Wolf, J. & zur Hausen, H. (1984) Analysis of proteins, helper dependence, and seroepidemiology of a new human parvovirus. *Virology* **134**, 64–71.
Giraud, C., Winocour, E. & Berns, K.I. (1994) Site-specific integration by adeno-associated virus is directed by a cellular DNA sequence. *Proc. Natl. Acad. Sci. USA* **91**, 10039–10043.
Gottlieb, J. & Muzyczka, N. (1988) *In vitro* excision of adeno-associated virus DNA from recombinant plasmids: isolation of an enzyme fraction from HeLa cells that cleaves DNA at Poly(G) sequences. *Mol. Cell. Biol.* **6**, 2513–2522.
Handa, H., Shiroki, K. & Shimojo, H. (1977) Establishment and characterization of KB cell lines latently infected with adeno-associated virus type 1. *Virology* **82**, 84–88.
Hermonat, P.L. (1994a) Adeno-associated virus inhibits human papillomavirus type 16: a viral interaction implicated in cervical cancer. *Cancer Res.* **54**, 2278–2281.
Hermonat, P.L. (1994b) Down-regulation of the human c-*fos* and c-*myc* proto-oncogene promoters by adeno-associated virus Rep78. *Cancer Lett.* **81**, 129–136.
Hermonat, P.L. & Muzyczka, N. (1984) Use of adeno-associated virus as a mammalian DNA cloning vector: transduction of neomycin resistance into mammalian tissue culture cells. *Proc. Natl. Acad. Sci. USA* **81**, 6466–6470.
Hermonat, P.L., Labow, M.A., Wright, R., Berns, K.I. & Muzyczka, N. (1984) Genetics of adeno-associated virus: isolation and preliminary characterization of adeno-associated virus type 2 mutants. *J. Virol.* **51**, 329–339.
Hirt, B. (1967) Selective extraction of polyoma DNA from infected mouse cell cultures. *J. Mol. Biol.* **26**, 365–369.
Hoggan, M.D., Blacklow, N.R. & Rowe, W.P. (1966) Studies of small DNA viruses found in various adenovirus preparations: physical, biological and immunological characteristics. *Proc. Natl. Acad. USA* **55**, 1467–1474.
Hong, G., Ward, P. & Berns, K.I. (1994) Intermediates of adeno-associated virus DNA replication *in vitro*. *J. Virol.* **68**, 2011–2015.
Im, D.S. & Muzyczka, N. (1990) The AAV origin binding proteins Rep68 is an ATP-dependent site-specific endonuclease with DNA helicase activity. *Cell* **61**, 447–457.
Kaplitt, M.G., Leone, P., Samuski, R.J., Xiao, X., Pfaff, D.W., O'Malley, K.L. & During, M.J. (1994) Long-term gene expression and phenotype correction using adeno-associated virus vectors in the mammalian brain. *Nature Genet.* **8**, 148–154.
Kotin, R.M. (1994) Prospects for the use of adeno-associated virus as a vector for human gene therapy. *Hum. Gene Ther.* **5**, 793–801.
Kotin, R.M. & Berns, K.I. (1989) Organization of adeno-associated virus DNA in latently infected Detroit 6 cells. *Virology* **170**, 460–467.
Kotin, R.M., Siniscalco, M., Samulski, R.J., Zhu, X., Hunter, L., Laughlin, C.A., McLaughlin, S. Muzyczka, N., Rocchi, M. & Berns, K.I. (1990) Site-specific integration by adeno-associated virus. *Proc. Natl. Acad. Sci. USA* **87**, 2211–2215.
Kotin, R.M., Menninger, J.C., Ward, D.C. & Berns, K.I. (1991) Mapping and direct visualization of a region-specific viral DNA integration on chromosome 19q13-qter. *Genomics* **10**, 831–841.
Kotin, R.M., Linden, R.M. & Berns, K.I. (1992) Characterization of a preferred site on human chromosome 19q for integration of adeno-associated virus DNA by non-homologous recombination. *EMBO J.* **11**, 5071–5078.
Kumar, S. & Leffak, M. (1991) Conserved chromatin structure in c-*myc* 5′-flanking DNA after viral transduction. *J. Mol. Biol.* **222**, 45–57.
Labow, M.A., Hermonat, P.L. & Berns, K.I. (1986) Positive and negative autoregulation of the adeno-associated virus type 2 genome. *J. Virol.* **60**, 515–524.
Labow, M.A., Graf, L.H. & Berns, K.I. (1987) Adeno-associated virus gene expression inhibits cellular transformation by heterologous genes. *Mol. Cell. Biol.* **7**, 1320–1325.

LaFace, D., Hermonat, P., Wakeland, E. & Peck, A. (1988) Gene transfer into hematopoietic progenitor cells mediated by an adeno-associated virus vector. *Virology* **162**, 483–486.
Laughlin, C.A., Myers, M.W., Risin, D.L. & Carter, B.J. (1978) Defective-interfering particles of the human parvovirus adeno-associated virus. *Virology* **94**, 162–174.
Laughlin, C.A., Tratschin, J.D., Coon, H. & Carter, B.J. (1983) Cloning of infectious adeno-associated virus genomes in bacterial plasmids. *Gene* **23**, 65–73.
Laughlin, C.A., Cardellichio, C.B. & Coon, H.C. (1986) Latent infection of KB cells with adeno-associated virus type 2. *J. Virol.* **60**, 515–524.
Lebkowski, J.S., McNally, M., Okarma, T.B. & Lerch, L.B. (1988) Adeno-associated virus: a vector system for efficient introduction and integration of DNA into a variety of mammalian cell types. *Mol. Cell. Biol.* **8**, 3988–3996.
Mayor, H.D., Drake, S., Stahman, J. & Mumford, D.M. (1976) Antibodies to adeno-associated satellite virus and herpes simplex in sera from cancer patients and normal adults. *Am. J. Obstet. Gynecol.* **126**, 100–104.
McCarty, D.M., Christensen, M. & Muzyczka, N. (1991) Sequences required for the coordinate induction of the AAV p19 and p40 promoters by the Rep protein. *J. Virol.* **65**, 2936–2945.
McCarty, D.M., Pereira, D.J., Zolotukhin, I., Zhou, X., Ryan, J.H. & Muzyczka, N. (1994) Identification of linear DNA sequences that specifically bind the adeno-associated virus Rep protein. *J. Virol.* **68**, 4988–4997.
McLaughlin, S.K., Collis, P., Hermonat, P.L. & Muzyczka, N. (1988) Adeno-associated virus general transduction vectors: analysis of proviral structures. *J. Virol.* **62**, 1963–1973.
Mendelson, E., Smith, M.G., Miller, I.L. & Carter, B.J. (1988a) Effect of a viral rep gene on transformation of cells by an adeno-associated virus vector. *Virology* **166**, 612–615.
Mendelson, E., Smith, M.G. & Carter, B.J. (1988b) Expression and rescue of a nonselected marker from an integrated AAV vector. *Virology* **166**, 154–165.
Mendelson, E., Grossman, Z., Mileguir, F., Rechavi, G. & Carter, B.J. (1992) Replication of adeno-associated virus type 2 in human lymphocytic cells and interaction with HIV-1. *Virology* **187**, 453–463.
Miller, D.G., Adam, M.A. & Miller, A.D. (1990) Gene transfer by retrovirus vectors occurs only in cells that are actively replicating at the time of the infection. *Mol. Cell. Biol.* **10**, 4329–4342.
Muro-Cacho, C.A., Samulski, R.J. & Kaplan, D. (1992) Gene transfer in human lymphocytes using a vector based on adeno-associated virus. *J. Immunother.* **11**, 231–237.
Muzyczka, N. (1992) Use of adeno-associated virus as a general transduction vector for mammalian cells. *Curr Top. Microbiol. Immunol.* **158**, 97–129.
Ni, T.-H., Zhou, X., McCarty, D., Zolotukhin, I. & Muzyczka, N. (1994) *In vitro* replication of adeno-associated virus DNA. *J. Virol.* **68**, 1128–1138.
Oelze, I., Rittner, K. & Sczakiel, G. (1994) Adeno-associated virus type 2 *rep* gene-mediated inhibition of basal gene expression of human immunodeficiency virus type 1 involves its negative regulatory functions. *J. Virol.* **68**, 1229–1233.
Philip, R., Brunette, E., Kilinski, L., Murugesh, D., McNally, M.A., Ucar, K., Rosenblatt, J., Okarma, T.B. & Lebkowski, J.S. (1994) Efficent and sustained expression in primary T lymphocytes and primary and cultured tumor cells mediated by adeno-associated virus plasmid DNA complexed to cationic liposomes. *Mol. Cell. Biol.* **14**, 2411–2418.
Podsakoff, G., Wong, K.K. & Chaterjee, S. (1994) Efficient gene transfer into nondividing cells by adeno-associated virus-based vectors. *J. Virol.* **68**, 5656–5666.
Redeman, B.E., Mendelson, E. & Carter, B.J. (1989) Adeno-associated virus Rep protein synthesis during productive infection. *J. Virol.* **63**, 873–882.
Roe, T., Reynolds, T.C., Yu, G. & Brown, P.O. (1993) Integration of murine leukemia virus DNA depends on mitosis. *EMBO J.* **12**, 2099–2108.
Ruffing, M., Zentgraf, H., Kleinschmidt, J.A. (1992) Assembly of viruslike particles by recom-

binant structural proteins of adeno-associated virus type 2 in insect cells. *J. Virol.* **66**, 6922–6930.

Russel, D.W., Miller, A.D. & Alexander, I.E. (1994) Adeno-associated virus vectors preferentially transduce cells in S phase. *Proc. Natl. Acad. Sci. USA* **91**, 8915–8919.

Samulski, R.J. (1993) Adeno-associated virus: integration at a specific chromosomal locus. *Curr. Opin. Genet. Dev.* **3**, 74–80.

Samulsi, R.J., Berns, K.I., Tan, M. & Muzyczka (1982) Cloning of adeno-associated virus into pBR322: rescue of intact virus from the recombinant plasmid in human cells. *Proc. Natl. Acad. Sci USA* **79**, 2077–2081.

Samulski, R.J., Chang, L.S. & Shenk, T. (1987) A recombinant plasmid from which an infectious adeno-associated virus genome can be excised *in vitro* and its use to study viral replication. *J. Virol.* **61**, 3096–3101.

Samulski, R.J., Chang, L.S. & Shenk, T. (1989) Helper-free stocks of recombinant adeno-associated viruses: Normal integration does not require viral gene expression. *J. Virol.* **63**, 3822–3828.

Samulski, R.J., Zhu, X., Xiao, X., Brook, J.D., Housman, D.E., Epstein, N. & Hunter, L.A. (1991) Targeted integration of adeno-associated virus (AAV) into human chromosome 19. *EMBO J.* **10**, 3941–3950.

Schlehöfer, J.R., Heilbronn, R., Georg-Fries, B. & zur Hausen, H. (1983) Inhibition of initiator-induced SV40 gene amplification in SV40-transformed chinese hamster cells by infection with a defective parvovirus. *Int. J. Cancer* **32**, 591–595.

Schlehöfer, J.R., Ehrlar, M. & zur Hausen, H. (1986) Vaccinia virus, herpes simplex virus, and carcinogens induce DNA amplification in a human cell line and support replication of a helpervirus dependent parvovirus. *Virology* **152**, 110–117.

Shelling, A. & Smith, M.G. (1994) Targeted integration of transfected and infected adeno-associated virus vectors containing the neomycin resistance gene. *Gene Ther.* **1**, 165–169.

Siegl, G., Bates, R.C., Berns, K.I., Carter, B.J., Kelly, D.C., Kurstak, E. & Tatersall, P. (1985) Characteristics and taxonomy of *Parvoviridae. Intervirology* **23**, 61–73.

Snyder, R.O., Im, D.S. & Muzyczka, N. (1990) *In vitro* resolution of covalently joined AAV chromosome ends. *J. Virol.* **64**, 6204–6213.

Sprecher-Goldberger, S., Thiry, L., Lefebvre, N., Debegel, D. & DeHalleux, F. (1971) Complement-fixation antibodies to adenovirus-associated viruses, adenoviruses, cytomegaloviruses and herpes simplex viruses in patients with tumors and in control individuals. *Am. J. Epidemiol.* **94**, 351–358.

Srivastava, A., Lusby, E.W. & Berns, K.I. (1983) Nucleotide sequence and organization of the adeno-associated virus 2 genome. *J. Virol.* **45**, 555–564.

Tenenbaum, L., Darling, J.L. & Hooghe-Peters, E.L. (1994) Adeno-associated virus (AAV) as a vector for gene transfer into glial cells of the human central nervous system. *Gene Ther.* **1** (suppl. 1), 80.

Tratschin, J.D., West, M.H., Sandbank, T. & Carter, B.J. (1984) A human parvovirus, adeno-associated virus, as a eucaryotic vector: transient expression and encapsidation of the procaryotic gene for chloramphenicol acetyltransferase. *Mol. Cell. Biol.* **4**, 2072–2081.

Tratschin, J.D., Miller, I.L., Smith, M.G. & Carter, B.J. (1985) Adeno-associated virus vector for high-frequency integration, expression, and rescue of genes in mammalian cells. *Mol. Cell. Biol.* **5**, 3251–3260.

Tratschin, J.D., Tal, Y. & Carter, B.J. (1986) Negative and positive regulation *in trans* of gene expression from adeno-associated virus vectors in mammalian cells by a viral *rep* gene product. *Mol. Cell. Biol.* **6**, 2884–2894.

Vincent, K.A., Moore, G.K. & Haigwood, N.L. (1990) Replication and packaging of HIV envelope genes in a novel adeno-associated virus vector system. *Vaccine* **90**, 353–359.

Walsh, C.E., Liu, J.M., Xiao, X., Young, N.M., Nienhuis, A.W. & Samulski, R.J. (1992) Regulated high level expression of a human gamma-globin gene introduced into erythroid cells by an adeno-associated virus vector. *Proc. Natl. Acad. Sci. USA* **89**, 7257–7261.

Walz, C. & Schlehöfer, J.R. (1992) Modification of some biological properties of HeLa cells containing adeno-associated virus DNA integrated into chromosome 17. *J. Virol.* **66**, 2990–3002.
Walz, C., Schlehöfer, J.R., Flentje, M., Rudat, V. & zur Hausen, H. (1992) Adeno-associated virus sensitizes HeLa cell tumors to gamma rays. *J. Virol.* **66**, 5651–5657.
Weitzman, M.D., Kyöstiö, S.R.M., Kotin, R.M. & Owens, R.A. (1994) Adeno-associated virus (AAV) Rep proteins mediate complex formation between AAV DNA and its integration site in human DNA. *Proc. Natl. Acad. Sci. USA* **91**, 5808–5812.
Winocour, E., Callaham, M.F. & Huberman, E. (1988) Perturbation of the cell cycle by adeno-associated virus. *Virology* **167**, 393–399.
Winocour, E., Puzis, L., Etkin, S., Koch, T., Danovitch, B., Mendelson, E., Shaulian, E., Karby, S. & Lavi, S. (1992) Modulation of the cellular phenotype by integrated adeno-associated virus. *Virology*, **190**, 316–329.
Yakobson, B., Koch, T. & Winocour, E. (1987) Replication of adeno-associated virus in synchronized cells without the addition of a helper virus. *J. Virol.* **61**, 972–981.
Yakobson, B., Hrynko, T.A., Peak, M.J. & Winocour, E. (1989) Replication of adeno-associated virus in cells irradiated with UV light at 254 nm. *J. Virol.* **63**, 1023–1030.
Yalkinoglu, A.O., Heilbronn, R., Bürkle, A., Schlehöfer, J.R. & zur Hausen, H. (1988) DNA amplification of adeno-associated virus as a response to cellular genotoxic stress. *Cancer Res.* **48**, 3123–3129.
Yang, Q., Chen, F. & Tremple, J.P. (1994) Characterization of cell lines that inducibly express the adeno-associated virus Rep proteins. *J. Virol.* **68**, 4847–4856.
Zhou, S.Z., Broxmeyer, H.E., Cooper, S., Harrington, M.A. & Srivastava, A. (1993) Adeno-associated virus 2-mediated gene transfer in murine hematopoietic progenitor cells. *Exp. Hematol.* **21**, 928–933.
Zhou, S.Z., Cooper, S., Kang, L.Y., Ruggieri, L., Heimfeld, S., Srivastava, A. & Broxmeyer, H.E. (1994) Adeno-associated virus 2-mediated high efficiency transfer into immature and mature subsets of hematopoietic progenitor cells in human umbilical cord blood. *J. Exp. Med.* **179**, 1867–1875.

CHAPTER 6

HERPES SIMPLEX VIRUS-BASED VECTORS

R. S. Coffin and D. S. Latchman

Department of Molecular Pathology, University College London Medical School, The Windeyer Building, 46 Cleveland Street, London W1P 6DB, UK

Table of Contents

6.1 Introduction

Herpes simplex viruses (HSV) 1 and 2 are large DNA viruses with genomes of around 150 kb which replicate in the nucleus of infected cells and infect sensory neurons. They are responsible for the symptoms of cold sores and genital herpes, respectively, and on rare occasions infection can spread to the central nervous system (CNS) causing an often fatal encephalopathy. HSV1 is endemic within the human population with approximately 80% of individuals testing positive for antibodies to HSV1, although a considerably lower percentage show symptoms of the disease. HSV2 is considerably less prevalent than HSV1. The genomes of HSV1 and HSV2 consist of a linear DNA molecule with a long and a short unique region each flanked by pairs of identical regions (Figure 1). Thus genes present in the repeated regions are present in two copies per genome and those in the unique regions in only one. Before replication the genome circularizes following which concatameric progeny DNA is produced before packaging of individual genomes into virus particles. Due to a high frequency of recombination mediated by the terminal regions of each of the repeats, the orienta-

Genetic Manipulation of the Nervous System
ISBN 0-12-437165-5

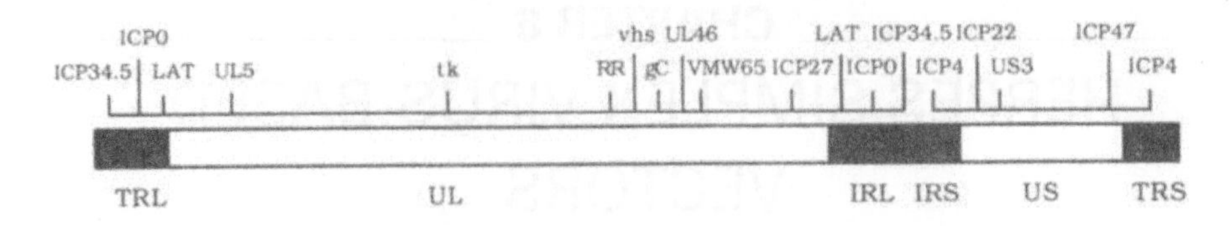

Figure 1 A schematic representation of the HSV1 genome. The positions of the terminal and internal short and long repeats (TRS, IRL, TRS, IRL), and the unique long and short regions (UL and US) are shown (see text), as are the positions of the genes mentioned in the text. ICP, infected cell protein; LAT, latency associated transcript; tk, thymidine kinase; RR, ribonucleotide reductase; vhs, virion host shutoff protein; gC, glycoprotein C.

tions of the long and short unique regions often change with respect to one another during replication, and thus herpes virus preparations exist as mixtures of the four possible DNA structural isoforms (see Roizman, 1979).

An important feature of the herpes virus life cycle is the ability of herpes viruses to remain latent within infected neuronal cells, often for the lifetime of the cell, but which can reactivate to a productive infection after suitable but as yet largely unknown molecular signals (reviewed by Latchman, 1990). Thus upon infection of a neuron at a peripheral site an infecting virion migrates to the cell body where it can either initiate a productive infection, or the virus enters latency. The initiation of a productive infection is dependent on the interaction of the herpes virion protein VMW65 with activating cellular transcription factors to produce a protein complex capable of trans-activating herpes immediate early genes (reviewed by O'Hare, 1993). Likewise it appears that the induction of latency is also dependent on cellular transcription factors but in this case repressing transcription factors which are only present in neurons (Lillycrop *et al.*, 1994). Thus the particular balance of repressing and activating transcription factors within an infected cell probably decides whether a productive or latent infection is established.

During latency the HSV genome is in a largely inactive chromatin-like structure (Deshmane and Fraser, 1989), with only a small region of the genome within the long repeat region (and thus present in two copies per genome) being transcribed (Stevens *et al.*, 1987). However the function of this unusual transcription unit, consisting of a large low abundance RNA and smaller RNAs which are highly abundant during latency and thought to be introns spliced from the larger species (Devi-Rao *et al.*, 1991; Farrell *et al.*, 1991), is as yet largely unknown. Suggestions for the function of these latency-associated transcripts (LATs) have included production of an encoded protein (although none has reliably been detected), an antisense role for the transcripts preventing expression of complementary immediate (IE) early genes, or the very fact of transcription allowing the genome to remain in a structure suitable to allow gene expression during the reactivation from latency. However, while no specific role has yet been assigned to the LATs, deletion of the LAT region from HSV1, while apparently having no effect on the establishment of latency, does appear to affect the efficiency of the reactivation process (e.g. Leib *et al.*, 1989; Trousdale *et al.*, 1991).

6.2 The potential of herpes simplex virus as a vector for the nervous system

It has often been suggested that HSV might be an ideal candidate vector virus to engineer the nervous system as it has evolved a life cycle in which infection is specifically targeted to neuronal cells, and it is capable of producing a latent infection which can be maintained for the lifetime of the cell. Thus the possibility exists of very long-term transgene expression (years) after once-only administration of such a vector system. Other viruses which can remain latent in infected cells by insertion of the virus genome into the host chromosome, such as retroviruses, require actively replicating DNA for insertion, and thus cannot infect neuronal cells in such a persistent manner. However, while the ability of HSV to infect terminally differentiated cells is useful, it is not unique as most other viruses which have been used as vectors for other cell types will infect and produce at least transient transgene expression in neurons. Particularly, it has been shown that disabled adenovirus vectors can express a transgene for at least 60 days in many neuronal cell types (see Chapters 2 and 3), and adeno-associated virus can give high level expression in a number of regions of the brain for at least 4 months (Kaplitt *et al.*, 1994; Chapter 5). The potential for longer term genetic correction using these viruses, however, is as yet unknown. Thus herpes viruses, while in many cases not necessarily providing specific advantages over other potential vector systems to engineer neurons, might with suitable development be of particular use where very long term transgene expression is necessary, and have the added advantage that with suitable deletions from the HSV genome very large DNA insertions could be made (possibly up to around 30 kb if all non-essential regions were deleted). Current adenovirus vectors will only allow insertions of 7–8 kb (although deletion of non-essential genes and expression of other genes in *trans* may well expand this) and adeno-associated virus vectors have a maximum possible insertion size of c. 4.5 kb.

6.3 Herpes simplex virus and other viral vectors compared

Wild-type HSV is highly pathogenic, and as a result direct inoculation of the nervous system has fatal consequences in test animals. It, like most other viral vectors, must therefore be disabled in some way.

HSV is also a large and structurally complex virus and interacts with the host cell at a number of levels, all of which alter the normal functioning of the cell to promote virus survival, replication and spread. These interactions occur both as a consequence of the delivery of particular viral proteins to the cell with the infecting virus particle (HSV particles are composed of approximately 40 virus-encoded proteins; for a review see Rickson, 1993) and as the result of protein synthesis following viral gene expression. Thus even if HSV is disabled such that viral gene expression is entirely prevented, and most cytopathic effects shown by HSV vectors have been atributed to HSV IE gene expression (see Johnson *et al.*, 1994), the effects of virion proteins entering the cell may also be important. However, while this is an aspect of the use of HSV as a vector which has so far been little studied, it does appear that at a gross level cells infected with virus particles from which viral gene expression is not possible (UV irradiated virus) appear to survive and remain healthy (Johnson *et al.*, 1992). While adenovirus is also a large and complex virus, it is considerably less so than HSV. Thus while considerable cytopathic effects and immune responses have been shown by adenovirus vectors, particularly first generation vectors in which viral gene expression has not been fully prevented, it is possible that the effects of virion proteins on host cell physiology might be easier to control. In this aspect of vector development adeno-associated virus would seem ideal as infecting virions are composed of only three protein types forming a simple capsid, and thus only minimal effects on the cell might be expected. However in this case the physiological effects of the possible contamination of vector stocks with the helper adenovirus necessary for growth may be of greater importance.

A further problem with HSV as a vector has been that while insertions of marker genes into the HSV genome give good short-term expression levels in neurons, gene expression is in many cases then shut off, presumably as the DNA forms the transcriptionally inactive and stable structure maintained during latency. It appears that in other vector systems (including plasmids which can be replicated and packaged into HSV particles; see later), as long as gene expression is initiated, expression will probably continue until the DNA is degraded within the cell. Thus although HSV has the potential advantage that it can remain latent within a cell, it appears that the formation of a DNA structure suitable to allow this to occur also produces the undesirable effect that unless the site of insertion into the HSV genome is carefully chosen, expression of the inserted gene is rapidly switched off. Not until this problem has been fully addressed, and the parameters which allow gene expression during latency have been more fully examined (see later, and Chapter 8), will the potential of HSV for very long term gene expression be able to be realized.

In view of these two fundamental properties of HSV (pathogenicity and transcriptional shut-down during latency), the remainder of this chapter describes the efforts which have been made both to reduce the pathogenicity and cytopathic effects of HSV1, and also to produce vectors capable of long-term gene expression *in vivo*.

6.4 Types of herpes simplex virus vector

The development of herpes vectors has until recently taken two routes. The first route has utilized deletions of essential genes and other genes involved in the production of cytopathic effects from the virus. The essential genes are complemented in culture allowing growth, but their absence from progeny virus renders them avirulent *in vivo*. The second approach, using defective viruses or amplicons, is based on a plasmid vector, allowing growth and genetic manipulation in bacteria, into which the potential transgene can be inserted. The presence of an HSV origin of replication and a packaging signal on the plasmid allows growth and packaging of the amplicon in cultured cells in the presence of a helper virus.

Herpes genomes are large (c. 150 kb), and thus cannot be manipulated as infectious plasmid clones. Rather, small regions of the genome are isolated as restriction fragments and cloned into plasmid vectors, where alterations (e.g. insertion of a promoter/transgene cassette, or deletion of a gene) can be made. As long as the altered sequences are flanked on each side by at least 1 kb of unaltered viral sequence the modified region can then be introduced into the virus genome by homologous recombination. Here, the plasmid DNA is mixed with DNA purified from HSV virions and introduced into susceptible cells by calcium phosphate transfection (Stow and Wilkie, 1976). As the purified HSV DNA is itself infectious under these conditions, virus plaques will form, and as replication has occurred in the presence of a plasmid containing homologous sequences (with an insertion or deletion) a small percentage of the plaques (approx. 1%) will have been generated from virus into which the altered sequence has recombined. Selection of the required plaques is either made by a phenotypic difference as compared to the wild type [e.g. marker gene activity, sensitivity to acyclovir if insertions are into the *tk* locus (giving resistance to the drug), or in the case of an essential gene deletion, growth only in a complementing cell line], or in the absence of such a selection by the random screening of plaques by Southern or dot blotting of isolated DNA. These techniques are used both for the insertion of transgenes into disabled viruses (and further disablement), and also for the development of disabled helper viruses for amplicon growth.

6.5 Disabled herpes viruses

Two types of disabled herpes viruses have been produced: (1) viruses with deletions in essential genes, and which thus require complementation for growth in culture; and (2) viruses with deletions in non-essential genes, not required for growth in culture, but which are required for growth in neurons *in vivo*. Viruses have also been produced in which deletions of both types have been made.

The first herpes virus vectors produced belonged to the second category above, with insertion of either an *Escherichia coli lacZ* gene (Ho and Mocarski, 1988) or a hypoxanthine phosphoribosyltransferase gene (HPRT; Palella *et al.*, 1988) into the thymidine kinase locus (*tk*) of HSV1. Both viruses gave activity for around 5 days in mouse dorsal root ganglia (DRGs; HPRT) or brains (*lacZ*). *Tk* mutants show reduced virulence in neurons, although they are lethal after high titre inoculation (Palella *et al.*, 1989). Thus they allowed the demonstration of the potential utility of HSV as a vector but provide insufficient safety for use in gene therapy.

More extensively disabled viruses were produced by deletions or insertions in the essential IE gene ICP4 (Dobson *et al.*, 1990; Chiocca *et al.*, 1990; Johnson *et al.*, 1992), which trans-activates early and late HSV genes. This provides an absolute block to virus replication and thus viruses with transgene insertions into this gene are incapable of producing a productive infection except in a complementing cell line (usually E5 cells; DeLuca *et al.*, 1985). However while no virus replication occurs and most virus proteins are not produced, the other IE genes (ICP0, ICP22, ICP27 and ICP47) and ribonucleotide reductase (*RR*; ICP6) are expressed. The IE genes have been shown alone or in combination to be responsible for considerable cytopathic effects in infected cells *in vitro* (Johnson *et al.*, 1994). Thus these mutants, while being relatively avirulent, produce localized necrosis around the site of inoculation in the CNS (Chiocca *et al.*, 1990), and although little untoward effect is seen in the peripheral nervous system (PNS; Dobson *et al.*, 1990), produce considerable cytopathic effects in cultured cells (Johnson *et al.*, 1992). Other IE genes have also been inactivated either individually (ICP0, ICP22, ICP27 and ICP47) or in combination (ICP4+ICP22, ICP4+ICP47; see Johnson *et al.*, 1992). Deletion of IE genes other than ICP4 or ICP27, however, does not entirely prevent virus replication, and none of these viruses is significantly less cytopathic in culture than deletions in ICP4 alone. Thus while insertion of *lacZ* into ICP0 instead of into ICP4 gives a larger area of staining in the brains of intracranially inoculated mice, the cytopathic effects produced are more marked and more extensive due to limited virus replication and spread (Chiocca *et al.*, 1990).

Like *tk*, deletion of other virus genes can give a phenotype allowing growth in culture but giving reduced virulence in neurons *in vivo*. These genes include US3 (a protein kinase gene; Leader and Purves, 1988), ICP22 (an IE gene required for growth in some cell types: Sears *et al.*, 1985), ICP345 (the so-called viral neurovirulence factor; Chou *et al.*, 1990), ICP6 (the large subunit of ribonucleotide reductase; Cameron *et al.*, 1988), UL5 (a component of the origin binding complex; Bloom and Stevens, 1994), UI46 (a protein which modifies VMW65; Zhang *et al.*, 1991) and VMW65 (here the gene is not deleted as it encodes an essential structural protein, but a small inactivating insertion abolishes its ability to trans-activate IE genes; Ace *et al.*, 1989). Such VMW65 mutants grow in culture at high multiplicity of infection (MOI) or in the presence of hexamethylene bisacetamide (McFarlane *et al.*, 1992). However, *in vivo* the failure to transactivate IE genes induces a latency-like infection (although at high inoculation titres some virus replication can be observed), and the lack of IE gene expression considerably reduces the cytopathic effects of the virus in infected cells, at least at low MOI. ICP34.5 deletions, which provide c. 10^6-fold greater LD_{50} than wild-type HSV in mice after intracranial inoculation (Chou *et al.*, 1990), or *VMW65* inactivation still allow the virus to enter the latent state (Steiner *et al.*, 1990; Robertson *et al.*, 1992), which may be important for very long term transgene expression.

Deletions in both essential and non-essential genes may provide viruses with better characteristics than deletions in essential or non-essential genes alone. Thus viruses have been produced with deletions in ICP4 together with VMW65 (Johnson *et al.*, 1994), which provide the complete replication block of ICP4 deletion mutants, together with reduced cytopathic effects (at least in cultured cells) as expression of the other IE genes is reduced. Further deletions were also made in *UL41*, the virion host shutoff protein (*vhs*), together with ICP4 or with ICP4 and VMW65. Vhs is a protein carried in the virus particle and responsible for shutting down host in favour of virus protein synthesis, and thus might be expected to induce considerable CPE in infected cells. However, although it can be shown that host protein synthesis is less affected in cells infected with *vhs* null mutants than in wild-type virus, cells infected with double or triple mutants in which UL41 has been deleted (ICP4+UL41 or ICP4+VMW65+UL41) show little difference in cytopathic effects to the ICP4 or ICP4+VMW65 mutants alone, at least in culture.

Thus from the above it might be expected that an optimal disabled virus may be produced by deletion of ICP4, ICP27 (and possibly other IE genes) and UL41, together with inactivation of VMW65. A further absolute block to replication in neurons may be provided by deletion of ICP34.5, and/or of other genes which reduce the efficiency of growth in neurons. However, while all these deletions may be desirable, growth in culture even in complementing cell lines may be considerably impaired (Johnson *et al.*, 1994), and production of high titre virus stocks may prove to be problematical with some of the more highly disabled mutants.

6.6 Promoters for transgene expression using disabled viruses

The earliest herpes vectors expressed HPRT or *lacZ* from an insertion into the *tk* locus of HSV1 using the *tk* promoter (Palella *et al.*, 1988), the ICP4 promoter (Ho and Mokarski, 1988) or the ICP8 promoter (Ho and Mokarski, 1988; an early gene), and while these gave expression in the short term (c. 4 days) in the PNS or CNS of mice, much longer term expression was not possible. This indicated that either the promoters or position of insertion were unsuitable for expression during latency. Further work has also shown that insertion of marker genes into areas of the genome not usually active during latency or under the control of a promoter not usually active during latency will not allow long-term gene expression. Thus insertion of an HSV glycoprotein C (gC) promoter/*lacZ* or IE cytomegalovirus promotor/*lacZ* cassette into the US3 gene (Fink *et al.*, 1992) gives expression in only the short term in the mouse CNS, as does insertion of *lacZ* into ICP0 under the control of the ICP0 promoter (Chiocca *et al.*, 1990), into ICP4 under the control of the ICP6 promoter (Chiocca *et al.*, 1990), or into gC under a number of viral or cellular promoters (Lokensgard *et al.*, 1994). However, somewhat surprisingly, expression of *lacZ* in a position antisense to ICP4 (and inactivating the gene) under the control of a Moloney murine leukaemia virus long terminal repeat (MMLV LTR) promoter, allows continued expression for at least 24 weeks in the PNS (mouse DRGs), although it gives more limited expression in the CNS (motor neurons of the hypoglossal nucleus) where no *lacZ* transcripts could be detected after 5 weeks (although expression had probably ceased considerably earlier; Dobson *et al.*, 1990).

An obvious method by which it might be possible to express transgenes in the long term using an HSV vector might be to utilize the natural ability of HSV to express the LATs during latency, either by direct insertion of transgenes into the region of the virus from which the LATs are expressed, or by expression of genes from LAT-derived

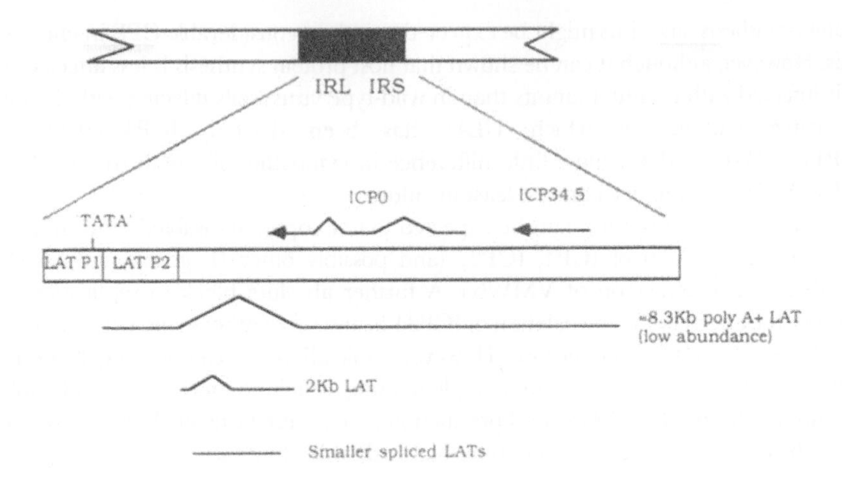

Figure 2 A schematic representation of the latently active region of HSV1. The positions of promoter elements (LAT P1 and LAT P2) are shown, together with the position of the TATA box and of the RNA transcripts within the region. Further details are given in the text.

promoters at ectopic sites within the genome. Both these approaches have been used with varying degrees of success.

The LATs are transcribed from the long repeat regions of the genome (Figure 2) and consist of a large low abundance c. 8.3 kb poly-adenylated transcript, a high abundance c. 2 kb, non-polyadenylated, nuclear transcript which may be spliced from the larger transcript, and transcripts of c. 1.5 kb which are further spliced products of the 2 kb species (see earlier). The 5′ terminus of the large transcript has been mapped to a position c. 600 bp upstream of the 5′ terminus of the 2 kb LAT and just downstream of a TATA box containing promoter (Dobson *et al.*, 1989). This promoter, while giving activity in most cell types, contains specific elements conferring enhanced activity in neurons (Batchelor and O'Hare, 1990, 1992; Zwaagstra *et al.*, 1990), and will here be referred to as LAT P1. A further TATA-less promoter element, which can act independantly from LAT P1, has recently been described in the region between the start of the large transcript and the start of the 2 kb LAT (Goins *et al.*, 1994), suggesting that at some point during the infection cycle the 2 kb LAT may be expressed separately from the large LAT. This promoter will here be referred to as LAT P2.

Early work identifying the LAT P1 promoter and demonstrating the possible utility of the promoter for transgene expression, inserted a rabbit β-globin RNA at a position just downstream of the LAT P1 TATA box (Dobson *et al.*, 1989). β-Globin RNA was detectable for at least 3 weeks in the DRGs of mice after foot-pad inoculation, although the level of RNA was considerably reduced during latency. About the same time Ho and Mocarski obtained *lacZ* expression for at least 8 weeks in the trigeminal ganglia of mice by insertion of the gene 136 bp upstream of the 5′ terminus of the 2 kb LAT (effectively after LAT P2; Ho and Mocarski, 1989). Here the staining was punctate within a small number of cells during latency, and staining could not be detected in vero cells in culture. This was, however, probably the first virus produced capable of expressing a transgene long term during latency (see later).

Further work to utilize LAT P1 by insertion of either a *lacZ* or nerve growth factor (NGF) gene just downstream of the TATA box (Margolis *et al.*, 1993) showed that while expression levels were high in mouse DRGs at early times after infection, by day 21 no transcripts could be detected. Similar results were obtained when a LAT P1/*lacZ* cassette was inserted into the gC locus of HSV (Margolis *et al.*, 1992; Dobson *et al.*, 1995), where although it was apparent that *lacZ* was expressed during the early stages of latency, LAT P1 alone was insufficient for long-term expression as staining steadily decreased over the course of a few weeks. Thus from the work of Ho and Mocarski and other work in which a GUS gene inserted c. 400 bp down-stream of the LAT P1 TATA box (i.e. within LAT P2; Wolfe *et al.*, 1992) was active in the CNS and PNS of mice for at least 4 months, it was indicated that regions downstream of the LAP P1 promoter were required for continued gene expression during latency. These regions could induce a change in the structure of the transcribed region compared to the remainder of the genome during latency, or act as further factor-binding promoter elements required for long-term expression but which are not contained within LAT P1, or a combination of the two. Demonstration of promoter activity by this region (LAT P2) by plasmid transfection experiments (Goins *et al.*, 1994) further suggested an important function in the continued transcription of the 2 kb and smaller LATs during latency. Moreover, when in this same study a LAT P2/*lacZ* cassette was inserted into the HSV gC locus stable expression (>300 days) was demonstrated in the DRGs of mice in a similar punctate fashion as seen by Ho and Mocarski after insertion of *lacZ* after LAT P2 in the LAT region. Here again staining could not be detected from replicating virus *in vitro*. However, the effect of deletion of LAT P2 on LAT expression, and thus its functional significance *in vivo*, is as yet unknown. Thus, while LAT P1/transgene fusions will only give short-term expression either within the LAT region or ectopically from another region of the genome, LAT P2 either alone at an ectopic site or in combination with LAT P1 in the LAT region, will give gene expression for an extended period of time, in the PNS at least, although due to the punctate nature of the staining expression levels are likely to be relatively low.

The early demonstration that an MMLV LTR/*lacZ* fusion could give long-term expression in the PNS when inserted into ICP4 (Dobson *et al.*, 1990), and that other promoters in ICP4 were only active in the short term (Chiocca *et al.*, 1990), suggested that something about the promoter and its proximity to the region of the genome actively transcribed during latency allowed continued latent activity. Thus a chimaeric promoter consisting of approximately 800 bp of the LAT P1 promoter, but not including the TATA box (and thus containing sequences necessary for neuronal expression but no transcription start site), followed by the MMLV LTR (containing possible elements necessary for long-term expression and a transcript start site) fused to *lacZ*, together with a number of other cellular and viral promoter/*lacZ* constructs, were separately inserted into the gC locus of HSV (Lokensgard *et al.*, 1994). Of these promoters only the MMLV LTR/LAT P1 fusion gave long-term activity in mouse DRGs, and unlike the punctate staining produced by the LAT P2/*lacZ* viruses above, activity could be shown at a high level in culture, during the acute phase of infection and undiminished during latency for at least 42 days after inoculation. Thus it appears that not only can the MMLV LTR substitute for the LAT P2 region in conferring long-term activity to LAT P1, but it can also, unlike LAT P2, allow expression levels to continue at levels close to those shown early in infection. However, although this promoter combination is the most promising so far for long-term gene expression from an HSV vector during latency, the absolute time limits for continued expression and its activity in the CNS have yet to be tested. Another factor which may be important for the continued expression observed using this promoter combination is that between the c. 800 pb LAT P1 sequence and the c. 450 pb MMLV LTR sequence used, approximately 800 bp of plasmid DNA was included due to the cloning procedure, and it is possible that this DNA could perform an important spacer function for continued activity by mimicking the overall structure of the LAT transcription unit.

6.7 Defective herpes viruses (amplicons)

An alternative method by which transgenes can be delivered to target cells using HSV has been developed following the observation that defective herpes virions accumulate in serially propagated virus stocks in culture, particularly if infection is performed at a high multiplicity (Frenkel, 1981). The DNA encapsidated in these virions consists of approximately 150 kb head-to-tail reiterations of short HSV sequences (as little as 3 kb) containing an HSV replication origin and a packaging signal, but which are themselves incapable of autonomous replication. Thus if a transgene were inserted into such a defective virus, not only would the vector be unable to replicate after delivery to target cells, but multiple gene copies would be contained in each virion, possibly resulting in high level transgene expression.

The first demonstration that such a system could be used as a vector was made in 1982 when Spaete and Frenkel cloned the 3.9 kb reiterated sequence from one such defective virus into a plasmid vector (Spaete and Frenkel, 1982). It was shown that this plasmid was amplified to high levels and packaged as tandem repeats after transfection into cultured cells in the presence of a wild-type helper virus. Subsequently a chicken ovalbumin gene was expressed at high levels using a similar system (Kwong and Frenkel, 1985). The determination of the precise sequences necessary for replication and packaging allowed the construction of an improved plasmid system containing only these sequences and gave high level expression of the chloramphenicol acetyl transferase gene under the control of a number of HSV promoters (Stow *et al.*, 1986).

While these plasmid-based systems can easily be manipulated and provide high level expression (at least in culture) from an HSV vector, they still require the presence of a helper virus for growth which will be present in the resulting defective virus stock. Thus to enable gene delivery to target cells without subsequent cell death the helper virus must also be disabled in some way. The first non-wild-type helper viruses used (tsK) had a temperature-sensitive mutation in the IE3 gene (ICP4) allowing growth at

31°C, at which temperature defective viruses were propagated, but which prevented growth at the non-permissive temperature of 37°C (Davison *et al.*, 1984). This system was used to prepare an HSV IE4/5 promoter/*LacZ*-containing defective virus stock which gave high level activity in cultured rat PNS neurons (Geller and Breakefield, 1988) and CNS neurons (Geller and Freese, 1990) and in a number of cultured human cell types (Boothman *et al.*, 1989). A similar plasmid also grown using tsK as a helper virus, with *lacZ* under CMV IE promoter control, gave *lacZ* expression *in vivo* in the rat brain for at least 2 weeks (Kaplitt *et al.*, 1991). Subsequently, these same or similar systems were used to express a number of physiologically active genes in a number of cultured neuronal cell types, inducing a physiological effect and demonstrating the utility of the approach for *in vitro* studies of gene function by HSV amplicon-mediated delivery (for further details see Chapter 7). These genes have included adenylate cyclase (Geller *et al.*, 1993), interleukin-2 (Eizenberg *et al.*, 1994), NGF (Geschwind *et al.*, 1994), the glucose transporter gene (Ho *et al.*, 1993), and growth-associated protein-43 (Verhaagen *et al.*, 1994), although in some cases the amplicon was propagated using an IE3-deleted helper virus in place of tsK (see below).

While tsK served as a helper virus suitable to demonstrate the potential of the vector system *in vitro*, and to allow experiments using physiologically active genes in cultured cells, the relatively high level reversion of tsK to a wild-type phenotype does not allow its effective use *in vivo*. Thus a helper virus (D30EBA) was developed with a deletion in the IE3 gene, which could be complemented in culture, and which gave a lower reversion frequency than tsK (Geller *et al.*, 1990). This system has been used to alter neuronal physiology *in vivo*, initially by expression of NGF from an amplicon vector (pHSVngf; Federoff *et al.*, 1992). In this case it was shown that the superior cervical ganglion in rats maintained tyrosine hydroxylase (TH) levels after axotomy (i.e. removing the natural supply of NGF necessary for TH expression) when treated with the pHSVngf virus, but not when treated with a similar virus expressing *lacZ*. More recently, and in the most effective use of an HSV vector to date, it was shown that an amplicon vector virus expressing TH could induce long-term behavioural recovery (at least 1 year) when delivered to the striatum of 6-hydroxydopamine-lesioned rats, a model for Parkinson's disease (During *et al.*, 1994). These experiments not only showed that gene expression could be induced in the brain at physiologically active levels and at an appropriate site for phenotypic recovery in this model of Parkinson's disease, but also that when using an amplicon vector, unlike many of the disabled viruses described earlier, gene expression levels are maintained in the long term. It does not appear that promoter choice is critical for long-term gene expression when using an amplicon system. This is unsurprising as amplicons contain little HSV DNA, and thus even when concatamerized would not be expected to take up the transcriptionally inactive and stable DNA structure maintained during latency, as this structure is probably dependent on particular features of the HSV sequence for formation. Rather, amplicon vectors would be expected to continue gene expression in a fashion similar to that shown by other viral vectors such as adenovirus or adeno-associated virus in which it would be expected that gene expression would continue until the input DNA is degraded within the cell.

Thus, amplicon vectors have a number of advantages over disabled HSV vectors, in that not only are they easier to manipulate as they can be grown as plasmids, but genes expressed in this way are not inactivated after short-term expression by the transcriptional shut-off which occurs with most promoters in the context of the virus genome as the virus enters latency. This gives the potential for the use of cell type-specific promoters for long-term expression which it is likely would only remain active in the short term if inserted at most sites within a disabled virus. For example, the pre-proencephalin promoter has been shown to give cell type-specific *lacZ* expression from an amplicon vector for at least 2 months in the rat brain (Kaplitt *et al.*, 1994). However, even with these advantages the safety of amplicon-based systems is still dependent on the particular helper virus used. As yet the IE3-deleted virus usually employed still gives relatively high levels of reversion to a wild-type phenotype, although further deleted viruses are being produced (see Chapter 7). Indeed any potentially safe, non-cytopathic disabled virus (such as those developed as vector viruses and described in the preceding sections) could be used as a helper virus providing improved safety for amplicon systems, although the problem of unreliable ratios of amplicon:helper virus would still remain.

6.8 Conclusions

Two types of HSV vector are available for gene transfer into neurons (disabled viruses and amplicon vectors), and both of these have been shown to be capable of delivering genes to a variety of cell types in the PNS and CNS of test animals. However, with both these systems problems of safety remain, either due to the cytopathic effects induced after inoculation of a disabled or helper virus, or, and more importantly, by the reversion of a disabled vector or helper virus to a wild-type phenotype. For example, approximately 10% of rats die after inoculation with the pHSVth/D30DBA combination used to correct the rat Parkinson's disease model above, due to reversion of the D30DBA helper virus to a wild-type phenotype at a frequency of 4×10^{-5} (During *et al.*, 1994). However most of these problems of reversion are caused by the use of complementing cell lines or disabled viruses with overlapping sequences (i.e. only partially deleted essential genes in the virus, or cell lines containing flanking as well as complementing sequences), allowing restoration of a wild-type phenotype by homologous recombination. However, these problems are being addressed both by the use of virus/cell line combinations with no overlapping sequences, and by the use of multiply deleted viruses for which repair of a number of genes would be necessary for pathogenicity. The likelihood of reversion of these viruses to wild type is essentially zero. Deletions in genes not necessary for growth in culture but necessary for virulence also further increase safety.

Thus, the safety considerations for the use of an amplicon or a disabled virus *in vivo* are essentially identical, as the safety of an amplicon is dependent on the helper virus used. However, the ease of manipulation, the possibility of higher levels of gene expression, and the potential for long-term gene expression may encourage the use of amplicons in many cases, particularly for *in vitro* studies of gene function or preliminary studies *in vivo*. It should be noted, however, that the use of amplicon vectors does not utilize the important ability of HSV to establish a life-long latent infection in neurons, as although expression can be maintained for at least 1 year in the rat brain using an amplicon vector, expression levels would be expected to slowly decline over time as amplicon DNA is degraded. With further development of promoters, disabled viruses may be able to maintain transgene expression throughout latency, which could be for many years. Disabled HSV vectors can also be used to express genes only in the short term, which may be important in some cases where continued activity would be undesirable, for example in the stimulation of nerve regrowth by growth factor expression. Here insertion of a transgene into a disabled vector at a position within the genome inactive during latency or under the control of a promoter not usually active during latency, results in transcriptional inactivation within around 1–2 weeks as the virus DNA forms the transcriptionally inactive and stable structure maintained during latency. This feature is not available with other virus vectors where the decrease in transgene expression over time is more gradual, and thus where short-erm gene delivery is less controllable.

References

Ace, C.I., McKee, T.A., Ryan, J.M., Cameron, J.M. & Preston, C.M. (1989) Construction and characterisation of a herpes simplex virus type 1 mutant unable to transinduce immediate early gene expression. *J. Virol.* **63**, 2260–2269.

Batchelor, A.H. & O'Hare, P. (1990) Regulation and cell-type specific activity of a promoter located upstream of the latency-associated transcript of herpes simplex virus type 1. *J. Virol.* **64**, 3269–3279.

Batchelor, A.H. & O'Hare, P. (1992) Localisation of *cis*-acting sequence requirements in the promoter of the latency-associated transcript of herpes simplex virus type 1 required for cell-type-specific activity. *J. Virol.* **66**, 3573–3582.

Bloom, D.C. & Stevens, J.G. (1994) Neuron-specific restriction of a herpes simplex virus recombinant maps to the UL5 gene. *J. Virol.* **68**, 3761–3772.

Boothman, D.A., Geller, A.I. & Pardee, A.B. (1989) Expression of the *E. coli lacZ* genes from a defective HSV-1 vector in various human normal, cancer-prone and tumor cells. *FEBS Lett.* **258**, 159–162.

Cameron, J.M., McDougal, I., Marsden, H.S., Preston, V.G., Ryan, D.M. & Subak-Sharpe, J.H. (1988) Ribonucleotide reductase encoded by herpes-simplex virus is a determinant of the pathogenicity of the virus in mice and a valid antiviral target. *J. Gen. Virol.* **69**, 2607–2612.

Chiocca, A.E., Choi, B.B., Cai, W., DeLuca, N.A., Schaffer, P.A., DeFiglia, M., Breakefield, X.O. & Martuza, R.L. (1990) Transfer and expression of the lacZ gene in rat brain neurons by herpes simplex virus mutants. *New Biol.* **2**, 739–736.

Chou, J., Kern, E.R., Whitley, R.J. & Roizman, B. (1990) Mapping of herpes simplex virus-1 neurovirulence to $\gamma_1$34.5, a gene nonessential for growth in culture. *Science* **250**, 1262–1266.

Davison, M.J., Preston, V.G. & McGeoch, D.J. (1984) Determination of the sequence alteration in the DNA of the herpes simplex virus type 1 temperature-sensitive mutant tsK. *J. Gen. Virol.* **65**, 859–863.

DeLuca, N.A., McCarthy, A.M. & Schaffer, P.A. (1985) Isolation and characterisation of deletion mutants of herpes simplex virus type 1 in the gene encoding immediate-early regulatory protein ICP4. *J. Virol.* **56**, 558–570.

Deshmane, S.L. & Fraser, N.W. (1989) During latency, herpes simplex virus type 1 DNA is associated with nucleosomes in a chromatin structure. *J. Virol.* **63**, 943–947.

Devi-Rao, G.B., Goodart, S.A., Hecht, L.M., Rochford, R., Rice, M.K. & Wagner, E.K. (1991) Relationships between polyadenylated and non-polyadenylated herpes simplex virus type 1 latency associated transcripts. *J. Virol.* **65**, 2179–2190.

Dobson, A.T.F., Sederati, F., Devi-Rao, G., Flanagan, W.M., Farrell, M.J., Stevens, J.G., Wagner, E.K. & Feldman, L.T. (1989) Identification of the latency associated transcript promoter by expression of rabbit β-globin mRNA in mouse sensory nerve ganglia latently infected with a recombinant herpes simplex virus. *J. Virol.* **63**, 3844–3851.

Dobson, A.T., Margolis, T.P., Sederati, F., Stevens, J.G. & Feldman, L.T. (1990) A latent, nonpathogenic HSV-1-derived vector stably expresses β-galactosidase in mouse neurons. *Neuron* **5**, 353–360.

Dobson, A.T., Margolis, T.P., Gomes, W.A. & Feldman, L.T. (1995) *In vivo* deletion analysis of the herpes simplex virus type 1 latency-associated transcript promoter. *J. Virol.* **69**, 2264–2270.

During, M.J., Naegele, J.R., O'Malley, K.L. & Geller, A.I. (1994) Long-term behavioral recovery in Parkinsonian rats by an HSV vector expressing tyrosine hydroxylase. *Science* **266**, 1399–1403.

Eizenberg, O., Kaplitt, M.G., Eitan, S., Pfaff, D.W., Hirschberg, D.L. & Schwartz, M. (1994) Linear dimeric interleukin-2 obtained by the use of a defective herpes simplex viral vector: conformation–activity relationship. *Mol. Br. Res.* **26**,156–162.

Farrell, M.J., Dobson, A.T. & Feldman, L.T. (1991) Herpes simplex latency associated transcript is a stable intron. *Proc. Natl. Acad. Sci. USA* **88**, 790–794.

Federoff, H.J., Geschwind, M.D., Geller, A.I. & Kessler, J.A. (1992) Expression of nerve growth factor *in vivo* from a defective herpes simplex virus 1 vector prevents effects of axotomy on sympathetic ganglia. *Proc. Natl. Acad. Sci. USA* **89**, 1636–1640.

Fink, D.J., Sternberg, L.R., Weber, P.C., Mata, M., Goins, W.F. & Glorioso, J.S. (1992) In vivo expression of β-galactosidase in hippocampal neurons by HSV-mediated gene transfer. *Hum. Gene Ther.* **3**, 11–19.

Frenkel, N. (1981) Defective interfering herpes viruses. In *The Human Herpes Viruses – An Interdisciplinary Perspective* (eds Nahmias, A.J., Dowdle, W.R. & Schcrazy, R.S.), pp. 91–120. New York, Elsevier.

Geller, A.I. & Breakefield, X.O. (1988) A defective HSV-1 vector expresses *Escherichia coli* β-galactosidase in cultured peripheral neurons. *Science* **241**, 1667–1669.

Geller, A.I. & Freese, A. (1990) Infection of cultured central nervous system neurons with a defective herpes simplex virus 1 vector results in stable expression of *Escherichia coli* β-galactosidase. *Proc. Natl. Acad. Sci. USA* **87**, 1149–1153.

Geller, A.I., Keyomarski, K., Bryan, J. & Pardee, A.B. (1990) An efficient deletion mutant packaging system for defective HSV-1 vectors; potential applications to neuronal physiology and human gene therapy. *Proc. Natl. Acad. Sci. USA* **87**, 8950–8954.

Geller, A.I., During, M.J., Haycock, J.W., Freese, A. & Neve, R. (1993) Long-term increases in neurotransmitter release from neuronal cells expressing a constitutively active adenylate cyclase from a herpes simplex virus type 1 vector. *Proc. Natl. Acad. Sci. USA* **90**, 7603–7607.

Geschwind, M.D., Kessler, J.A., Geller, A.I. & Federoff, H.J. (1994) Transfer of the nerve growth factor gene into cell lines and cultured neurons using a defective herpes simplex vector. *Mol. Brain Res.* **24**, 327–335.

Goins, W.F., Sternberg, L.R., Croen, K.D., Krause, P.R., Hendricks, R.L. Fink, D.J., Straus, S.E., Levine, M. & Glorioso, J.C. (1994) A novel latency-associated promoter is contained within the herpes simplex virus type 1 UL flanking repeats. *J. Virol.* **68**, 2239–2252.

Ho, D.Y. & Mocarski, E.S. (1988) β-Galactosidase as a marker in the peripheral and neural tissues of the herpes simplex virus-infected mouse. *Virology* **167**, 279–283.

Ho, D.Y. & Mocarski, E.S. (1989) Herpes simplex virus latent RNA (LAT) is not required for latent infection in the mouse. *Proc. Natl. Acad. Sci. USA* **86**, 7596–7600.

Ho, D.Y., Mocarski, E.S. & Sapolsky, R.M. (1993) Altering central nervous system physiology with a defective herpes simplex virus vector expressing the glucose transporter gene. *Proc. Natl. Acad. Sci. USA* **90**, 3655–3659.

Johnson, P.A., Miyanohara, A., Levine, F., Cahill, T. & Freidman, T. (1992) Cytotoxicity of a replication-defective mutant of herpes simplex virus type 1, *J. Virol.* **66**, 2952–2965.

Johnson, P.A., Wang, M.J. & Freidman, T. (1994) Improved cell survival by the reduction of immediate-early gene expression in replication-defective mutants of herpes simplex virus type 1 but not by a mutation in the virion host shutoff function. *J. Virol.* **68**, 6347–6362.

Kaplitt, M.G., Pfaus, J.G., Kleopoulos, S.P., Hanlon, B.A., Rabkin, S.D. & Pfaff, D.W. (1991) Expression of a functional foreign gene in adult mammalian brain following in vivo transfer via a herpes simplex virus type 1 defective viral vector. *Mol. Cell. Neurosci.* **2**, 320–330.

Kaplitt, M.G., Kwong, A.D., Kleopoulos, S.P., Mobbs, C.V., Rabkin, S.D. & Pfaff, D.W. (1994) Preproencephalin promoter yields region-specific and long-term expression in adult brain after direct in vivo gene transfer via a defective herpes simplex virus vector. *Proc. Natl. Acad. Sci. USA* **91**, 8979–8983.

Kwong, A.D. & Frenkel, N. (1985) The herpes simplex virus amplicon IV. Efficient expression of a chimeric chicken ovalbumin gene amplified within defective virus genomes. *Virology* **142**, 421–425.

Latchman, D.S. (1990) Molecular biology of herpes virus latency. *J. Exp. Pathol.* **71**, 133–141.

Leader, D.P. & Purves, F.C. (1988) The herpesvirus protein kinase: a new departure in protein phosphorylation? *Trans. Biochem. Sci.* **13**, 244–246.

Leib, D.A., Bogard, C.L., Kosz-Vnenchak, M., Hicks, K.A., Coen, D.M., Knipe, D.M. & Schaffer, P.A. (1989) A deletion mutant of the latency-associated transcript of herpes simplex virus type 1 reactivates from the latent state with reduced frequency. *J. Virol.* **63**, 2893–2900.

Lillycrop, K.A., Howard, M.K., Estridge, J.K. & Latchman, D.S. (1994) Inhibition of herpes simplex virus infection by ectopic expression of neuronal splice variants of Oct-2 transcription factor. *Nucl. Acid Res.* **22**, 815–820.

Lokensgard, J.R., Bloom, D.C., Dobson, A.T. & Feldman, L.T. (1994) Long-term promoter activity during herpes simplex virus latency. *J. Virol.* **68**, 7148–7158.

Margolis, T.P., Sederati, F., Dobson, A.T., Feldman, L.T. & Stevens, J.G. (1992) Pathways of viral gene expression during acute neuronal infection with HSV-1. *Virology* **189**, 150–160.

Margolis, T.P., Bloom, D.C., Dobson, A.T., Feldman, L.T. & Stevens, J.G. (1993) Decreased reporter gene expression during latent infection with HSV LAT promotor constructs. *Virology*. **197**, 585–592.

McFarlane, M., Daksis, J.I. & Preston, C.M. (1992) Hexamethylene bisacetamide stimulates herpes-simplex virus immediate early gene-expression in the absence of trans-induction by VMW65. *J. Gen. Virol.* **73**, 285–292.

O'Hare, P. (1993) The virion transactivator of herpes simplex virus. *Semin. Virol.* **4**, 145–155.

Palella, T.D., Silverman, L.J., Schroll, C.T., Homa, F.L., Levine, M. & Kelley, W.M. (1988) Herpes simplex virus-mediated human hypoxanthine-guanine phosphoribosyl transferase gene transfer into neuronal cells. *Mol. Cell. Biol.* **8**, 457–460.

Palella, T.D., Hidaka, Y., Silverman, L.J., Levine, M., Glorioso, J.C. & Kelley, W.M. (1989) Expression of human HPRT mRNA in brains of mice infected with a recombinant herpes simplex virus-1 vector. *Gene* **80**, 137–144.

Rickson, F.J. (1993) Structure and assembly of herpes viruses. *Semin. Virol.* **4**, 135–144.

Robertson, L.M., Maclean, A.R. & Brown, S.M. (1992) Peripheral replication and latency reactivation kinetics of the non-neurovirulent herpes simplex virus type 1 variant 1716. *J. Gen. Virol.* **73**, 967–970.

Roizman, B. (1979) The structure and isomerisation of herpes simplex virus genomes. *Cell* **16**, 481–494.

Sears, A.E., Halliburton, I.W., Meignier, B., Silver, S. & Roizman, B. (1985) Herpes simplex virus 1 mutant deleted in the α22 gene: growth and gene expression in permissive and restrictive cells and establishment of latency in mice. *J. Virol.* **55**, 338–346.

Spaete, R.R. & Frenkel, N. (1982) The herpes simplex virus amplicon: a new eukaryotic defective-virus cloning amplifying vector. *Cell* **30**, 295–304.

Steiner, I., Spivack, J.G., Deshmane, S.L., Ace, C.I., Preston, C.M. & Fraser, N.W. (1990) A herpes simplex virus type 1 mutant containing a nontransducing VMW65 protein establishes latent infection *in vivo* in the absence of viral replication and reactivates efficiently from explanted trigeminal ganglia. *J. Virol.* **64**, 1630–1638.

Stevens, J.G., Wagner, E.K., Devi-Rao, G.B., Cook, M.L. & Feldman, L.T. (1987) RNA complementary to a herpesvirus α gene mRNA is prominent in latently infected neurons. *Science* **235**, 1056–1059.

Stow, N.D. & Wilkie, N.M. (1976) An improved technique for obtaining enhanced infectivity with herpes simplex virus type 1 DNA. *J. Gen. Virol.* **33**, 447–458.

Stow, N.D., Murray, N.D. & Stow, E.C. (1986) Cis-acting signals involved in the replication and packaging of herpes simplex virus type-1 DNA. *Cancer Cells* **4**, 497–507.

Trousdale, M.D., Steiner, I., Spivack, J.G., Deshmane, S.L., Brown, S.M., Maclean, A.R., Subak-Sharpe, J.H. & Fraser, N.W. (1991) *In vivo* and *in vitro* reactivation impairment of a herpes simplex virus type 1 latency-associated transcript variant in a rabbit eye model. *J. Virol.* **65**, 6989–6993.

Verhaagen, J., Hermens, W.T.J.M.C., Oestreicher, A.B., Gispen, W.H., Rabkin, S.D., Pfaff, D.W. & Kaplitt, M.G. (1994) Expression of the growth associated protein B-50/GAP43 via adefective herpes-simplex virus vector results in profound morphological changes in non-neuronal cells. *Mol. Br. Res.* **26**, 26–36.

Wolfe, J.H., Deshmane, S.L. & Fraser, N.W. (1992) Herpesvirus vector gene transfer and expression of β-glucuronidase in the central and peripheral nervous system of MPS VII mice. *Nature Genet.* **1**, 379–384.

Zhang, Y., Sirko, D.A. & McKnight, J.L.C. (1991) Role of herpes simplex virus type 1 UL46 and UL47 in αTIF-mediated transcriptional induction: characterisation of three viral deletion mutants. *J. Virol.* **65**, 829–841.

Zwaagstra, J.C.H., Ghiasi, H., Slanina, S.N., Nesburn, A.B., Wheatly, S.C., Lillycrop, K., Wood, J., Latchman, D.S., Patel, K. & Wechsler, S.L. (1990) Activity of herpes simplex virus type 1 latency-associated transcript (LAT) promotor in neuron-derived cells: evidence for neuron specificity and for a large LAT transcript. *J. Virol.* **64**, 5019–5028.

CHAPTER 7

GENE DELIVERY USING HERPES SIMPLEX VIRUS TYPE 1 PLASMID VECTORS

Karen L. O'Malley* and Alfred I. Geller†

*Department of Anatomy and Neurobiology, Washington University School of Medicine, St Louis, MO 63110, and †Division of Endocrinology, Children's Hospital, Boston, MA 02115, and Program in Neuroscience, Harvard Medical School, Cambridge, MA, USA

Table of Contents

7.1 Introduction

Over the past decade, the development of virus vector systems has led to new approaches for studying gene function in neurons as well as novel strategies for gene therapy. Vector systems that have been used to deliver genes into cells in the nervous system include retroviruses, adenovirus vectors, defective herpes simplex virus type 1 (HSV-1) vectors, recombinant HSV-1 vectors, and most recently adeno-associated virus vectors. This chapter will summarize current results using defective HSV-1 vectors, or HSV-1 plasmid vectors, to directly modify neuronal physiology resulting in long-term biochemical and behavioral changes. Other viral vectors are discussed in the appropriate chapters of this book. The specific topics to be covered here include

Genetic Manipulation of the Nervous System
ISBN 0-12-437165-5

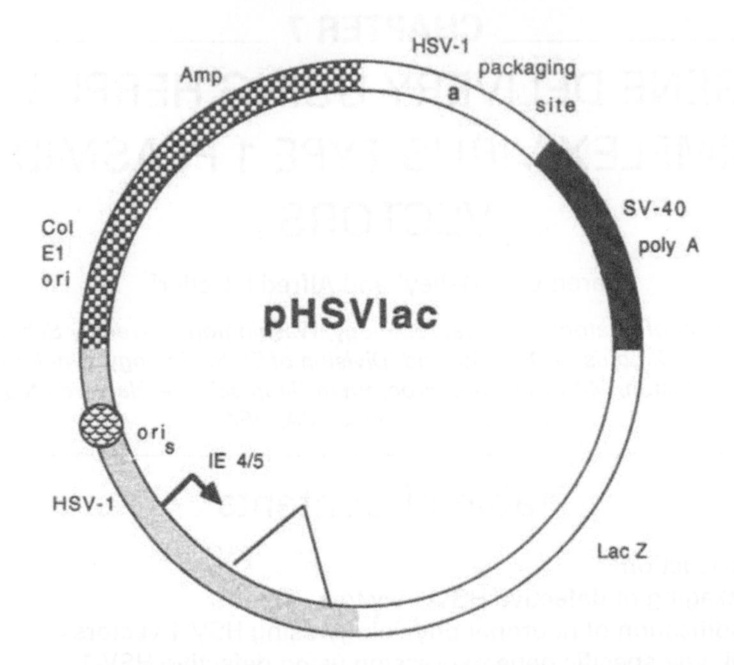

Figure 1 Schematic diagram of pHSVlac, the prototype defective HSV-1 vector. pHSVlac contains three kinds of genetic elements: (1) an HSV-1 origin of DNA replication (ori$_S$, small circle filled with waves) and the packaging site (**a** sequence, clear segment at top right) are required for packaging HSV-1 vectors into HSV-1 particles; (2) a transcription unit composed of the HSV-1 IE4/5 promoter (arrow), the intervening sequence following that promoter (triangle), the *E. coli lacZ* gene (clear segment at bottom right), and the SV-40 early region polyadenylation site (black segment); (3) sequences from pBR322 (checkerboard segment) are required to propagate pHSVlac in *E. coli*.

improvements in the packaging system, use of HSV-1 vectors to perturb neuronal physiology, the potential for targeting gene expression to specific cell types by the use of cell type-specific promoters in HSV-1 vectors, and finally use of this system in gene therapy paradigms.

7.2 Packaging of defective HSV-1 vectors

The prototype HSV-1 vector, pHSVlac (Figure 1; Geller and Breakefield, 1988; Geller and Freese, 1990), contains three kinds of genetic elements. These include sequences for the propagation of pHSVlac in *Escherichia coli*, HSV-1 sequences for the propagation of pHSVlac in an HSV-1 virus stock, and a transcription unit consisting of the HSV-1 immediate early (IE) 4/5 promoter, the intervening sequence following the promoter, the *E. coli lacZ* gene, and the SV40 early region polyadenylation site.

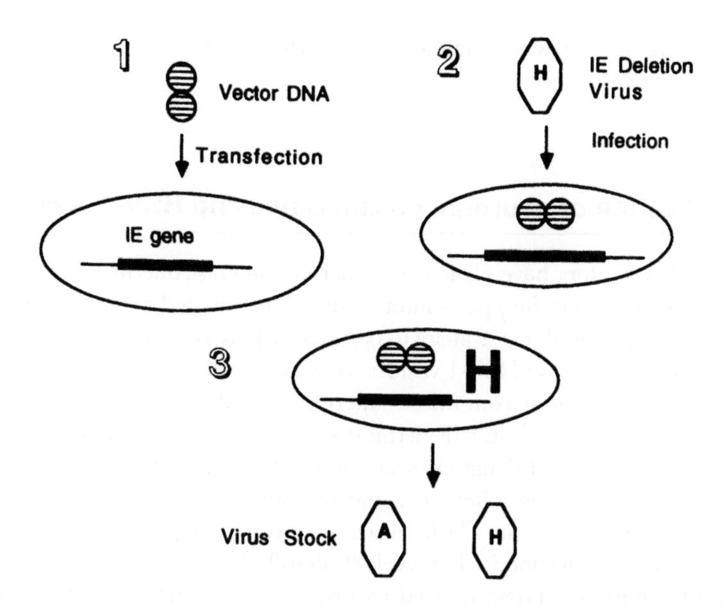

Figure 2 Schematic diagram of the packaging procedure for defective HSV-1 vectors using a deletion mutant of HSV-1. (1) Using a transfection procedure such as calcium phosphate or lipofectamine, vector DNA is delivered into a fibroblast cell line which expresses an HSV-1 IE gene. (2) The cells are then infected with a mutant of HSV-1 harboring a deletion in the same IE gene. (3) The IE gene in the cell line complements the HSV-1 deletion mutant resulting in a productive lytic infection and the packaging of both vector and helper into HSV-1 particles.

The first deletion mutant packaging system (Geller *et al.*, 1990) used HSV-1 strain 17 D30EBA (Patterson and Everett, 1990) and M64A cells (Davidson and Stow, 1985). D30EBA contains a deletion in the IE3 gene, the major regulatory protein of HSV-1, and is grown in M64A cells which express the IE3 gene. To package vector DNA into HSV-1 particles (Figure 2), the complementing cell line is transfected with vector DNA and then superinfected with the helper virus. Unfortunately, in this packaging system, recombination between the IE3 gene in the M64A cells and the helper virus (D30EBA) produces wild-type HSV-1 at a frequency of $\sim 10^{-5}$. Moreover, a small fraction ($\sim$10%) of the rats injected with these virus stocks die within 2 weeks after gene transfer because the wild-type HSV-1 present in the these virus stocks can cause HSV-1 encephalitis (During *et al.*, 1994).

In order to overcome some of these problems, we developed two new packaging systems (Lim *et al.*, 1994). These are: (1) KOS strain d120, which harbors a deletion in the IE3 gene, and E5 cells (DeLuca *et al.*, 1985); and (2) KOS strain 5 *dl*1.2 (McCarthy *et al.*, 1989), which harbors a deletion in the IE2 gene, and 2.2 cells (Smith *et al.*, 1992). Both of these packaging systems have very limited sequence homology between the IE gene in the cell line and the mutant virus, and consequently show a reversion frequency to wild-type HSV-1 of $<10^{-7}$. This represents an improvement of at least 100-fold compared to the D30EBA packaging system. When gene transfer is performed using these new packaging systems virtually all of the rats survive until deliberately sacrificed.

7.3 Modification of neuronal physiology using HSV-1 vectors

Defective HSV-1 vectors have elicited considerable interest due to their ability to transfer and express genes into post-mitotic, fully differentiated neurons. A number of recent studies have used this strategy to perturb various aspects of neuronal physiology. For example, defective HSV-1 vectors were used to deliver nerve growth factor (NGF; vector pHSVngf) into axotomized superior cervical ganglion cells deprived of their target-derived source of NGF (Federoff *et al.*, 1992). In these studies, axotomized ganglia transduced with pHSVngf virus not only exhibited no decline in expression of the neurotransmitter biosynthetic enzyme tyrosine hydroxylase (TH) but rather manifested modest increases in TH levels compared to axotomized controls which exhibited a significant decline in TH levels (Federoff *et al.*, 1990). Thus, the *de novo* expression of a normally target-derived requisite gene product could be achieved using this system. In addition, infection of cultured striatal or basal forebrain cells with pHSVngf results in induction of choline acetyltransferase (Geschwind *et al.*, 1994). Recently, the HSV-1 vector system was used to transduce the NGF receptor, trkA, into neonatal nodose neurons and embryonic spinal motor neurons (Xu *et al.*, 1994). These cell types are normally non-responsive to NGF. However, introduction of the trkA receptor resulted in their conversion into NGF-responsive neurons. Similarly, when an HSV-1 vector containing the p75 low affinity NGF receptor was transduced into fibroblast cells expressing the trkA receptor, a new high affinity NGF binding site was created consistent with data suggesting a functional interaction occurs between the p75NGF receptor and the trkA tyrosine kinase receptors (Battleman *et al.*, 1993). Furthermore, expression of the p75NGF receptor in cultured cortical cells resulted in a high level of NGF binding to these cells.

In addition to growth factor transduction, HSV-1 vectors have been used to introduce signal transduction enzymes into various cell types *in vitro* and *in vivo*. For example, the HSV-1 system was used to express the catalytic domain of yeast adenylate cyclase (cyr; vector pHSVcyr) in a neuronal cell line. Because this form of the enzyme is no longer regulated but instead is constitutively active, PC12 cells infected with pHSVcyr exhibited increased cAMP levels, protein kinase A activity, protein phosphorylation and neurotransmitter release (Geller *et al.*, 1993). Similar results were seen when sympathetic neurons were infected suggesting that pHSVcyr directed a long-term activation of the cAMP pathway resulting in a long-term increase in neurotransmitter release (Geller *et al.*, 1993).

HSV-1 vector-mediated expression of a glucose transporter (Ho *et al.*, 1993) and the GluR6 subunit of the kainate receptor family (Bergold *et al.*, 1993) have also yielded changes in neuronal physiology. A major attribute of the system is apparent in the latter example. Microapplication of only 100 HSV/GluR6 virus particles into

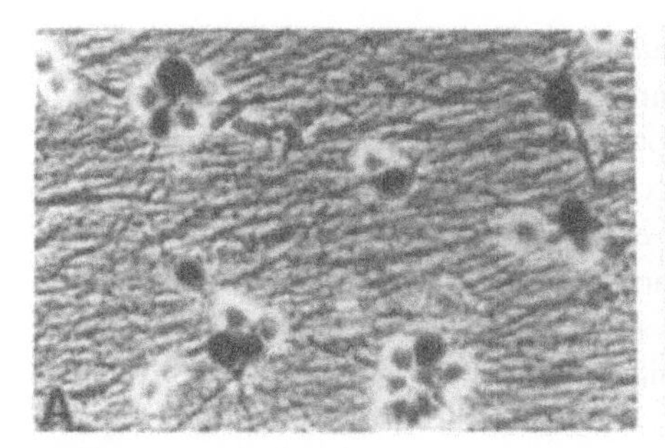

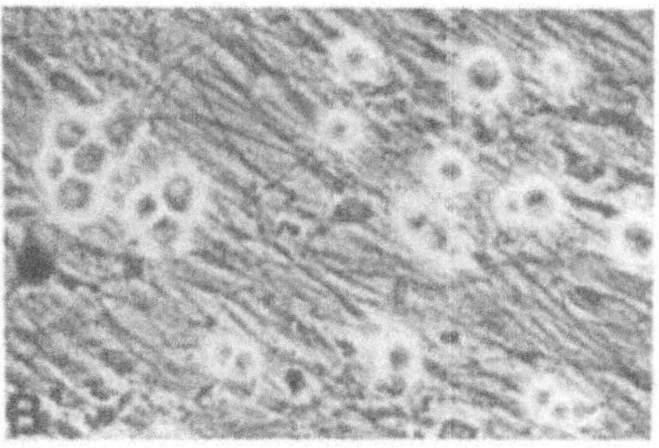

Figure 3 Infection of neuronal cultures with pTHlac-663 leads to the cell type-specific expression of β-galactosidase. Superior cervical ganglia (SCG; TH expressing) and dorsal root ganglia (DRG; TH non-expressing) were prepared and maintained for 7 days *in vitro*. Cultures were infected with pTHlac-663 and X-gal cytochemistry was performed 1 day after infection. Phase contrast photomicrographs of (A) SCG and (B) DRG. X-gal-positive neurons were predominantly observed in SCG cultures, but were largely absent in DRG cultures.

organotypic hippocampal slice cultures resulted in a large loss of CA3 and hilar neurons due to the specific introduction of the virus. Thus, a limited number of infectious particles of vector can have a marked physiological effect. Taken together these types of studies emphasize the power and utility of the defective HSV-1 vector system for studying various aspects of neuronal function.

7.4 Cell type-specific gene expression using defective HSV-1 vectors

Because HSV-1 vectors can infect a wide variety of cell types, use of a constitutive promoter in the vector results in recombinant gene expression in multiple cell types. Restricting gene expression to a particular cell type would have important implications not only for gene therapy but also for studying promoter function and for targeting a particular genetic perturbation to a specific cell type. Recent reports have suggested that cell type-specific promoters may function properly in defective HSV-1 vectors. For example, a human neurofilament L promoter directed neuronal-specific expression of β-galactosidase in PC12 cells and in cultures of both sympathetic and cortical cells (Federoff *et al.*, 1990). Additionally, the rat preproenkephalin promoter conferred cell type-specific expression of β-galactosidase for up to 2 months after stereotactic injection into specific areas of the brain (Kaplitt *et al.*, 1994).

We have utilized this system to study the cell type-specific expression of the rat TH promoter in cultured rat neurons. As shown in Figure 3, infection of TH-expressing superior cervical ganglia neurons versus the non-TH-expressing dorsal root ganglia neurons with an HSV-1 vector which used 663 bp fragment of the TH promoter to regulate expression of the *E. coli lacZ* gene resulted in a 17-fold difference in reporter gene expression in these two neuronal populations. Constructs containing only 278 or 181 bp of the TH promoter exhibited only a 5-fold difference in β-galactosidase expression when tested in this paradigm (not shown). These data suggest that HSV-1 vector-mediated gene transfer can be successfully employed to dissect the genetic elements responsible for cell type-specific expression of a given gene in cultured neurons (Oh *et al.*, 1995). Using this approach, the mechanisms of TH regulation in cultured neurons by various growth factors such as CNTF and cholinergic development factor/leukemia inhibitory factor might also be studied. Recently, we have extended these studies *in vivo*. For these experiments, a 6.8 kb fragment of rat TH promoter was used to direct *lacZ* gene expression after stereotactic injection into the midbrain of adult rats. Expression of β-galactosidase was detected by X-gal histochemistry at both 4 days and 6 weeks after gene transfer (Song *et al.*, 1994). Although some expression in inappropriate cell types can be observed, this study and that of Kaplitt *et al.* (1994) demonstrate that HSV-1 vectors containing cell type-specific promoters can be used to target expression of recombinant gene products to specific cell types. Collectively, these studies raise the possibility that promoter analysis can be performed using HSV-1 vectors, and that for *in vivo* promoter studies this approach may provide a more rapid and economical alternative to the construction of transgenic animals. It seems likely that strategies which use cell type-specific promoters to restrict expression of recombinant genes to chosen cell types will be of increasing importance in the utilization of HSV-1 vectors both for gene therapy and for studying neuronal physiology.

7.5 The use of HSV-1 vectors in gene therapy paradigms

Amelioration of disorders that affect the central nervous system (CNS) presents a special challenge due to the inaccessibility of most brain structures, the post-mitotic nature of most neurons, and the complex nature of most neurological disorders. Thus, it is not surprising that most gene therapy protocols have focused on accessible, peripheral targets. However, one CNS disorder that might be amenable to gene therapy approaches is Parkinson's disease (PD). PD, a neurodegenerative disorder, is characterized by the progressive loss of the dopaminergic neurons in the substantia nigra pars compacta (SNc) which project to the corpus striatum (Yahr and Bergmann, 1987). The principal therapy for PD is the oral administration of levo-dopa (L-dopa; Yahr *et al.*, 1969) which is converted to dopamine (DA) by endogenous striatal aromatic amino acid decarboxylase (AADC, EC4.1.1.28; Tashiro *et al.*, 1989; Li *et al.*, 1992). This treatment is initially effective, but then loses efficacy over a period of several years (Yahr *et al.*, 1969). Various methods to improve the delivery of L-dopa have been explored. Basic research efforts have used precursor loading of L-dopa (Cotzias *et al.*, 1967; Yahr *et al.*, 1969; Martin, 1971; Rossor *et al.*, 1980; Yahr and Bergmann, 1987), DA agonists (such as bromocryptine; Yahr and Bergmann, 1987) and implantable DA delivery systems (either polymeric or pump systems; Hargraves and Freed, 1987), with varying effectiveness. Typically, like oral L-dopa, these treatments are effective initially but then gradually lose their efficacy.

The development of efficient gene transfer techniques has led to strategies for gene therapy whereby *in vitro* modification of candidate cells with subsequent reimplantation into specific brain regions is combined. For example, previously we inserted the rat TH cDNA into a retroviral vector which was then used to develop TH-expressing fibroblast cell lines. Such cells had TH enzyme activity, produced L-dopa and appeared effective in treating a rat model of PD (Wolff *et al.*, 1988). While the power of this approach is obvious, significant scientific and clinically relevant questions remain. These include efficient introduction of DNA, host rejection, tumor formation, disruption of local neuronal circuitry, migration of transplanted cells, and others.

An alternative therapeutic strategy is to convert a fraction of the striatal cells into L-dopa-producing cells by direct gene transfer of TH using a defective HSV-1 vector. Potential advantages of this approach include production of L-dopa at the required site of action, so that diffusion over significant distances is not necessary, and alleviation of potential problems with either graft rejection or tumor formation. To test this strategy we inserted a human TH cDNA into an HSV-1 vector (pHSVth). The capability of pHSVth to direct production and release of L-dopa in cultured CNS cells was investigated. As described (Geller *et al.*, 1995), following infection of cultured striatal cells, pHSVth directed the expression of TH RNA and TH protein for up to 1 week. Moreover, release of L-dopa as well as low levels of DA was detected in cultures infected with pHSVth. These results suggested that pHSVth might be effective in animal models of PD.

To test the hypothesis that pHSVth could direct behavioral recovery in a rodent model of PD, pHSVth or pHSVlac virus, or vehicle alone, was delivered by stereotactic injection into the partially denervated striatum of 6-hydroxydopamine (6-OHDA)-lesioned rats. As shown in Figure 4, when injected into the 6-OHDA-lesioned rat model of PD, the pHSVth virus directed approximately 60–70% behavioral recovery, which could be observed in animals for up to 1 year (During *et al.*, 1994). Further analysis of pHSVth-treated animals demonstrated increases in TH enzyme activity and striatal DA levels, the presence of TH immunoreactive cells, the expression of human TH RNA, and the persistence of pHSVth DNA. Thus, these studies hold promise for the efficacy of this gene therapy approach to PD.

Several features of HSV-1 vectors may assist in furthering this approach. First, as described above, the ability to use cell type-specific promoters to target expression to particular types of striatal cells may be advantageous. Moreover, inducible promoters, which can be regulated by the addition of a small molecule, may also be useful for modulating the level of foreign gene expression. Second, since HSV-1 vectors have the potential to express more than one gene, the possibility exists to co-express both TH and AADC in the same cell to direct more efficient production of DA. Currently, there is conflicting evidence as to whether *in situ* L-dopa or DA production would be more effective in rodent models of PD (Horellou *et al.*, 1990; Kang *et al.*, 1994). The HSV-1 system should allow both models to be tested.

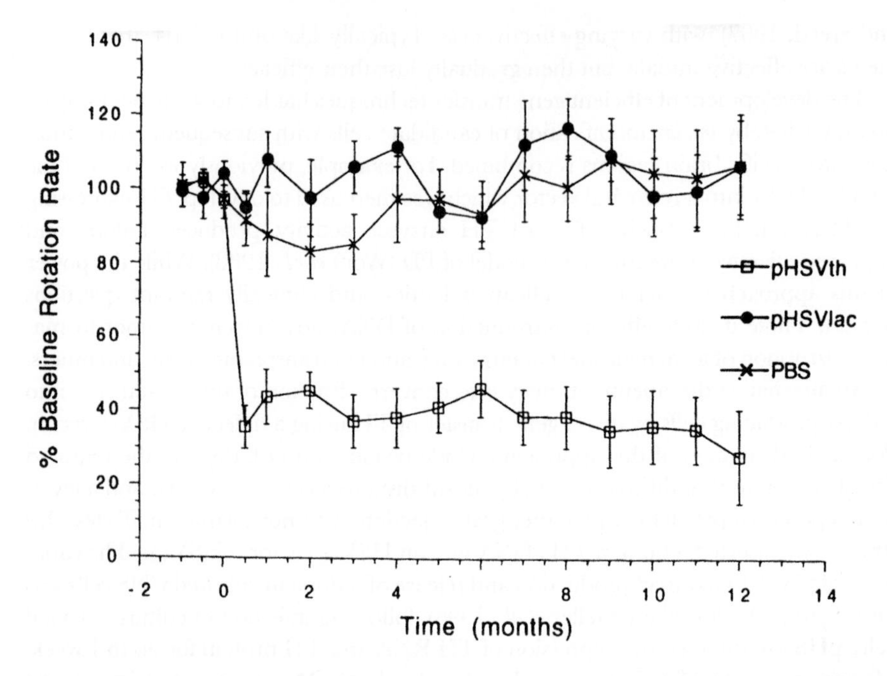

Figure 4 Rotation rates following stereotactic injection of pHSVth, pHSVlac or phosphate-buffered saline (PBS) into the partially denervated striatum. Rats lesioned with 6-OHDA were tested at least 3 times at weekly intervals for the apomorphine-induced rotation rate. At time 0 gene transfer was performed. The apomorphine-induced rotation rate was determined every week for the first 3 months after gene transfer and then every other week. The difference in rotation rates between pHSVth and the control groups was statistically significant (6 months, $p<0.01$; 1 year, $p<0.05$).

7.6 A genetic intervention strategy, using HSV-1 vectors, to analyze neuronal physiology *in vivo*

It may be possible to use HSV-1 vectors to alter the physiology of specific groups of neurons in the adult rat brain (Geller *et al.*, 1991). The tools of modern neuroscience, including behavioral tests, might then be used to analyze the role of these neurons in brain functions.

Our strategy has three components. First, expression of the catalytic domain of a signal transduction enzyme results in an unregulated enzyme that is always active, thereby activating a specific signal transduction pathway and causing long-term changes in neuronal physiology, such as increased neurotransmitter release. For example, as described above, expression of the catalytic domain of adenylate cyclase in cultured neurons activates the cAMP pathway, resulting in a long-term increase in neurotransmitter release under specific conditions (Geller *et al.*, 1993).

Second, a cell type-specific promoter is used to limit expression of a signal transduction enzyme to a particular type of neuron. As described above, HSV-1 vectors containing the preproenkephalin promoter (Kaplitt *et al.*, 1994) or the TH promoter (Oh *et al.*, 1995; Song *et al.*, 1995) can direct cell type-specific expression in the adult rat brain.

Third, stereotactic injection of the vector into a specific area in the adult brain limits gene transfer to cells in a specific area. For example, injection of vectors into the striatum limits gene expression to striatal cells and striatal projection neurons (During *et al.*, 1994).

This genetic intervention strategy (Geller *et al.*, 1991) is designed to cause changes in neuronal physiology to a specific type of neuron, in a specific area of the adult brain. For example, a vector which uses the TH promoter to regulate expression of the catalytic domain of an adenylate cyclase might be used to affect the function of a specific group of catecholaminergic neurons in the adult rat brain. Resulting changes in behavior could then be attributed to a particular type of neuron and a particular signal transduction pathway.

7.7 Concluding remarks

In this Decade of the Brain, efforts have been accelerated to understand the basic functions of CNS systems. Such efforts have been advanced by new technological approaches allowing old questions to be asked in new ways. Genetic intervention in the nervous system using defective HSV-1 vectors represents one such approach. Specifically, the ability to direct recombinant gene expression to post-mitotic cell types allows for the direct manipulation and/or perturbation of CNS systems. Such perturbations may prove useful for analyzing the role of specific types of neurons in brain functions. Moreover, these types of approaches lay the groundwork for the development of gene therapy systems via directed gene expression in animal models of human CNS disorders. Ultimately they may be applicable to neurological disorders such as Parkinson's or Alzheimer's disease.

Currently, several different virus systems are being evaluated in various short-term and long-term paradigms. Here, we have emphasized the potential of defective HSV-1 vectors in expressing recombinant genes in neural cells in the brain. Although improvements in the defective HSV-1 vector system are clearly warranted, results from many different laboratories suggest this type of approach can be useful in addressing specific questions of gene function. Moreover, the long-term behavioral recovery observed following direct gene transfer in an animal model of Parkinson's disease offers the hope that this type of technology might eventually offer amelioration of human CNS diseases.

Acknowledgements

Supported by American Parkinson Disease Association and AG10827 (NIH, AG and KO); 50081 (NIH, KO); American Health Assistance Foundation, Burroughs Wellcome Fund, and the National Parkinson Foundation (AG).

References

Battleman, D.S., Geller, A.I. & Chao, M.V. (1993) HSV-1 vector mediated gene transfer of the human nerve growth factor receptor p75hNGFR defines high affinity NGF binding. *J. Neurosci.* **13**, 941–951.

Bergold, P.J., Casaccia-Bonnefil, P., Xiu-Liu, Z. & Federoff, H.J. (1993) Transsynaptic neuronal loss induced in hippocampal slice cultures by a herpes simplex virus vector expressing the GluR6 subunit of the kainate receptor. *Proc. Natl. Acad. Sci. USA* **90**, 6165–6169.

Cotzias, G.C., Van Woert, M.H. & Schiffer, L.M. (1967) Aromatic amino acids and modification of parkinsonism. *N. Engl. J. Med.* **276**, 374–379.

Daubner, S.C. & Fitzpatrick, P.F. (1993) Alleviation of catecholamine inhibition of tyrosine hydroxylase by phosphorylation at serine 40. *Adv. Exp. Med. Biol.* **338**, 87–92.

Davidson, I. & Stow, N.D. (1985) Expression of an immediate early polypeptide and activation of a viral origin of DNA replication in cells containing a fragment of herpes simplex virus DNA. *Virology* **141**, 77–88.

DeLuca, N.A., McCarthy, A.M. & Schaeffer, P.A. (1985) Isolation and characterization of deletion mutants of herpes simplex virus type 1 in the gene encoding immediate-early regulatory protein ICP4. *J. Virol.* **56**, 558–570.

During, M.J., Naegele, J., O'Malley, K. & Geller, A.I. (1994) Long-term behavioral recovery in Parkinsonian rats by an HSV-1 vector expressing tyrosine hydroxylase. *Science* **266**, 1399–1403.

Federoff, H.J., Geller, A.I. and Lu, B. (1990) Neuronal specific expression of the human neurofilament L promoter in a HSV-1 vector. *Abs. Soc. Neurosci.* **154**, 2.

Federoff, H.J., Geschwind, M.D., Geller, A.I. & Kessler, J.A. (1992) Expression of NGF *in vivo*, from a defective HSV-1 vector, prevents effects of axotomy on sympathetic neurons. *Proc. Natl. Acad. Sci. USA* **89**, 1636–1640.

Geller, A.I. & Breakefield, X.O. (1988) A defective HSV-1 vector expresses *Escherichia coli* β-galactosidase in cultured peripheral neurons. *Science* **241**, 1667–1669.

Geller, A.I. & Freese, A. (1990) Infection of cultured central nervous system neurons with a defective Herpes Simplex Virus 1 vector results in stable expression of *Escherichia coli* β-galactosidase. *Proc. Natl. Acad. Sci. USA* **87**, 1149–1153.

Geller, A.I., Keyomarski, K., Bryan, J. & Pardee, A.B. (1990) An efficient deletion mutant packaging system for defective HSV-1 vectors; potential applications to neuronal physiology and human gene therapy. *Proc. Natl. Acad. Sci. USA* **87**, 8950–8954.

Geller, A.I., During, M.J. & Neve, R.L. (1991) Molecular analysis of neuronal physiology by gene transfer into neurons with herpes simplex virus vectors. *Trends Neurosci.* **14**, 428–432.

Geller, A.I., During, M.J., Haycock, J.W., Freese, A. & Neve, R.L. (1994) Long-term increases in neurotransmitter release from neuronal cells expressing a constitutively active adenylate cyclase from a HSV-1 vector. *Proc. Natl. Acad. Sci. USA* **90**, 7603–7607.

Geller, A.I., Freese, A., During, M.J. & O'Malley, K.L. (1995) A HSV-1 vector expressing tyrosine hydroxylase causes production and release of L-Dopa from cultured rat striatal cells. *J. Neurochem.* **64**, 487–496.

Geschwind, M.D., Kessler, J.A., Geller, A.I. & Federoff, H.J. (1994) Transfer of the nerve growth factor gene into cell lines and cultured neurons using a defective herpes simplex virus vector. Transfer of the NGF gene into cells by a HSV-1 vector. *Mol. Brain Res.* **24**, 327–335.

Hargraves, R. & Freed, W.J. (1987) Chronic intrastriatal dopamine infusion in rats with unilateral lesions of the substantia nigra. *Life Sci.* **40**, 959–966.

Ho, D.Y., Mocarski, E.S. & Sapolsky, R.M. (1993) Altering central nervous system physiology with a defective HSV-1 vector expressing the glucose transporter gene. *Proc. Natl. Acad. Sci. USA* **90**, 3655–3659.

Horellou, P., Brundin, P., Kalen, P., Mallet, J. & Bjorklund, A. (1990) In vivo release of DOPA and dopamine from genetically engineered cells grafted to the denervated rat striatum. *Neuron* **5**, 393–402.

Kang, U.J., Fisher, L.J., Jinnah, H.A., Rosenberg, T.H., Joh, T., Friedmann, F.H., Gage, F.H. (1991) Intracerebral grafting of AADC-expressing primary fibroblasts for site-specific delivery of dopamine. *Abstr. Soc. Neurosci.* **17**, 424.12.

Kaplitt, M.G., Kwong, A.D., Kleopoulos, S.P., Mobbs, C.V., Rabkin, S.D. & Pfaff, D.W. (1994) Preproenkephalin promoter yields region-specific and long-term expression in adult brain after direct *in vivo* gene transfer via a defective herpes simplex viral vector. *Proc. Natl. Acad. Sci. USA* **91**, 8979–8983.

Li, X.M., Jourio, A.V., Paterson, I.A., Walz, W., Zhu, M.Y. & Boulton, A.A. (1992) Gene expression of aromatic L-amino acid decarboxylase in cultured rat glial cells. *J. Neurochem.* **59**, 1172–1175.

Lim, F., Starr, P., Song, S., Hartley, D., Lang, P., Wang, Y. & Geller, A.I. (1994) Methods for using a defective HSV-1 vector system. In *DNA Cloning: Mammalian Systems* (ed. Glover, D.).

Martin, W.E. (1971) Adverse reactions during treatment of Parkinson's disease with levodopa. *J. Am. Med. Assoc.* **216**, 1979–1983.

McCarthy, A.M., McMahan, L. & Schaffer, P.A. (1989) Herpes simplex virus type 1 ICP27 deletion mutants exhibit altered patterns of transcription and are DNA deficient. *J. Virol.* **63**, 18–27.

Oh, Y.J., Wong, S.C., Moffat, M., Ullrey, D., Geller, A.I. & O'Malley, K.L. (1995) A herpes simplex virus-1 one vector containing the rat tyrosine hydroxylase promoter directs cell-type specific expression of β-galactosidase in cultured rat peripheral neurons. In press.

Patterson, T. & Everett, R.D. (1990) A prominent serine-rich region in Vmw175, the major transcriptional regulator protein of Herpes Simplex Virus type 1, is not essential for virus growth in tissue culture. *J. Gen. Virol.* **71**, 1775–1783.

Rossor, M.N., Watkins, J., Brown, M.J., Reid, J.L. and Dollery, C.T. (1980) Plasma levadopa, dopamine and therapeutic response following levadopa therapy of parkinsonian patients. *J. Neurol. Sci.* **46**, 385–392.

Smith, I.L., Hardwicke, M.A. & Sandri-Goldin, R.M. (1992) Evidence that the HSV immediate early protein ICP27 acts post-transcriptionally during infection to regulate gene expression. *Virology* **186**, 74–86.

Song, S., Wang, Y., Hartley, D., Ullrey, D., Bak, S.Y., Ashe, O., Bryan, J., Haycock, J., During, M., Neve, R.L., O'Malley, K.L. & Geller, A.I. (1994) Cell type specific expression from the tyrosine hydroxylase promoter may account for the rotational behavior induced by activation of the PKC pathway in SNc neurons with a HSV-1 vector. *Abstr. Soc. Neurosci.* **20**, 586.10.

Tashiro, Y., Kaneko, T., Sugimoto, T., Nagatsu, I., Kikuchi, H. & Mizuno, N. (1989) Striatal neurons with aromatic L-amino acid decarboxylase-like immunoreactivity in the rat. *Neurosci. Lett.* **100**, 29–34.

Wolff, J.A., Fisher, L.J., Xu, L., Jinnah, H.A., Langlais, P.J., Iuvone, M.P., O'Malley, K.L., Rosenberg, M.B., Shimohara, S., Friedmann, T., Gage, F.H. (1989) Grafting fibroblasts genetically modified to produce L-Dopa in a rat model of Parkinson's Disease. *Proc. Natl. Acad. Sci. USA* **86**, 9011–9014.

Xu, H., Federoff, H., Maragos, J., Parada, L.F. & Kessler, J.A. (1994) Viral transduction of *trkA* into cultured nodose and spinal motor neurons conveys NGF responsiveness. *Devel. Biol.* **163**, 152–161.

Yahr, M.D. & Bergmann, K.J. (eds) (1987) *Parkinson's Disease.* New York, Raven Press.

Yahr, M.D., Duvoisin, R.C., Schear, M.J., Barrett, R.E. & Hoehn, M.M. (1969) Treatment of parkinsonism with levodopa. *Arch. Neurol.* **21**, 343–354.

CHAPTER 8

PROBLEMS IN THE USE OF HERPES SIMPLEX VIRUS AS A VECTOR

Lawrence T. Feldman

Department of Microbiology and Immunology, UCLA School of Medicine, Los Angeles, CA 90024, USA

Table of Contents

8.1 Introduction

In this chapter I will review the use of herpes simplex virus type I (HSV1) as a vector for expression of foreign genes. The use of HSV1 as a vector has been investigated by a number of groups and these studies have included both replicating and non-replicating viruses (Breakefield and DeLuca, 1991; Johnson *et al.*, 1992; Miyamohara *et al.*, 1992; Davar *et al.*, 1994) and helper-dependent viral DNA constructs termed amplicons (Federoff *et al.*, 1992; Battleman *et al.*, 1993). The ability of HSV1 to establish a long-term latent infection in neurons has made it an attractive candidate as a vector for the delivery and expression of foreign genes in the nervous system. Latent infections are characterized by the absence of viral gene expression of the lytic cycle. During latency RNA accumulates from a single transcription unit and these RNAs have been designated the latency-associated transcripts (LAT). Since the LAT RNAs are not cytopathic, and can be deleted, the latent state is an ideal one to express foreign genes with minimal effect on the host cell.

It has also been established that the latent viral genome persists as multiple episomal copies in each neuron (Rock and Fraser, 1983, 1985). In this state the viral DNA can be reactivated to allow transcription of the lytic genes and viral amplification.

Genetic Manipulation of the Nervous System
ISBN 0-12-437165-5

However, it has been demonstrated that non-replicating viral mutants establish latent infections efficiently (Katz *et al.*, 1990; Sedarati *et al.*, 1993; Ramakrishnan *et al.*, 1994), which enhances the potential of HSV1 as a vector since non-replicating viruses do not reactivate (Johnson *et al.*, 1992; Sederati *et al.*, 1993). Thus, one of the main goals of using HSV1 as a vector is to create a modified virus which retains its capacity to infect post-mitotic cells such as neurons, where it establishes a latent infection for the life of the animal, while crippling its ability to express viral genes which could harm the host either during the initial infection or during reactivation.

A second advantage of HSV as a vector is its high capacity to accept foreign DNA; inserts of over 10 kb are easily obtained, and even greater amounts of DNA can be incorporated into the viral genome if necessary.

While these two aspects of the virus continue to hold promise, there are both theoretical and practical problems which have prevented the extensive use of HSV as a vector. These problems are each discussed separately below.

8.2 Cytopathic effects

Infection of the animal with wild-type HSV1 causes a local infection of the epithelial cells and subsequent infection of the neuronal cells innervating that tissue. In the neuron the virus takes one of two pathways, either productively infecting the neuron and producing more virus and eventually killing the cell, or entering a relatively quiescent latent state as described above. Thus, even though the goal of using HSV1 as a vector is to express genes during latency, the capacity of the virus to enter into a productive infection of neurons must be eliminated or greatly reduced.

The use of HSV1 as a vector became a practical goal with the discovery that the *ICP4* gene could be deleted and the virus grown as a host range mutant (DeLuca *et al.*, 1985). The ICP4 protein is a transcriptional activator which is required for both early and late viral gene transcription. Thus the *ICP4* viruses are unable to replicate to any degree in the absence of the *ICP4* gene product. DeLuca *et al.* (1985) generated the host range E5 cells expressing *ICP4* and established the basic backbone of a modified herpes virus which is unable to replicate in the host. Although other steps remain to be taken, this single viral mutation so reduced the otherwise extremely cytopathic effect of the virus that it made the use of HSV1 as a vector possible.

In further studies, DeLuca has deleted the *ICP27* gene from the *ICP4*$^-$ virus. The *ICP27* gene has been shown to adversely affect cellular RNA splicing. This double mutant does not revert back to wild-type virus with any measure of frequency and is propogated on an *ICP4*:*ICP27* cell line. The double mutant also expresses somewhat lower levels of early proteins relative to the single *ICP4*$^-$ mutant.

To reduce further the cytotoxicity of the virus, DeLuca and colleagues introduced other mutations into the *ICP4*:*ICP27* virus background. *UL41*, which encodes the virion host shut-off protein, was disrupted, as well as *ICP6*, the gene for a protein kinase made in the absence of ICP4. The resulting virus is 10- to 30-fold less cytotoxic than the *ICP4:ICP27* double mutant as measured by the ability of cells infected at different multiplicities to form colonies (N. DeLuca, personal communication).

Another group working in the area of reducing cytopathic effects of HSV1 is Friedmann's laboratory. Johnson and Freidmann have used modified HSV1 derivatives to examine the effects of immediate early (IE) gene expression on cytopathic effects *in vitro*. Infection of either BHK or CV-1 cells with an *ICP4*$^-$ virus caused considerable cytopathic effect at a variety of multiplicities of infection (Johnson *et al.*, 1992). Co-transfection of plasmids encoding *ICP0, ICP22* and *ICP27* with a neomycin plasmid all cause decreased colony formation in the presence of G418. This suggested that these IE genes are cytopathic. Based on these results, they constructed a double mutant in which the *ICP4* gene was deleted and in which the VP-16 transactivation function was altered. This latter mutation prevents the VP-16 virion protein from interacting with Oct-1 to stimulate IE gene transcription. This double mutant was greatly reduced in its capacity to produce cytopathic effects upon infection of either BHK or CV-1 cells. Interestingly, in the same study it was found that mutation of the virion host shutoff (VHS) function did not significantly contribute to cytopathic effects. These experiments illustrate the importance of halting IE gene expression when using HSV as a vector in cell culture. It is likely that such expression may also be important in infection of neurons in the animal, however many neurons do not constitutively express Oct-1 and the level of IE gene expression appears to be directly related to the presence of amplified DNA templates (Kosz-Vnenchak *et al.*, 1993). Therefore the cytopathic effect of IE gene expression in neurons may be less than in cell culture.

8.3 The problem of long-term expression

One of the interesting problems which has arisen is the problem of long-term expression. We and others discovered an unusual behavior of viral and cellular promoters inserted into the virus (reviewed by Glorioso *et al.*, 1992). These promoters were functional in tissue culture infections and in infections of neurons for a few days. However, as the virus entered a latent state, expression from these promoters decreased over time so that by 21 days post-infection no RNA signals from the reporter gene could be observed by *in situ* hybridization (Margolis *et al.*, 1993). A puzzling aspect of this is that in wild-type infections the LAT promoter remains functional throughout latency, and LAT can be observed by *in situ* hybridization at any time after the latent infection has been established (Stevens *et al.*, 1987; Dobson *et al.*, 1989). Still, use of the LAT promoter to drive these reporter genes failed to give long-term expression (Margolis *et al.*, 1992).

In one of the earliest viral vectors constructed, we placed the murine leukemia virus long terminal repeat (LTR) (Price *et al.*, 1987) driving the *lacZ* gene into the *ICP4* gene. This created a defective virus, 8117–43, of the kind that DeLuca and colleagues had previously constructed, a host range virus which must be grown on ICP4-expressing

E5 cells (DeLuca *et al.*, 1985). Interestingly, this virus expressed *lacZ* RNA by *in situ* hybridization 30 days after infection of mouse dorsal root ganglia, although expression was not observed in the hypoglossal nucleus (Dobson *et al.*, 1990). Despite this success, all other viruses subsequently constructed by us and many others have failed to yield RNA of sufficient quantity to be detected by either *in situ* hybridization or by Northern blot analysis. Lower levels of RNA expression have been observed by RT-PCR, however this result is subject to alternative explanations of promoter-driven expression, such as low level reactivation from the latent state or read-through expression from other promoters. In addition, expression at such low levels is unlikely to be useful in many experimental settings. In our view, expression must be detectable by *in situ* hybridization for the vector to have a reasonable chance of success.

In our laboratory, we addressed three essential questions. One, why did the LTR construct work when others did not? Second, why did the normal LAT promoter continue to drive LAT expression and not other genes? Third, why did the LTR construct work in dorsal root ganglia and not in the hypoglossal nucleus?

In considering why the LTR promoter was able to remain active in 8117–43, we thought it possible that the presence upstream of the LAT promoter/enhancer may have contributed to neuronal expression. To test this idea, the LTR-*lacZ* construct was inserted into the virus at the glycoprotein C locus (*gC*), away from the repeat region and the LAT promoter. This virus did not express RNA during latency, suggesting that the LAT promoter did contribute to LTR function in neurons (Lokensgard *et al.*, 1994). Since we also knew that the LAT promoter-*lacZ* constructs also did not remain active when inserted into the virus at *gC*, we reasoned that the two DNA fragments might work in concert. We then placed the upstream part of the LAT promoter, excluding the TATA box but including the enhancer region, next to the LTR-*lacZ* construct. A virus containing this combination continued to express RNA during latency (Lokensgard *et al.*, 1994). This indicated that functional promoter expression during latency may require two elements, and that the upstream LAT promoter and the LTR together provide these functions.

As to the second question, of why the LAT promoter continues to drive LAT during latency but not *lacZ*, we have found that a region 3′ of the transcription start site of the LAT transcription unit is apparently required for long-term expression. We therefore refer to this region of the LAT transcription unit as the LTE region. The reason why LAT continues to be made during latency but the *lacZ* gene at *gC* shuts off is that the promoter at LAT is connected to the LTE region, but is absent in the promoter construct driving *lacZ* at *gC*. Addition of a DNA fragment comprising the LTE to the *lacZ* construct restores long-term expression of the *lacZ* gene from the LAT promoter (J.R. Lokensgard and L.T. Feldman, unpublished).

In latent infections of the brain stem, Fraser and colleagues have shown that the viral DNA is present in the form of nucleosomes (Deshmane and Fraser, 1989). Because the formation of nucleosomes often leads to a reduction in transcription, one possible model by which the LTR and LTE DNA fragments could function is to correctly order the nucleosomes locally around the TATA box so that transcription can continue on a long-term basis.

8.4 Targeted expression

In another set of experiments we set out to examine levels of transcription from different viral and cellular promoters. These experiments addressed the third question of why the LTR in 8117–43 was functional in dorsal root ganglia but not in the hypoglossal nucleus.

In these experiments expression was measured at 4 days post-infection to eliminate the problem of long-term expression. Viruses containing the *lacZ* gene driven by the LTR and the methallothionein promoter gave moderate levels of expression in the dorsal root ganglia. This expression was substantially increased by addition of the upstream region of the LAT promoter (Lokensgard *et al.*, 1994). Thus the addition of, essentially, DNA binding sites, increased promoter expression dramatically. This was further demonstrated by recent work with *in vivo* expression from a LAT promoter deletion series. As specific regions of the upstream part of the LAT promoter were removed, particularly between −330 and −250, expression in neurons decreased (Dobson *et al.*, 1995). It is therefore likely that targeted expression is nothing more than making a good match between transcription factor binding sites on the promoter and high levels of the same transcription factor in a given cell type.

Thus a reasonable explanation for the lack of expression of 8117–43 in the hypoglossal nucleus may have been a lack of transcription factors which could bind to either the LTR or the LAT enhancer. We have recently observed that 8117–43 continues to express on a long-term basis in hippocampal neurons of the central nervous system, but at a much lower level than that found in the dorsal root ganglia (Bloom *et al.*, 1995). Again, the explanation could be that the neurons of the hippocampus do not contain transcription factors suitable for high level expression from the LAT enhancer region.

We therefore have a picture of long-term expression in which either the LTR or the 3′ LAT region can effect long-term expression, and secondly the level of expression observed appears to derive mainly from the upstream enhancer region in the construct and its ability to bind to transcription factors present in the neurons (see Figure 1). Thus in neurons of the dorsal root ganglia, where the factors can bind to the LAT enhancer, expression is high, and in other neurons it is lower or not observed. Nevertheless, we appear to be in a position to put a promoter–enhancer construct together which could afford high levels of long-term expression in the neuron of interest.

Recently, Geller and colleagues have succeeded in establishing low but measurable levels of tyrosine hydroxylase from an amplicon-based HSV (During *et al.*, 1994). Expression from this virus produced long-term recovery in a rat model of Parkinson's disease. This interesting experiment is treated in greater depth in Chapter 7.

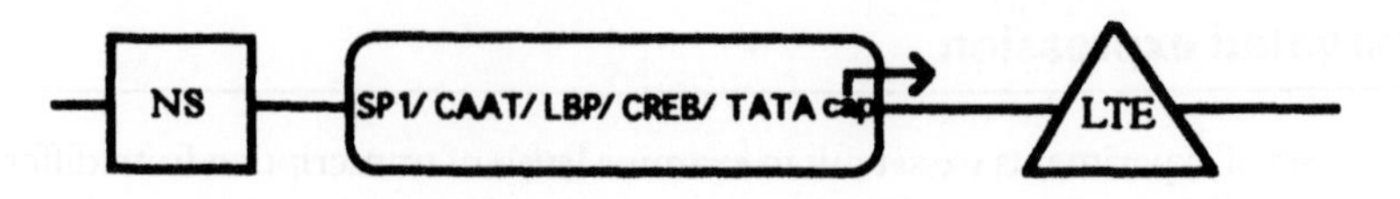

Figure 1 Separation of LAT promoter functional properties. NS, DNA binding sites for neuronal specific transcription factors; LTE, long-term expression element.

8.5 Neural tracing studies

One of the interesting current uses of herpes viruses as vectors exploits their neurotropic and neuroinvasive properties to trace neural circuits. Enquist and his colleagues have used modified strains of pseudorabies virus (PRV) to determine the dendritic architecture of neurons involved in the baroreceptor vagal reflex. Expression of the β-galactosidase gene from a recombinant virus allowed them to successfully define the dendritic morphology of motorneurons and interneurons that innervate the heart (Standish *et al.*, 1995). This group also used two different strains of PRV to double label second- and third-order neurons. Basically, the two viruses were separately put into either the stomach or the esophagus, or stomach and cecum of rats. Double labeling of the same neurons did not occur in the primary motor nuclei for the different viscera but did occur in some neurons of the solitary tract, ventrolateral medulla, parapyramidal area, paragigantocellular reticular formation, raphe magnus and the A5, A6 and sub-A6 catecholaminergic areas – all known afferent sources to the above primary motor nuclei. This double labeling provides anatomical evidence that some second- and third-order afferents can simultaneously modulate final common paths to different viscera and thereby serve an integratory or coordinating function (Miselis *et al.*, 1995).

8.6 Expression in non-neuronal cells

Miyanohara *et al.* (1992) report the transient expression of β-galactosidase, hepatitis B virus surface antigen and canine factor IX in *ICP4*$^-$ defective viruses. When these viruses were inoculated into mouse liver by direct injection or by the portal vein, these reporter genes under control of the human CMV promoter expressed at a high level but for only a few days. Lower level expression, but for up to 3 weeks, was obtained by expression under the LAT promoter. These results indicate the potential use of the liver as a site for infection with HSV vectors, and underscore the importance of the requirement for long-term expression.

Acknowledgments

I wish to thank Neal DeLuca, Paul Johnson and Lynn Enquist for communicating their unpublished work. I also wish to acknowledge my long-term collaborator on viral vectors, Dr Jack G. Stevens, as well as Dr Nigel Maidment, both at UCLA. Collaborative studies in both of these labs has been most helpful in gaining insight into the problems of long-term expression in the PNS and CNS. I also wish to thank Dr David Bloom in Jack's lab, for his help over the years on many of these problems. Finally, I would like to acknowledge the outstanding work principally of two members of my laboratory. Dr Anthony Dobson was instrumental in finding the LAT promoter and in constructing the first vector, 8117–43. Dr James Lokensgard constructed all of the viruses which led to our understanding of long-term expression as well as enhancer-mediated expression, as depicted in Figure 1 of this text. This work was supported by PHS grants NS-30420 to J.G.S. and A128338 to L.T.F.

References

Battleman, D.S., Geller, A.I. & Chao, M.V. (1993) HSV-1 vector-mediated gene transfer of the human nerve growth factor receptor p75 (hNGFR) defines high affinity NGF binding. *J. Neurosci.* **13**, 941–951.

Bloom, D.S., Maidment, N.T., Tan, A., Dissette, V.B., Feldman, L.T. & Stevens, J.G. (1995) Long term expression of a reporter gene from latent herpes simplex virus in the rat hippocampus. *Molec. Brain Res.* (in press).

Breakefield, X.O. & DeLuca, N.A. (1991) Herpes simplex virus for gene delivery to neurons. *New Biol.* **3**, 303–218.

Davar, G., Kramer, M.F., Garber, D., Roca, A.L., Andersen, J.K., Bebrin, W., Coen, D.M., Kosz-Vnenchak, M., Knipe, D.M. & Breakefield, X.O. (1994) Comparative efficacy of expression of genes delivered to mouse sensory neurons with herpes virus vectors. *J. Comp. Neurol.* **339**, 3–11.

DeLuca, N.A., McCarthy, A.M. & Schaffer, P.A. (1985) Isolation and characterization of deletion mutants of herpes simplex virus type 1 in the gene encoding immediate-early regulatory protein ICP4. *J. Virol.* **63**, 943–947.

Deshmane, S.L. & Fraser, N.W. (1989) During latency, herpes simplex virus type 1 DNA is associated with nucleosomes in a chromatin structure. *J. Virol.* **63**, 943–947.

Dobson, A.T., Sederati, F., Devi, R.G., Flanagan, W.M., Farrell, M.J., Stevens, J.G., Wagner, E.K. & Feldman, L.T. (1989) Identification of the latency-associated transcript promoter by expression of rabbit beta-globin mRNA in mouse sensory nerve ganglia latently infected with a recombinant herpes simplex virus. *J. Virol.* **63**, 3844–3851.

Dobson, A.T., Margolis, T.P., Sedarati, F., Stevens, J.G. & Feldman, L.T. (1990) A latent, non-pathogenic HSV-1-derived vector stably expresses beta-galactosidase in mouse neurons. *Neuron* **5**, 353–360.

Dobson, A.T., Margolis, T.P., Gomes, W.A. & Feldman, L.T. (1995) *In vivo* deletion analysis of the herpes simplex virus type 1 LAT promoter. *J. Virol.* **69**, 2264–2270.

During, M.J., Naegele, J.R., O'Malley, K.L. & Geller, A.I. (1994) Long-term behavioral recovery in parkinsonian rats by an HSV vector expressing tyrosine hydroxylase. *Science* **266**, 1399–1403.

Federoff, H.J., Geschwind, M.D., Geller, A.I. & Kessler, J.A. (1992) Expression of nerve growth factor *in vivo* from a defective herpes simplex virus 1 vector prevents effects of axotomy on sympathetic ganglia. *Proc. Natl. Acad. Sci. USA* **89**, 1630–40.

Glorioso, J.C., Goins, W.F. & Fink, D.J. (1992) Herpes simplex virus-based vectors. *Sem. Virol.* **3**, 265–276.

Johnson, P.A., Miyanohara, A., Levine, F., Cahill, T. & Friedmann, T. (1992) Cytotoxicity of a replication-defective mutant of herpes simplex virus type 1. *J. Virol.* **66**, 2952–2965.

Katz, J.P., Bodin, E.T. & Coen, D. (1990) Quantitative polymerase chain reaction analysis of herpes simplex virus DNA in ganglia of mice infected with replication incompetent mutants. *J. Virol.* **64**, 4288–4295.

Kosz-Vnenchak, M., Jacobson, J., Coen, D.M. & Knipe, D.M. (1993) Evidence for a novel regulatory pathway for herpes simplex virus gene expression in trigeminal ganglia. *J. Virol.* **63**, 5383–5393.

Lokensgard, J.R., Bloom, D.C., Dobson, A.T. & Feldman, L.T. (1994) Long-term promoter activity during herpes simplex virus latency. *J. Virol.* **68**, 7148–7158.

Margolis, T.P., Sedarati, F., Dobson, A.T., Feldman, L.T. & Stevens, J.G. (1992) Pathways of viral gene expression during acute and neuronal infection with HSV-1. *Virology* **189**, 150–160.

Margolis, T.P., Bloom, D.C., Dobson, A.T., Feldman, L.T. & Stevens, J.G. (1993) LAT promoter activity decreases dramatically during the latent phase of gangliaonic infection with HSV. *Virology* **197**, 585–592.

Miselis, R.R., Ynag, M., Robbins, A., Whealy, M. & Enquist, L. (1995) Transsynaptic double-labelling within the visceral neuraxis with uniquely marked strains of PRV. (in prep.)

Miyanohara, A., Johnson, P.A., Elam, R.L., Dai, Y., Witztum J.L., Verma, I.M. & Friedmann, T. (1992) Direct gene transfer to the liver with herpes simplex virus type 1 vectors: transient production of physiologically relevant levels of circulating factor IX. *New Biol.* **4**, 238–246.

Price, J., Turner, D. & Cepko, C. (1987) Lineage analysis in the vertebrate nervous system by retrovirus-mediated gene transfer. *Proc. Natl. Acad. Sci. USA* **84**, 156–160.

Ramakrishnan, R., Fink, D.J., Jiang, G., Desai, P., Glorioso, J.C. & Levine, M. (1994) Competetive quantitative PCR analysis of herpes simplex virus type 1 DNA and latency-associated transcript RNA in latently infected cells of the rat brain. *J. Virol.* **68**, 1864–1873.

Rock, D.L. & Fraser, N.W. (1983) Detection of HSV-1 genome in central nervous system of latently infected mice. *Nature* **302**, 523–525.

Rock, D.L. & Fraser, N.W. (1985) Latent herpes simplex virus type 1 DNA contains two copies of the virion DNA joint region. *J. Virol.* **55**, 849–852.

Sedarati, F., Margolis, T.P. & Stevens, J.G. (1993) Latent infection can be established with drastically restricted transcription and replication of the HSV-1 genome. *Virology* **192**, 687–691.

Standish, A., Enquist, L.W., Miselis, R.R. & Schwaber J.S. Dendritic morphology of cardiac related medullary neurons defined using circuit-specific infection by PRV-BaBlu, a recombinant virus expressing β-galactosidase. (in prep.)

Stevens, J.G., Wagner, E.K., Devi, R.G.B., Cook, M.L. & Feldman, L.T. (1987) RNA complementary to a herpesvirus alpha gene mRNA is prominent in latently infected neurons. *Science* **235**, 1056–1059.

Wolfe, J.H., Deshmane, S.L. & Fraser, N.W. (1992) Herpesvirus vector gene transfer and expression of beta-glucuronidase in the central nervous system of MPS VII mice. *Nature Genet.* **1**, 379–384.

CHAPTER 9

GENE DELIVERY TO THE NERVOUS SYSTEM BY DIRECT INJECTION OF RETROVIRAL VECTOR

Ryoichiro Kageyama,[*] Makoto Ishibashi[*†] and Koki Moriyoshi[*]

[*]*Institute for Immunology and* [†]*Department of Anatomy, Kyoto University Faculty of Medicine, Sakyo-ku, Kyoto 606, Japan*

Table of Contents

9.1 Introduction

The analysis of the functions of a gene in developmental processes requires the study of gain-of-function mutations, which reveal the effects of persistent expression of the gene, as well as of loss-of-function mutations, which show what the gene is necessary for. To make these mutations, transgenic and gene targeting methods are widely used (Hogan *et al.*, 1986; Joyner, 1993). Although these methods offer a powerful approach to problems of mammalian development, they require special equipment and are technically demanding. Furthermore, if over-expression or null mutation of the gene is lethal, mutant embryos do not survive to full term, hampering the analysis of mutant phenotypes.

One alternative and very successful method is genetic manipulation by a recombinant retrovirus (Mann *et al.*, 1983; Cone and Mulligan, 1984; Miller *et al.*, 1985). A retroviral vector is an infectious vehicle which efficiently transfers exogenous genes into mitotic cells. After infection, a single DNA copy of the RNA viral genes is stably

Genetic Manipulation of the Nervous System
ISBN 0-12-437165-5

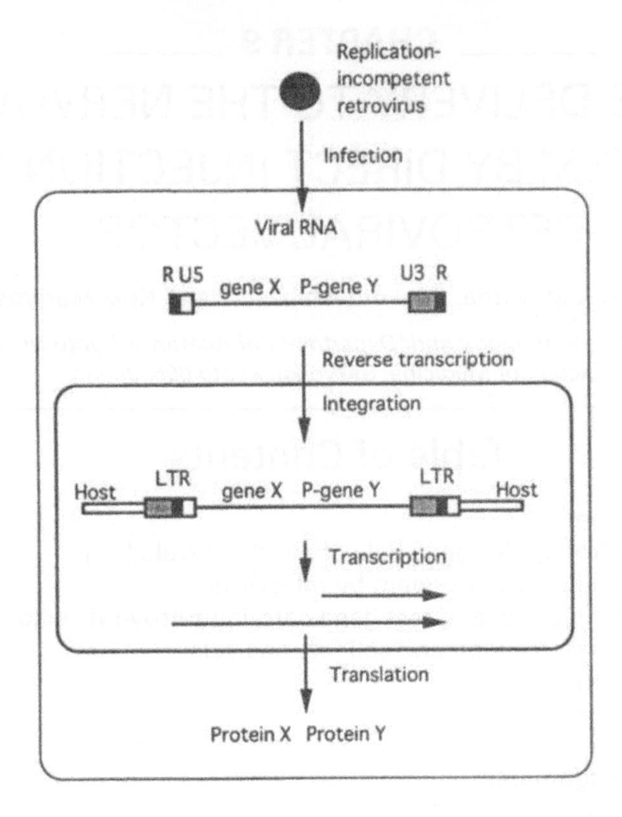

Figure 1 Replication-incompetent retrovirus-mediated gene expression. After infection, a single DNA copy of the RNA viral genome produced by reverse transcriptase is integrated into the host genome. The DNA copy is precisely transmitted to their progeny. Exogenous genes X and Y are transcribed from the 5′-LTR promoter and the internal promoter (P), respectively. Because replication-incompetent retrovirus does not contain genes for viral capsid proteins, infected cells do not produce viral particles. Thus, this viral vector cannot spread horizontally to neighbouring unrelated cells.

integrated into the genome of host cells (Figure 1; Weiss *et al.*, 1985). The DNA copy (the provirus form) contains the whole intact vector sequence, and the integrated genes are precisely transmitted to their progeny. If the exogenous genes are under the control of constitutive promoters, they will be permanently expressed. This approach enables us to examine the behaviour of groups of mutated (retrovirus-infected) cells in a wild-type environment (non-infected cells), thus facilitating the analysis of mutant phenotypes. Whereas adenovirus vector also efficiently transfers exogenous genes, the adenovirus-mediated gene expression is transient because the adenoviral genome is not integrated into the host genome (Yang, 1992; Brody and Crystal, 1994). Thus, the fact that the retrovirus-infected cells and their progeny will permanently express the exogenous genes makes a retroviral vector more advantageous for the analysis of the long-term behaviour of infected cells. But, because the integration of retroviral genes into the host cell genome requires the cell to go through S phase, retrovirus is not the choice when genes are to be introduced into post-mitotic cells.

9.2 General strategy for gene delivery to the developing mammalian nervous system by retrovirus

9.2.1 Target cells of the developing nervous system

Recombinant retrovirus has been successfully used to deliver the exogenous genes to the developing mammalian nervous system (Sanes *et al.*, 1986; Price *et al.*, 1987; Turner and Cepko, 1987; Luskin *et al.*, 1988; Walsh and Cepko 1988; Price and Thurlow, 1988; Galileo *et al.*, 1990; Turner *et al.*, 1990). In mammals, the neural tube derived from the embryonic ectoderm gives rise to the central nervous system (CNS; Jacobson, 1991). Here, the neural precursor cells are dividing in the ventricular zone, which is next to the lumen. When they determine their fate to become neurons, the precursor cells stop cell division, migrate out of the ventricular zone to the outer regions, and undergo terminal differentiation (see Figure 4). Neural precursor cells also give rise to glial cells, which are capable of proliferation. Thus, neural precursor cells and dividing glial cells, but not neurons which are post-mitotic, are the major target of retroviral vectors.

9.2.2 Retrovirus carrying a cell-marking gene

In most cases, a retroviral vector has been used for cell lineage analysis to mark neural precursor cells with *lacZ*, which enables visualization of the infected cells and their progeny with X-gal staining. Because X-gal staining occurs in somata, the cell type of *lacZ*-expressing cells (neural precursor cells, neurons or glial cells) can be determined. It has been shown that neural precursor cells infected with *lacZ*-transducing retrovirus differentiate to neurons and glial cells (Turner and Cepko, 1987; Luskin *et al.*, 1988; Walsh and Cepko, 1988). If a nuclear localization signal sequence is fused to *lacZ*, blue staining with X-gal occurs in nuclei but not in somata, facilitating immunohistochemical study with antibodies reactive to cytoplasmic antigens (Galileo *et al.*, 1990).

Recently, another gene, human placental alkaline phosphatase gene (*hPLAP*), has been developed for a cell-marking method (Fields-Berry *et al.*, 1992; Halliday and Cepko, 1992). The membrane-associated nature of hPLAP is advantageous in determining cell types and other morphological features such as projection patterns of neurons.

The gene for jellyfish green fluorescent protein (GFP) is another tool that shows unique potential for monitoring living cells (Chalfie *et al.*, 1994). Cells expressing GFP can be easily identified (green fluorescence) by irradiation of long-wave UV without fixing cells. Thus, this gene would provide a novel technique that enables real-time analysis of cell fate, growth and differentiation.

In one application of the above-mentioned cell-marking techniques, we have recently developed a retrovirus that transduces another gene (a test gene) in addition to a cell-marking gene (Ishibashi *et al.*, 1994). This virus is useful for the *in vivo* functional analysis of the test gene because the behaviour and fate of cells that permanently express the test gene can be monitored by simply examining the stained cells.

9.2.3 Direct injection of retrovirus

To infect neural precursor cells with retrovirus, virus solution should be injected into the ventricles of the embryonic CNS. There are several methods available for injection of virus into mouse and rat developing nervous system. One of the easiest methods (but difficult to obtain constant results) is where, after ventral laparotomy of pregnant animals, the virus is injected into embryos without opening the uterus (*in utero*; Cockroft, 1990). Injections are aimed at the head of embryos localized by transillumination of the uterus. But, this is in most part a blind injection, and thus it is difficult to correctly target neural precursor cells. Another method, which is more difficult but more reproducible, is virus injection into the brain vesicles of embryos exposed outside of the uterus (*ex utero*; Muneoka *et al.*, 1990). This method allows direct access of virus to the ventricles of the developing CNS under the microscope, and thus neural precursor cells present in the wall of the brain vesicles are reproducibly infected with retrovirus. A third method concerns neural retina, where neurogenesis continues post-natally. In this case, virus can be injected into the eyes of newborn mice to infect neural precursor cells (Turner and Cepko, 1987). When virus is injected into the space between the pigment epithelium and the retina, the virus spreads in the back of the eye and infects neural retinal precursor cells, which will differentiate into rods, bipolar cells, amacrine cells and Müller glial cells.

In the following sections, we describe: (1) the cell-marking gene- and test gene-carrying recombinant retroviral vector that we have recently developed; and (2) the method of virus injection into the developing CNS of *ex utero* embryos. We also describe the applications of this retrovirus injection method to study gain-of-function and loss-of-function mutations.

9.3 Cell-marking gene- and test gene-carrying retroviral vector

9.3.1 Structure of the retroviral vector

Because retroviral genomes are made of RNA, the provirus form (a DNA copy produced by reverse transcriptase) cloned in a plasmid is used for genetic manipulation. The replication-competent retroviral genomes contain ψ, *gag*, *pol* and *env* flanked by the repeat sequence at both ends. The ψ sequence is required for encapsidation of viral genomes, but the *gag*, *pol* and *env* genes that encode the capsid, reverse transcriptase, integrase and envelope proteins can be deleted for replication-incompetent retrovirus. Thus, the *gag, pol* and *env* genes are exchanged with other genes in most recombinant retroviruses. There are various types of replication-incompetent retroviral vectors available, but those containing the *neo* gene are convenient for virus production because virus-producing cell lines are selected by drug resistance. In addition, vectors containing the *lacZ* gene are useful for virus titration and monitoring cell fate because the infected cells and their progeny are easily identified by X-gal staining. Thus, retroviruses transducing these two genes, *neo* and *lacZ*, are often used.

When the effects of a gene on cell differentiation are investigated by retrovirus, three different genes, *neo, lacZ* and the gene to be tested, should be expressed from one retroviral vector. Because the *neo* and *lacZ* genes can be fused (Friedrich and Soriano, 1991), two promoters will be necessary to express the three genes. We constructed pLXSG vector, which expresses the *lacZ–neo* fusion gene from the internal SV40 early promoter (Ishibashi *et al.*, 1994), by exchanging the *lacZ–neo* gene with the *neo* gene of the retroviral vector pLXSN (Miller and Rosman, 1989; Figure 2). The third gene (a test gene) can be cloned into the *Xho*I site which is located downstream of the 5′-long terminal repeat (LTR). This gene will be expressed from the 5′-LTR promoter. Because of the limitation of the genome size that the retrovirus can transmit (less than 10 kb in total between the two LTR sequences), the test gene should be less than 3–4 kb.

9.3.2 Preparation of retroviral stock

To make infectious retroviral particles, the vector DNA should be transfected into packaging cells, which provide the viral structural proteins encoded by *gag, pol* and *env*. There are two types of packaging cell lines, the ecotropic- and amphotropic-type producer lines. Ecotropic retrovirus is infectious only to rodent cells, while amphotropic retrovirus has a very broad host range including humans and therefore needs more caution in handling. We usually use ψ2 packaging cells (Mann *et al.*, 1983) because they produce ecotropic retrovirus. Before transfection, the vector DNA is linearized by a restriction enzyme that digests only the outside of the viral portion (such as *Sca*I of pLXSG) because the linearized form seems to be integrated more efficiently than the closed circular form. The transfected packaging cells are then selected for drug resistance (G418). The RNA produced by transcription of the transfected viral DNA in the cells is subsequently packaged into viral particles by recognizing the ψ packaging sequence of the viral RNA.

Cells that are selected in the presence of G418 do not necessarily produce retrovirus. Thus, each colony should be tested for virus production. G418-resistant cells are grown in medium without G418, and after 2 days the recovered medium is added to rodent cells with polybrene for titration of virus. Polybrene helps virus adsorption to cells. After 2–3 days, the rodent cells are stained with X-gal and blue cells are counted. Virus-producing cells that give the highest titres should be expanded, and the recovered medium containing the retrovirus is subsequently concentrated. There are several methods for virus concentration, but we usually use the limited filtration method with ultrafiltration concentrators such as Centricon 100 (Amicon). Virus solu-

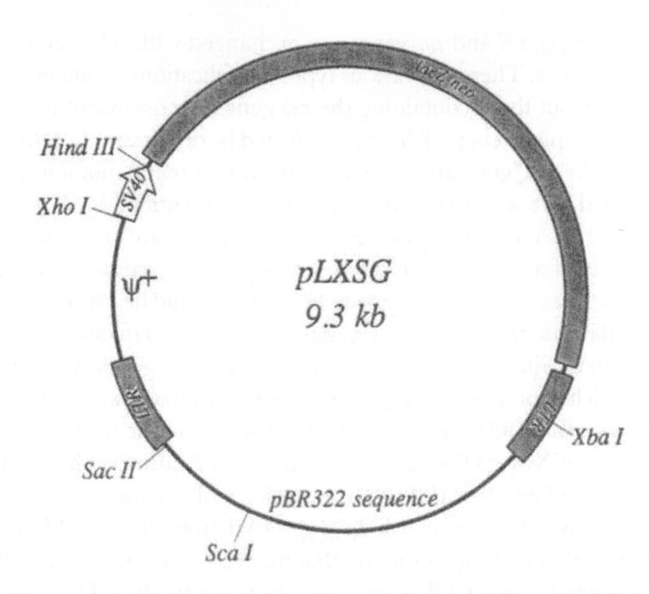

Figure 2 Structure of the retroviral vector pLXSG (Ishibashi *et al.*, 1994). The provirus DNA cloned into the plasmid pBR322 contains Moloney sarcoma virus LTR at the 5′-terminus, Moloney leukaemia virus LTR at the 3′-terminus and the *lacZ–neo* fusion gene which is under the control of the internal SV40 early promoter. It also has the ψ packaging signal sequence as well as a part of the *gag* gene (ψ^+). The test gene can be cloned into the *Xho*I site located between the 5′-LTR and the internal SV40 promoter. The DNA can be linearized by *Sca*I located in the plasmid portion before transfection into packaging cells (unless the test gene contains *Sca*I sites).

tion can be concentrated about 100-fold. The final titre should be $\sim 1\times10^6$ colony forming units (cfu) ml^{-1} or more.

It is very important to confirm that a test gene is transduced by retrovirus. Functional assays are preferable, but reverse transcriptase-mediated PCR, Northern or Western blot analysis may be sufficient.

When retrovirus is used for cell lineage analysis, it is necessary to check that virus stock does not contain helper virus that can spread the viral vector horizontally to neighbouring unrelated cells.

9.4 *Ex utero* manipulation and cell fate analysis

9.4.1 Virus injection to the nervous system of *ex utero* embryos

Embryos can be exposed outside by a longitudinal incision of the uterine myometrium opposite to the placenta. These exposed embryos develop normally *ex utero* (outside of uterus but intraperitoneally) as long as their connection to the uterus and placenta is

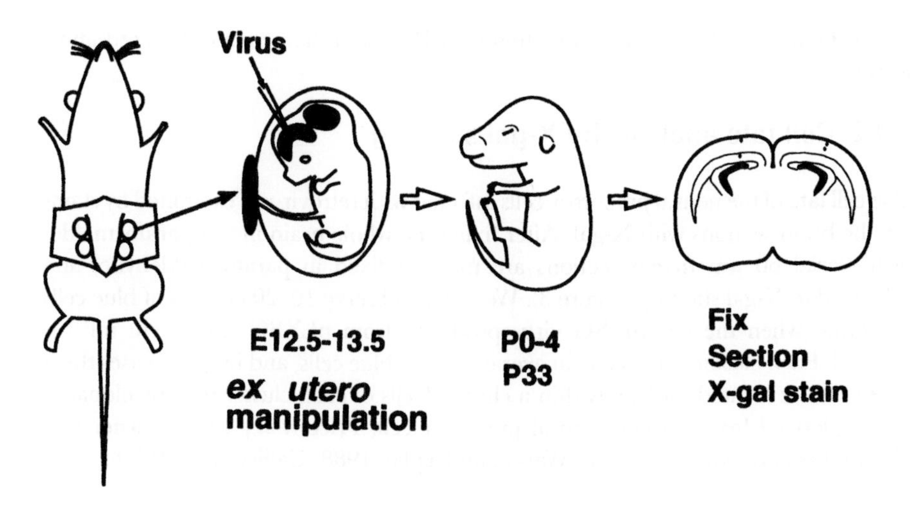

Figure 3 *Ex utero* manipulation and cell fate analysis. Pregnant mice on E12.5 or 13.5 (the vaginal plug is detected on E0) are anaesthesized by injecting Nembutal at a dose of 1.2 $\mu l\ g^{-1}$ body weight. After midline laparotomy, the uterine wall is incised to expose embryos with extra-embryonic membranes. Using a heat-pulled glass micropipette, 1 μl of retrovirus solution with 0.2 $mg\ ml^{-1}$ polybrene and 0.05% trypan blue is injected into the lateral ventricle. Embryos are allowed to develop *ex utero*. After delivery by Cesarean section on E19.5, neonates are fostered. Brains are harvested post-natally. After fixation, brains are embedded in OCT compound and frozen. Serial 60 μm sections are cut using a cryostat, refixed and stained with X-gal (5-bromo-4-chloro-3-indolyl β-galactopyranoside).

intact. Extra-embryonic membranes should not be damaged during incision. Embryos of days 12.5 (E12.5) and 13.5 are the most suitable for virus injection because they contain abundant neural precursor cells. Furthermore, the ventricles of the developing CNS can be observed from the outside at this stage. Thus, it is possible to inject virus solution precisely into the ventricles under the microscope (Figure 3). We usually inject around 1 μl of virus solution containing 0.2 $mg\ ml^{-1}$ polybrene and 0.05% trypan blue by using a heat-pulled glass micropipette. The outer diameter of the micropipette should be less than 50 μm. After injection, the blue dye (trypan blue) expands throughout the injected ventricles and sometimes to the next ventricles. This confirms that the virus has been injected to the correct places. At most, four embryos can be manipulated per pregnant mouse, and the rest should be removed from the uterus at the same time. During this procedure, embryos should be submerged in phosphate-buffered saline (PBS). After the procedure, the abdominal wall and skin are sutured separately without closing the uterus. The survival rate of the manipulated embryos is around 50% in the authors' laboratory. Embryos that are younger than E12.5 may be difficult to manipulate and the survival rate is much lower.

Embryos can be recovered at any time, but to analyse the effects of an exogenous gene on neural differentiation, it is better to allow embryos to develop to post-natal stages because neurogenesis continues after birth. Because the uterus is open, embryos should be delivered by Cesarean section on E19.5, and the neonates fostered afterwards.

9.4.2 Cell fate analysis by X-gal staining

The cell fate of the neural precursor cells infected with retrovirus is examined by staining the brain sections with X-gal. After fixing the whole brain in 2% paraformaldehyde, serial 60 μm frozen sections are made, refixed in paraformaldehyde, and subjected to X-gal staining (Figure 3). We usually observe 10–20 clusters of blue cells per brain when the control SG virus produced from pLXSG (1×10^6 cfu ml^{-1}) is injected. Each cluster consists of an average of five blue cells, and in many cases these blue cells are located radially within a cluster. Cells of each cluster may be clonal in origin (derived from a single neural precursor cell; Price *et al.*, 1987; Turner and Cepko, 1987; Luskin *et al.*, 1988; Walsh and Cepko, 1988; Galileo *et al.*, 1990).

9.5 Applications

9.5.1 Gain-of-function analysis

The method described above is applicable to analysis of the *in vivo* functions of a gene in neural differentiation. Particularly, it is suitable for gain-of-function analysis in specific subsets of differentiating cells. Here, we present results obtained recently in our laboratory.

Our laboratory has been focusing on helix–loop–helix (HLH) transcription factors involved in mammalian neurogenesis (Akazawa *et al.*, 1992; Sasai *et al.*, 1992; Ishibashi *et al.*, 1993; Takebayashi *et al.*, 1994; Sakagami *et al.*, 1994). Among them, HES-1, a mammalian HLH factor structurally related to the product of *Drosophila* pair-rule gene *hairy*, is expressed at high levels throughout the ventricular zone of the developing CNS, which contains neural precursor cells (Sasai *et al.*, 1992). HES-1 expression decreases as neural differentiation proceeds (Figure 4). Because of this negative correlation between HES-1 expression and neurogenesis, we tested whether HES-1 negatively regulates neural differentiation in the CNS (Ishibashi *et al.*, 1994).

To examine this hypothesis, we made two recombinant retroviruses, SG and SG-HES1. SG is a control retrovirus produced from pLXSG. SG-HES1 virus is made by inserting the HES-1 cDNA into the *Xho*I site of pLXSG. To determine whether HES-1 is expressed in SG-HES1 virus-infected cells, we carried out luciferase analysis. HES-1 represses transcription by binding to the N box sequences (CACNAG; Sasai *et al.*, 1992). The luciferase reporter gene under the control of the promoter containing the N box elements was transfected, and subsequently SG and SG-HES1 viruses were infected. The luciferase activity was significantly reduced when SG-HES1 virus was infected but not when SG virus was injected. This assay thus confirmed that SG-HES1 virus directed expression of functional HES-1 protein.

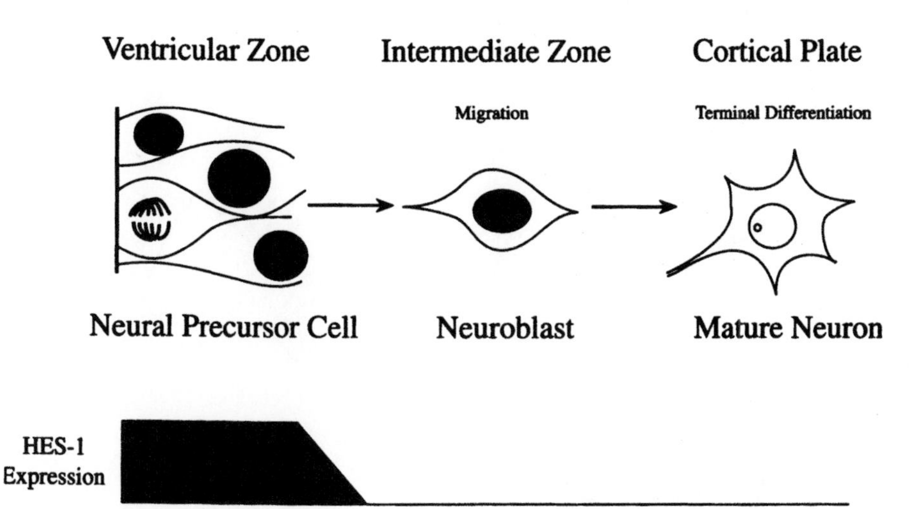

Figure 4 Neural differentiation and HES-1 expression. In mammalian developing CNS, neural precursor cells divide in the ventricular zone, which is next to the lumen. When they determine their fate to become neurons, they stop cell division, migrate out of the ventricular zone towards the outer layers, and undergo terminal differentiation. Helix–loop–helix factor HES-1, a mammalian homologue of the product of *Drosophila hairy*, is expressed throughout the ventricular zone. As neural differentiation proceeds, HES-1 expression decreases (Sasai *et al.*, 1992), and HES-1 transcript is not detected in differentiated neural cells.

We next injected the retroviruses into the lateral ventricles of *ex utero* E12.5 and E13.5 mouse embryos, as described above. The infected brains were examined postnatally by X-gal staining. We found many blue cells in the ventricular zone as well as in the outer layers of the brains infected with SG virus (Figure 5A). Many of the cells detected in the outer layers extended processes. These results suggest that cells infected with SG virus migrated out of the ventricular zone into the outer layers and underwent neural differentiation. In contrast, almost all of the blue cells detected in the brains infected with SG-HES1 virus were present in the ventricular zone (Figure 5B). Furthermore, they were round in appearance and none of them extended any processes. These results suggest that continuous expression of HES-1 prevents neural differentiation in the CNS. Thus, HES-1 is a negative regulator of mammalian neurogenesis, and down-regulation of HES-1 expression is necessary for neuronal and glial differentiation.

9.5.2 Loss-of-function analysis

A retroviral vector is also very useful for loss-of-function analysis by expressing antisense transcripts. For antisense analyses, oligonucleotides are widely used. However, antisense oligonucleotides are, in most cases, inefficient for *in vivo* analysis partly because they are easily degraded and partly because the absorption rate of oligonu-

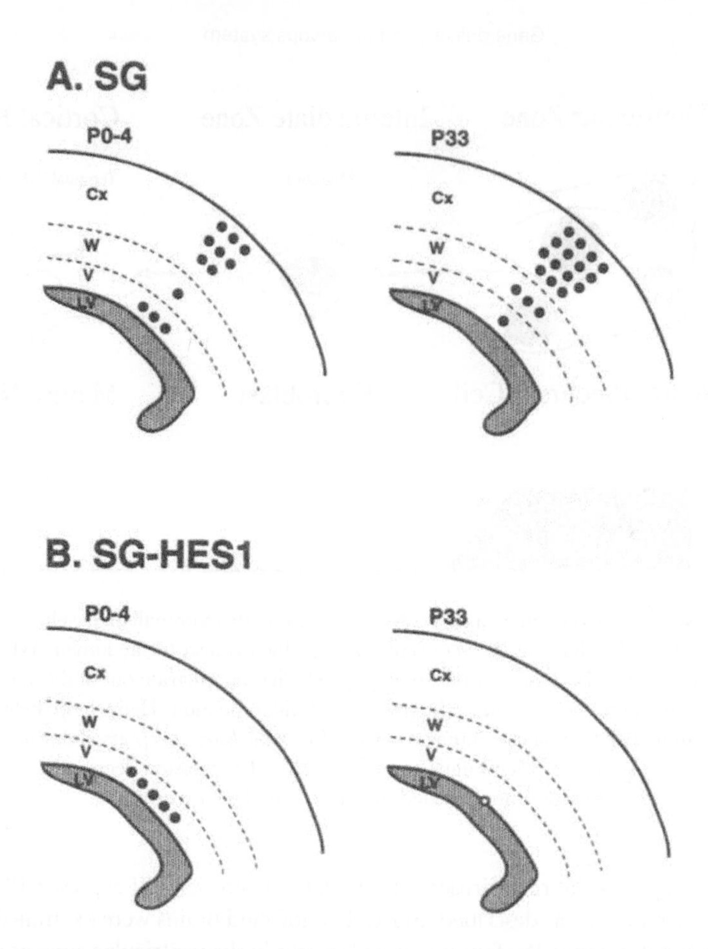

Figure 5 X-gal analysis of brains infected with SG virus (A) and SG-HES1 virus (B) (Ishibashi *et al.*, 1994). (A) Distribution of X-gal-stained clones (dots) in SG virus-infected brains. Brains were infected with SG virus (produced from pLXSG) on E12.5 or E13.5. Each clone consists of 1–10 labelled cells. On post-natal days 0–4 (P0–4, ~25% of the identified clones remained in the ventricular/subventricular zone (V), 8% in the white matter (W) and 67% reached the cortex (Cx). Many of the labelled cells found in the cortex extended processes, suggesting that these cells differentiate into neurons or glial cells. On P33, almost 80% of the labelled cells migrated out of the ventricular zone and reached the cortex. Only ~5% of the labelled cells remained in the ventricular/subventricular zone. These results show that neural precursor cells infected with the control SG virus undergo normal neural differentiation. LV, lateral ventricle. (B) Distribution of X-gal-stained clones in SG-HES1 virus-infected brains. Brains were infected with SG-HES1 virus (produced from pLXSG-HES1) on E12.5 or E13.5. Each clone mostly consists of one labelled cell. On P0-4, all labelled cells are present in the ventricular/subventricular zone, and none are found in the outer layers. All of the labelled cells are round in appearance and have no mature processes. On P33, the X-gal-stained cells decrease to ~10% compared with the number on P0-4, suggesting that most of the HES-1-expressing cells die during the post-natal 4–5 weeks. Furthermore, the labelled cells are present only in the ependymal layer, the epithelial lining of the ventricles. These results show that persistent expression of HES-1 prevents neural differentiation in the CNS.

cleotides into cells is quite low. Thus, *in vivo* antisense analysis requires repeated injection of oligonucleotides. For example, repeated injection of antisense N-methyl-D-aspartate (NMDA) receptor channel oligonucleotides into the lateral cerebral ventricle is required to obtain a 30–40% reduction of NMDA receptor expression in adult rat (Wahlestedt *et al.*, 1993). Thus, antisense analysis with embryos by simply injecting a dose of antisense oligonucleotides may not work. However, repeated injection into embryos is very difficult. This difficulty may be overcome by the retroviral vector directing antisense expression, and there are several studies reporting such an analysis. For example, Schwann cells infected with a retrovirus that produces myelin-associated glycoprotein (MAG) antisense RNA express low levels of MAG and do not myelinate axons (Owens and Bunge, 1991). Another example concerns neuroblasts infected with a retrovirus that expresses antisense integrin RNA (Galileo *et al.*, 1992). Neuroblasts infected with a control retrovirus migrate along radial glia to form the tectal plate, whereas neuroblasts infected with the retrovirus expressing antisense integrin RNA do not. Thus, the retrovirus method is undoubtedly more reliable and promising than the oligonucleotide method. Furthermore, this antisense retrovirus method has advantages to the gene targeting method because the former enables one to inactivate the expression of genes in specific subsets of differentiating cells, in contrast to the latter method that in general affects the entire embryo.

In the case of pLXSG, a test gene is cloned into the *Xho*I site so that the antisense strand may be transcribed from the 5′-LTR. After injection of this virus into the developing CNS, brains are stained with X-gal, which will reveal the effects of antisense expression on neural differentiation.

Another useful approach to loss-of-function mutation is retrovirus-mediated ribozyme expression (Weerasinghe *et al.*, 1991; Shore *et al.*, 1993). Although the ribozyme may not completely eliminate the target mRNA, it can efficiently reduce the mRNA level. Thus, ribozyme-transducing retrovirus also has a great potential for loss-of-function analysis, although reduction of mRNA levels does not necessarily result in reduction of protein levels in some cases (Baier *et al.*, 1994).

9.6 Conclusions

Recently, more and more genes that are expressed in the developing nervous system have been isolated. Once these genes are isolated, elucidation of their functions in neural development is the next main objective. In this chapter, we present the method of one such functional assay: manipulation of gene expression in subsets of differentiating neural cells by retrovirus. This method allows the manipulated cells to stay in physiological conditions because the other surrounding cells are normal. This is a sharp contrast to the transgenic and gene targeting methods that more or less affect the entire embryo. Thus, if the gene activation or inactivation is crucial to survival of embryos, transgenic and gene targeting manipulation would often lead to embryonic lethality, making the analysis more difficult. The method that we have presented does not cause embryonic lethality because only the manipulated cells are affected, making this analysis much easier. Furthermore, this method is relatively easy and does not require any sophisticated equipment. Thus, this retrovirus technique will give substantial potential for studying the molecular mechanisms of mammalian neural development.

Acknowledgements

We thank Drs S. Nakanishi and K. Shiota for encouragement of this study. This work was supported by a research grant from the Ministry of Education, Science, and Culture of Japan, the Sankyo Foundation, and the Inamori Foundation.

References

Akazawa, C., Sasai, Y., Nakanishi, S. & Kageyama, R. (1992) Molecular characterization of a rat negative regulator with a basic helix–loop–helix structure predominantly expressed in the developing nervous system. *J. Biol. Chem.* **267**, 21879–21885.

Baier, G., Coggeshall, K.M., Baier-Bitterlich, G., Giampa, L., Telford, D., Herbert, E., Shih, W. & Altman, A. (1994) Construction and characterization of lck- and fyn-specific tRNA: ribozyme chimeras. *Mol. Immunol.* **31**, 923–932.

Brody, S.L. & Crystal, R.G. (1994) Adenovirus-mediated *in vivo* gene transfer. *Ann. NY Acad. Sci.* **716**, 90–103.

Chalfie, M., Tu, Y., Euskirchen, G., Ward, W.W. & Prasher, D.C. (1994) Green fluorescent protein as a marker for gene expression. *Science* **263**, 802–805.

Cockroft, D.L. (1990) Dissection and culture of postimplantation embryos. In *Postimplantation Mammalian Embryos: A Practical Approach* (eds Copp, A.J. & Cockroft, D.L.). Oxford, Oxford University Press.

Cone, R.D. & Mulligan, R.C. (1984) High-efficiency gene transfer into mammalian cells: generation of helper-free recombinant retrovirus with broad mammalian host range. *Proc. Natl. Acad. Sci. USA* **81**, 6349–6353.

Fields-Berry, S.C., Halliday, A.L. & Cepko, C.L. (1992) A recombinant retrovirus encoding alkaline phosphatase confirms clonal boundary assignment in lineage analysis of murine retina. *Proc. Natl. Acad. Sci. USA* **89**, 693–697.

Friedrich, G. & Soriano, P. (1991) Promoter traps in embryonic stem cells: a genetic screen to identify and mutate developmental genes in mice. *Genes Devel.* **5**, 1513–1523.

Galileo, D.S., Gray, G.E., Owens, G.C., Majors, J. & Sanes, J.R. (1990) Neurons and glia arise from a common progenitor in chicken optic tectum: demonstration with two retroviruses and cell type-specific antibodies. *Proc. Natl. Acad. Sci. USA* **87**, 458–462.

Galileo, D.S., Majors, J., Horwitz, A.F. & Sanes, J.R. (1992) Retrovirally introduced antisense integrin RNA inhibits neuroblast migration *in vivo*. *Neuron* **9**, 1117–1131.

Halliday, A.L. & Cepko, C.L. (1992) Generation and migration of cells in the developing striatum. *Neuron* **9**, 15–26.

Hogan, B., Costantini, F. & Lacy, E. (1986) In *Manipulating the Mouse Embryo: A Laboratory Manual.* Cold Spring Harbor Press, Cold Spring Harbor, NY.

Ishibashi, M., Sasai, Y., Nakanishi, S. & Kageyama, R. (1993) Molecular characterization of HES-2, a mammalian helix–loop–helix factor structurally related to *Drosophila hairy* and *Enhancer of split*. *Eur. J. Biol.* **215**, 645–652.

Ishibashi, M., Moriyoshi, K., Sasai, Y., Shiota, K., Nakanishi, S. & Kageyama, R. (1994) Persistent expression of helix–loop–helix factor HES-1 prevents mammalian neural differentiation in the central nervous system. *EMBO J.* **13**, 1799–1805.

Jacobson, M. (1991) *Developmental Neurobiology*, 3rd edn. New York, Plenum Press.

Joyner, A.L. (1993) In *Gene Targeting: A Practical Approach*. Oxford University Press, Oxford.

Luskin, M.B., Pearlman, A.L. & Sanes, J.R. (1988) Cell lineage in the cerebral cortex of the mouse studied *in vivo* and *in vitro* with a recombinant retrovirus. *Neuron* **1**, 635–647.

Mann, R., Mulligan, R.C. & Baltimore, D. (1983) Construction of a retrovirus package mutant and its use to produce helper-free defective retrovirus. *Cell* **33**, 153–159.

Miller, A.D. & Rosman, G.J. (1989) Improved retroviral vectors for gene transfer and expression. *BioTechniques* **7**, 980–990.

Miller, A.D., Law, M.-F. & Verma, I.M. (1985) Generation of helper-free amphotropic retroviruses that transduce a dominant-acting, methotrexate-resistant dihydrofolate reductase gene. *Mol. Cell. Biol.* **5**, 431–437.

Muneoka, K., Wanek, N., Trevino, C. & Bryant, S.V. (1990) *Ex utero* surgery. In *Postimplantation Mammalian Embryos: A Practical Approach* (Copp, A.J. and Cockroft, D.L. eds) Oxford University Press, Oxford.

Owens, G.C. & Bunge, R.P. (1991) Schwann cells infected with a recombinant retrovirus expressing myelin-associated glycoprotein antisense RNA do not form myelin. *Neuron* **7**, 565–575.

Price, J. & Thurlow, L. (1988) Cell lineage in the rat cerebral cortex: a study using retroviral-mediated gene transfer. *Development* **104**, 473–482.

Price, J., Turner, D. & Cepko, C. (1987) Lineage analysis in the vertebrate nervous system by retrovirus-mediated gene transfer. *Proc. Natl. Acad. Sci. USA* **84**, 156–160.

Sakagami, T., Sakurada, K., Sakai, Y., Watanabe, T., Nakanishi, S. & Kageyama, R. (1994) Structure and chromosomal locus of the mouse gene encoding a cerebellar Purkinje cell-specific helix–loop–helix factor HES-3. *Biochem. Biophys. Res. Commun.* **203**, 594–601.

Sanes, J.R., Rubenstein, J.L.R. & Nicolas, J.-F. (1986) Use of a recombinant retrovirus to study post-implantation cell lineage in mouse embryos. *EMBO J.* **5**, 3133–3142.

Sasai, Y., Kageyama, R., Tagawa, Y., Shigemoto, R. & Nakanishi, S. (1992) Two mammalian helix–loop–helix factors structurally related to *Drosophila hairy* and *Enhancer of split*. *Genes Devel.* **6**, 2620–2634.

Shore, S.K., Nabissa, P.M. & Reddy, E.P. (1993) Ribozyme-mediated cleavage of the BCRABL oncogene transcript: *in vitro* cleavage of RNA and in vivo loss of P210 protein-kinase activity. *Oncogene* **8**, 3183–3188.

Takebayashi, K., Sasai, Y., Sakai, Y., Watanabe, T., Nakanishi, S. & Kageyama, R. (1994) Structure, chromosomal locus, and promoter analysis of the gene encoding the mouse helix–loop–helix factor HES-1. *J. Biol. Chem.* **269**, 5150–5156.

Turner, D.L. & Cepko, C.L. (1987) A common progenitor for neurons and glia persists in rat retina late in development. *Nature* **328**, 131–136.

Turner, D.L., Snyder, E.Y. & Cepko, C.L. (1990) Lineage-independent determination of cell type in the embryonic mouse retina. *Neuron* **4**, 833–845.

Wahlestedt, C., Golanov, E., Yamamoto, S., Yee, F., Ericson, H., Yoo, H., Inturrisi, C.E. & Reis, D.J. (1993) Antisense oligodeoxynucleotides to NMDA-R1 receptor channel protect cortical neurons from excitotoxicity and reduce focal ischaemic infarctions. *Nature* **363**, 260–263.

Walsh, C. & Cepko, C.L. (1988) Clonally related cortical cells show several migration patterns. *Science* **241**, 1342–1345.

Weerasinghe, M., Liem, S.E., Asad, S., Read, S.E. & Joshi, S. (1991) Resistance to human immunodeficiency virus type (HIV-1) infection in human $CD4^+$ lymphocyte-derived cell lines conferred by using retroviral vectors expressing an HIV-1 RNA-specific ribozyme. *J. Virol.* **65**, 5531–5534.

Weiss, R., Teich, N., Varmus, H. & Coffin, J. (1985) In *RNA Tumor Viruses*. Cold Spring Harbor Laboratory, Cold Spring Harbor, NY.

Yang, N.S. (1992) Gene transfer into mammalian somatic cells *in vivo*. *Crit. Rev. Biotech.* **12**, 335–356.

CHAPTER 10

GENE DELIVERY TO THE NERVOUS SYSTEM USING RETROVIRAL VECTORS

Miguel Sena-Esteves,[1] Manish Aghi,[1] Peter A. Pechan, Edward M. Kaye[1,2] and Xandra O. Breakefield[1]

[1]Molecular Neurogenetics Unit, Neurology Service, Massachusetts General Hospital and Neuroscience Program, Harvard Medical School, Boston, MA, and [2]Tufts University School of Medicine, Floating Hospital for Children, Boston, MA USA

Table of Contents

10.1 Introduction

Retrovirus vectors derived from Moloney murine leukemia virus (MMLV), originally developed by Mulligan and Berg (1981), have been the most utilized vector for gene delivery to the nervous system. They have been extensively modified through the years to increase their safety, titer, cell selectivity and control of transgene expression. The low toxicity of retrovirus vectors, as compared with adenovirus and herpes simplex virus vectors, and their high efficiency of gene delivery to a wide range of cell types has made them the first choice for gene delivery to mitotic cells. Their current limitations for gene delivery to the nervous system lie in their limited gene capacity (about 8 kb), and in the fact that their integration into the host cell genome requires the disassembly of the nuclear membrane during mitosis, as most neural cells are post-mitotic after birth. The low titers and instability of these vectors that previously restricted direct gene delivery *in vivo* now appear to have been overcome by pseudotyping of the virus particles (see Section 10.6).

Genetic Manipulation of the Nervous System
ISBN 0-12-437165-5

Many ingenious applications of gene delivery have been effected using retrovirus vectors. In terms of basic neuroscience, they have served to track cell lineages in development of the nervous system by delivery of histochemical marker proteins, and to elucidate the function of particular proteins in development, both by inhibiting their synthesis or by causing atypical expression. Retrovirus vectors bearing marker genes have also been used to identify the few mitotic cells in the post-natal nervous system. Their main use, however, has been in development of therapeutic strategies for diseases of the nervous system. In one mode, cells capable of division in culture are genetically modified using retrovirus vectors and then grafted into the nervous system. Modifications have included delivery of temperature-sensitive oncogenes to primary cells in culture to conditionally immortalize them for use as 'replacement' neural cells, and of genes for trophic factors and other neuroactive molecules to cells so that they can serve as 'biological minipumps' *in vivo*. In therapeutic strategies for brain tumors, retrovirus vectors are used to deliver genes selectively into mitotic tumor cells by grafting packaging cell lines into the tumor mass.

10.2 Vectors and packaging cell lines

Retrovirus vectors have been used for over 10 years in delivery of genes to cells in culture and *in vivo* (for review see Gilboa *et al.*, 1986; Mulligan, 1993). The basic features of retroviruses pertinent to their use as vectors will be briefly reviewed here (see Section 10.6 for new developments). Most retrovirus vectors derive from MMLV, an RNA virus which can infect many types of cells. Particles enter the cells by binding to glycoproteins on the plasma membrane of the host cell, followed by fusion of the virus envelope with the cell membrane. There is some cell specificity of virus entry conferred by the interaction between envelope elements in the virus particles and the cell membrane. Two general classes of retrovirus particles have been described: ecotropic, which infect murine cells most efficiently, and amphotropic particles, which can infect cells of many species. If a cell expresses retrovirus glycoproteins on its surface, due to the presence of retrovirus sequences in its genome, it will resist subsequent infection by particles of the same class. This can be a problem for some rodent cells which harbor endogenous retrovirus sequences, thus making cells resistant to infection with these vectors.

Following entry into cells, the retrovirus particle forms a pre-integration complex and viral RNA is converted to double-stranded DNA through the action of reverse transcriptase encoded by the virus. Following circularization via the long terminal repeats (LTRs) the viral DNA enters the cell nucleus as a nucleosome complex when the nuclear membrane is disassembled at mitosis. The viral genome integrates into the host cell genome such that the LTRs end up flanking the retrovirus sequences, with the 3′-LTR of the infecting virus being duplicated to become both the 5′- and 3′-LTRs of the integrated virus (Figure 1a). Sites of integration in the host cell genome appear to be fairly random, but there may be some preferred sites (Shih *et al.*, 1988).

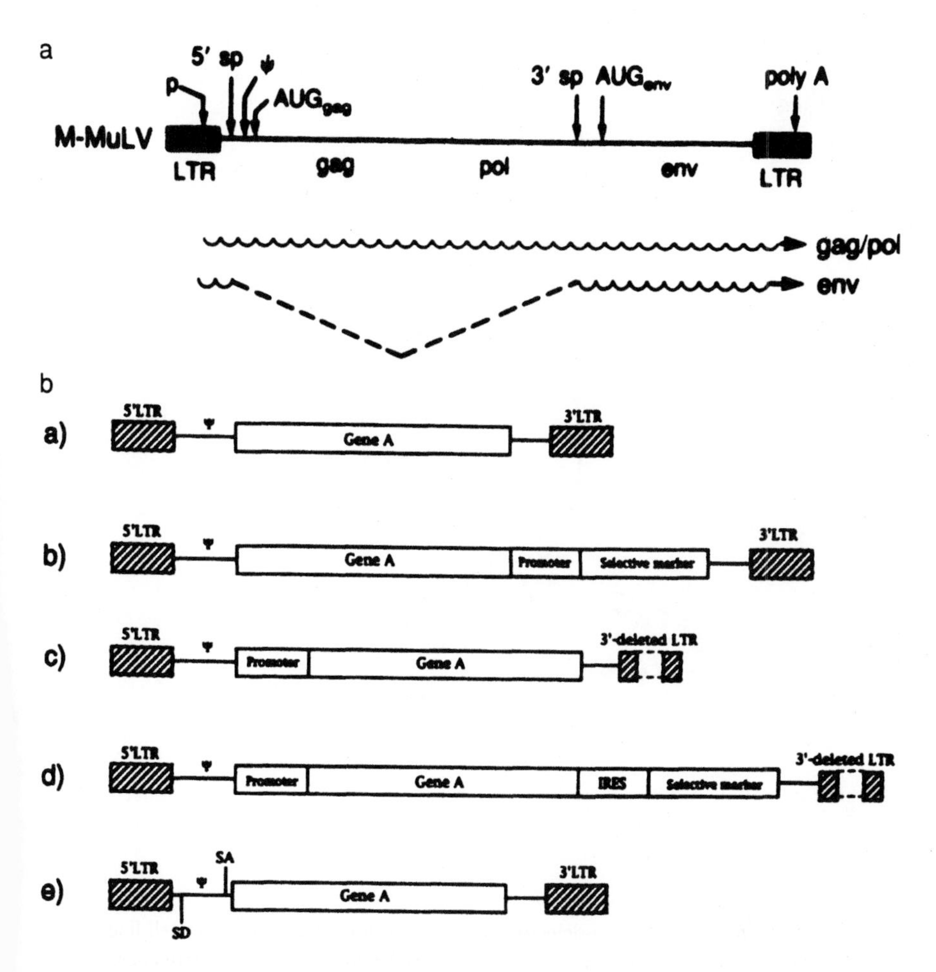

Figure 1 Retrovirus vectors. (A) MMLV generates two type of transcripts (wavy lines): unspliced (*gag/pol*) and spliced (*env*) RNA, both under regulation of the 5′-LTR (left black box).p, promoter; sp, RNA splice sites; ψ, packaging signal; AUG, start site of translation; polyA, polyA addition site. Reprinted with permission from Gilboa *et al.* (1986). (B) The retroviral vectors currently used for gene therapy can be divided in monocistronic (a, c, e) and polycistronic (b, d, f, g). The vector initially developed by Mulligan's group (a) (Cepko *et al.*, 1984) was modified to include a selective marker for mammalian cells (b). To avoid the problem of inactivation of the LTR promoter *in vivo*, vectors were developed to include another promoter. However, interference between promoters can decrease the efficiency of these vectors. To avoid this problem new vectors are designed in such a way that the 5′-LTR is inactive after integration by insertion of a small deletion in the 3′-LTR (c). Inclusion of an IRES sequence between the transgenes allows translation of two polypeptides from the same message (d and g). If genes are small enough three of them can be inserted with IRES sequences between them under control of the same promoter. More efficient transcription and packaging was achieved by extension of *gag* sequences and inclusion of a splice acceptor site from the *env* gene (e).

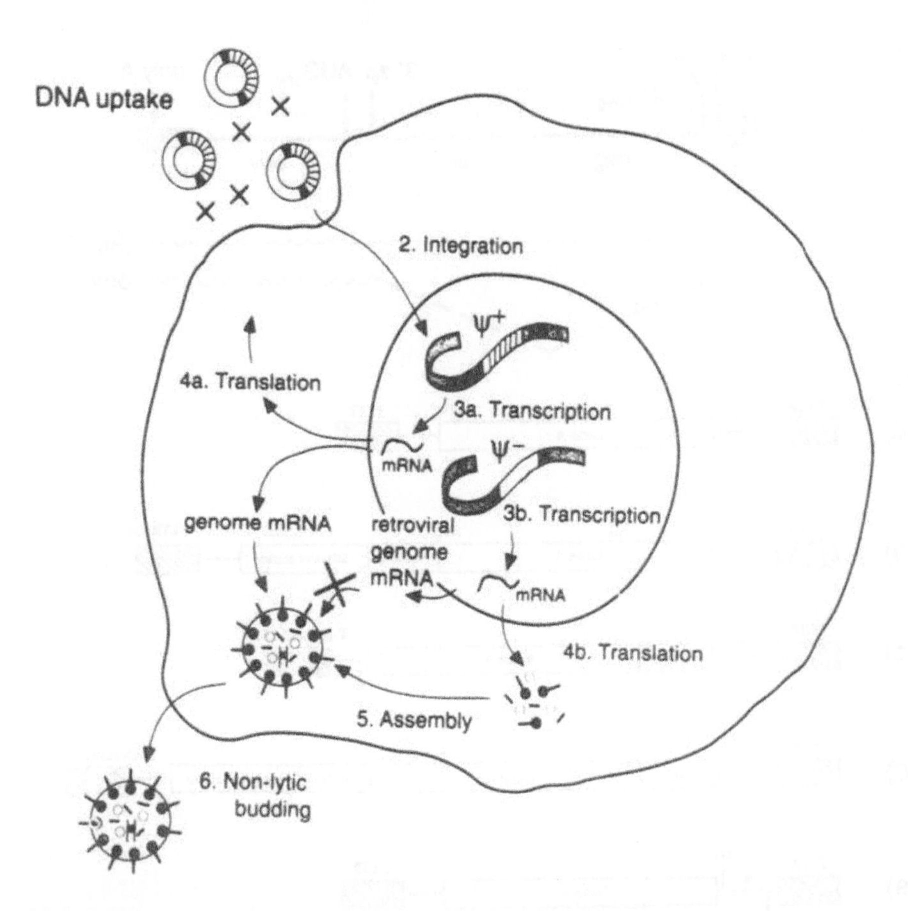

Figure 2 Packaging of replication-defective retroviral vectors. The packaging cell line contains wild-type retroviral sequences mutated at the ψ packaging signal. These sequences code for viral proteins that make up the viral particles but cannot themselves be packaged. DNA plasmid constructs corresponding to retroviral vectors containing the transgene(s) of interest are transfected into the packaging cell line. Successful transfectants are cloned under neomycin selection and express an RNA transcript encoding the transgenes which can be packaged because of an intact ψ signal ($ψ^+$). Modified and reprinted with permission from Breakefield and Geller (1987).

Integrated sequences are transcribed to yield mRNAs using the promoter/enhancer in the 5′-LTR and the polyadenylation site in the 3′-LTR. Packaging of the full length RNA transcript into retrovirus particles is dependent on the presence of the ψ sequence. The translational products of the unspliced mRNA are gag, structural components of the virus particle, and *pol*, which includes the reverse transcriptase. Excision of an intron generates a shorter mRNA species which is translated to yield *env*, the virus coat protein. *Env* is embedded in the lipid bilayer of the virus envelope and determines the host range of the virus particle. Cells infected with wild-type retrovirus can continuously produce retrovirus particles with essentially no compromise of host cell functions.

Retrovirus vectors need to retain the retrovirus LTR and ψ sequences, but the rest of the virus genome can be replaced with transgene sequences. For recombinant vector constructions, DNA equivalents of the virus sequence are cloned into plasmids (Cepko *et al.*, 1984). Plasmid constructs are transfected into cultured cells which have been genetically modified to serve as packaging cell lines (Figure 2). Most current packaging cell lines are derived from the murine fibroblast line, 3T3 (Mann *et al.*, 1983), and are termed ecotropic or amphotropic depending on which type of *env* protein they express. In the most basic version of a packaging cell line, the cells are stably transfected with retrovirus sequences which are wild type except for a mutation in the ψ sequence. This allows for expression of all the retrovirus proteins, but not for packaging of the retrovirus RNA into particles – hence cells bud off empty particles (which can be monitored by the reverse transcriptase activity in the medium). Following transfection and stable integration of the plasmid construct, vector RNA is packaged into particles by virtue of the ψ sequence it bears. In this basic system there is a relatively high frequency of recombination between helper virus and vector virus sequences, such that wild-type particles can be generated. Several groups have now designed packaging cells in which the helper genome is divided into two or more segments bearing additional mutations and integrated at separate sites in the host cell genome, such that multiple recombination events would be needed to reconstitute the wild-type genome (Danos and Mulligan, 1988, Markovitz *et al.*, 1988; Miller, 1990; Pear *et al.*, 1993). In these packaging lines the chance of generating wild-type virus is essentially zero. Retrovirus vectors are harvested from the medium of packaging cells at typical titers of 10^5–10^6 infectious units ml^{-1}.

Some typical vector designs are illustrated in Figure 1b. Two transgenes can be incorporated into the same vector, either by transcribing the first via the 5′-LTR promoter and the second via an internal promoter, or by creating a common transcript under the same promoter which includes separate sites for initiation of translation mediated by an internal ribosome entry site (IRES) between transgene coding sequences (Ghattas *et al.*, 1991; Adam *et al.*, 1991). The packaging efficiency of the viral RNA can be increased by extending the length of *gag* sequence prior to the transgene sequences to include splice donor and acceptor sites, as in the pMFG vector (Ohashi *et al.*, 1992). In order to prevent transcriptional override by the 5′-LTR and thus allow the internal promoter to retain its integrity, a small deletion can be placed in the 3′-LTR of the construct, which becomes the 5′-LTR of the integrated sequences, and serves to inactivate its promoter/enhancer functions (Yee *et al.*, 1987). Alternatively the transgene and its promoter can be inserted in the 3′-LTR, termed double-copy vectors (Hantzopoulos *et al.*, 1989).

In the most common form of retrovirus vectors the MMLV LTR serves as the promoter/enhancer element to regulate expression of at least one transgene. This is a strong constitutive promoter in most cells, but like most virus promoters tends to be down-regulated following grafting *in vivo*. Down-regulation of the MMLV LTR has been described in immortalized fibroblasts and neonatal astrocytes grafted into the brain (Frim *et al.*, 1993b; Cunningham *et al.*, 1994). However, long-term expression from this same promoter can occur in some cells *in vivo*, e.g. tumor cells in the brain (Tamiya *et al.*, 1995) and blood cells derived from genetically modified hematopoietic stem cells (Ohashi *et al.*, 1992). Use of the LTR from the mouse mammary tumor virus provides glucocorticoid inducibility to gene expression (Buetti and Kuhnel, 1986). Other virus promoter/enhancer elements utilized in retrovirus vectors include those derived from cytomegalovirus (CMV), Rous sarcoma virus (RSV) and SV40. The T7 polymerase promoter in combination with a nuclear targeted T7 polymerase (Lieber *et al.*, 1993), and the mammalian RNA polymerase type III in combination with a promoter for mammalian tRNA (Sullenger *et al.*, 1990) have been used to achieve high level transgene expression in cultured cells. Synthetic regulatory elements derived from prokaryotic inducible promoters are being developed to effect specific, drug-inducible expression of transgenes, e.g. derived from the bacterial *lac* and *tet* operons (Gossen and Bujard, 1992; Hannan *et al.*, 1993).

Scharfman *et al.* (1991) showed that stable, *albeit* low level, transgene expression could be achieved following peripheral grafting of fibroblasts by using the 'housekeeping' promoter for dihydrofolate reductase, which is active in most mammalian cell types. Recently the phosphoglycerate kinase (PGK) promoter, which is constitutively active at higher levels in most cells (Adra *et al.*, 1987; Hawley *et al.*, 1989) has been found to remain active in grafted fibroblasts over extended periods (Moullier *et al.*, 1993b). Several vector cassettes are available which incorporate this promoter (Hawley *et al.*, 1994). Virus promoters can also remain on *in vivo* when coupled to enhancer elements associated with high level gene expression in mammalian cells. Verma's group demonstrated that the enhancer element from the muscle-specific creatinine kinase promoter can effect continued expression of the CMV promoter in differentiated muscle cells *in vivo* (Dai *et al.*, 1992). Wilson *et al.* (1990) have also shown that coupling of the CMV enhancer element with the β-actin promoter can yield high level, stable expression in many cell types after grafting.

10.3 Basic neuroscience studies

Retrovirus vectors have been used to elucidate cell lineages and molecular mechanisms involved in neuronal development and function, as well as to assess the fate of cells grafted into different environments and to create replacement neural cells to supplement missing functions.

Direct injection of retrovirus particles bearing marker genes has proved very useful in tracing the clonal lineage of neural precursor cells in development, which have integrated retrovirally encoded sequences during terminal cell divisions (Cepko, 1988; Sanes *et al.*, 1986). Precursor cells are labelled with the *lacZ* marker gene and their progeny are tracked over time and characterized with respect to cell type and routes of migration, e.g. in the retina (Turner *et al.*, 1990), the cerebral cortex (Krushel *et al.*, 1993), the cerebellum (Ryder and Cepko, 1994) and the perineurium (Bunge *et al.*,

1989). In post-natal animals it is more difficult to label cells due to the low titer and instability of this vector, as well as the non-mitotic state of most cells. However this approach has provided a means to identify the few dividing cell types present. Direct injection of the vector into the brain, or grafting of packaging cells to maximize the time window of infection, can effect gene delivery to a few endothelial cells, reactive astrocytes, microglia and other as yet identified cell types (Short *et al.*, 1990b; D. Landis, personal communication). Division of some of these cells may be stimulated during the injury response to the injection procedure. Retrovirus gene delivery to adult brain has also helped to elucidate the neural stem cell population in the subependyma near the lateral ventricle which continues to generate glia and neurons in adult life (Morshead *et al.*, 1994; Luskin and McDermott, 1994; for further details of neuronal precursor cells in the adult brain see Chapter 13). When packaging cell lines are grafted into tumors in the brain there is extensive and selective gene transfer to tumor cells, which forms the basis of retrovirus therapy for brain tumors (see Section 10.5).

The function of particular neural proteins has been elucidated by retrovirus-encoded antisense RNA which interferes with expression of the protein. Antisense technology is still an incomplete art and one can expect to achieve only partial inhibition (50–95%) of translation of the targeted mRNA. Still in many cases this can result in phenotypic changes in cells. Schwann cells infected in culture with a retrovirus vector bearing an antisense RNA for the myelin-associated proteins, MAG or PO, showed a reduction in their ability to myelinate neuronal processes (Owens and Bunge, 1991; Owens and Boyd, 1991). Neuroblasts from the optic tectum, infected with retrovirus vectors bearing an antisense construct for the neuroadhesion molecule, β_1-integrin, as well as the *lacZ* gene, were no longer able to migrate normally along radial glia when grafted back into the developing optic tectum (Galileo *et al.*, 1992).

Different primary cell types have also been labeled in culture with a marker gene and then grafted back into different environments to follow their fate. Primary Schwann cells, labeled in culture with the *lacZ* gene, retain the ability to become normal components of peripheral nerve sheaths (Langford and Owens, 1990; Federoff *et al.*, 1992). However, after extended passage in culture, these cells can become 'immortalized' and then produce a neurofibroma when grafted back into the environment of the nerve fiber (Langford *et al.*, 1988; Eccleston *et al.*, 1991). Several groups have also grafted cultured Schwann cells into the CNS and found that they undergo limited proliferation and migrate out along white matter tracts (Kromer and Cornbrooks, 1985; Guenard *et al.*, 1993; Brook *et al.*, 1994). In developmental paradigms marker genes can be used to follow the fate of neuroblasts from different regions of the developing nervous system grafted into various regions of embryos at different times in development to elucidate when cells become committed to a differentiated pathway and whether environmental factors can influence their choice, e.g. neuroblasts from the quail neural tube become neurons or glia when grafted into the CNS, and melanocytes when grafted into the periphery (Stocker *et al.*, 1993). Such labeling also allows one to monitor the fate of different cultured cells grafted into the brain, such as transformed cells, which in the case of embryonic carcinoma cells differentiate into neuron-like cells (Wojcik *et al.*, 1993) or in the case of glioma cells form large tumors (Lampson *et al.*, 1993; Scharf *et al.*, 1994). Immortalized fibroblast lines, e.g. Rat I (Shimohama *et al.*, 1993), form small fibromas which grafted in the adult brain; while primary skin fibroblasts (Wolff *et al.*, 1989), neonatal astrocytes (Cunningham *et al.*, 1991b), myoblasts (Jiao *et al.*, 1994) and other primary cell types tend to form a stable, non-proliferative, non-migratory grouping of cells at the site of implantation.

Retrovirus vectors have proved to be an efficient way of immortalizing embryonic neural cells in culture by introduction of oncogenes while cells are still undergoing mitoses. Introduction of the SV40 large T antigen (T ag) or other oncogenes into neural cells can result in immortalization without transformation (Cepko, 1988; Frederickson *et al.*, 1988). A number of neural lines have been generated by retroviral delivery of sequences encoding the SV40 T ag gene containing one or more temperature-sensitive mutations. These lines proliferate at the permissive temperature (usually 34°C) in culture, but manifest growth arrest and differentiated properties of neurons and glia at the non-permissive temperature (37–39°C) in culture or when grafted back into animals. Following grafting, they appear to respond to and interact with endogenous neural cells, e.g. astrocytic lines migrate in the brain (Snyder *et al.*, 1992; Goodman *et al.*, 1993; Okoye *et al.*, 1994), as do normal astrocytes (Zhou *et al.*, 1990), and neuronal lines extend processes, appear to form synaptic connections (Bredesen *et al.*, 1990; Almazan and McKay, 1992; Snyder *et al.*, 1992; Onifer *et al.*, 1993) and produce neurotransmitters (Anton *et al.*, 1994).

10.4 Biological delivery of neuroactive substances

Retrovirus vectors have also been used to deliver potentially therapeutic genes efficiently to defined cell populations in culture, which can then be grafted into the brain (Figure 3; Breakefield and Geller, 1987; Fisher and Gage, 1993; Doering, 1994). A wide variety of different cell types have been used for this purpose, including primary cells, e.g. fibroblasts, astrocytes, muscle cells and endothelial cells, and continuous cell lines, e.g. immortalized fibroblasts, endocrine cells and neural cells. Encapsulation of cells in biocompatible alginate–polylysine spheres or acrylonitrile/vinylchloride copolymer capsules allows them to retain their viability and capacity to continuously secrete neurotrophins for at least 2 months *in vivo*, at the same time reducing immune rejection and facilitating their removal (Maysinger *et al.*, 1993; Hoffman *et al.*, 1993). A combination of gene transfer and intracerebral grafting appears to be an effective approach toward the protection of neurons from a variety of different types of injury (Gage *et al.*, 1987; Isacson, 1993). Neuroactive substances which have proven amenable to this delivery mode include neurotrophic factors, neuropeptides and biosynthetic enzymes for neurotransmitters. A similar approach has also been used for enzyme replacement therapy in the peripheral and central nervous system, e.g. lyso-

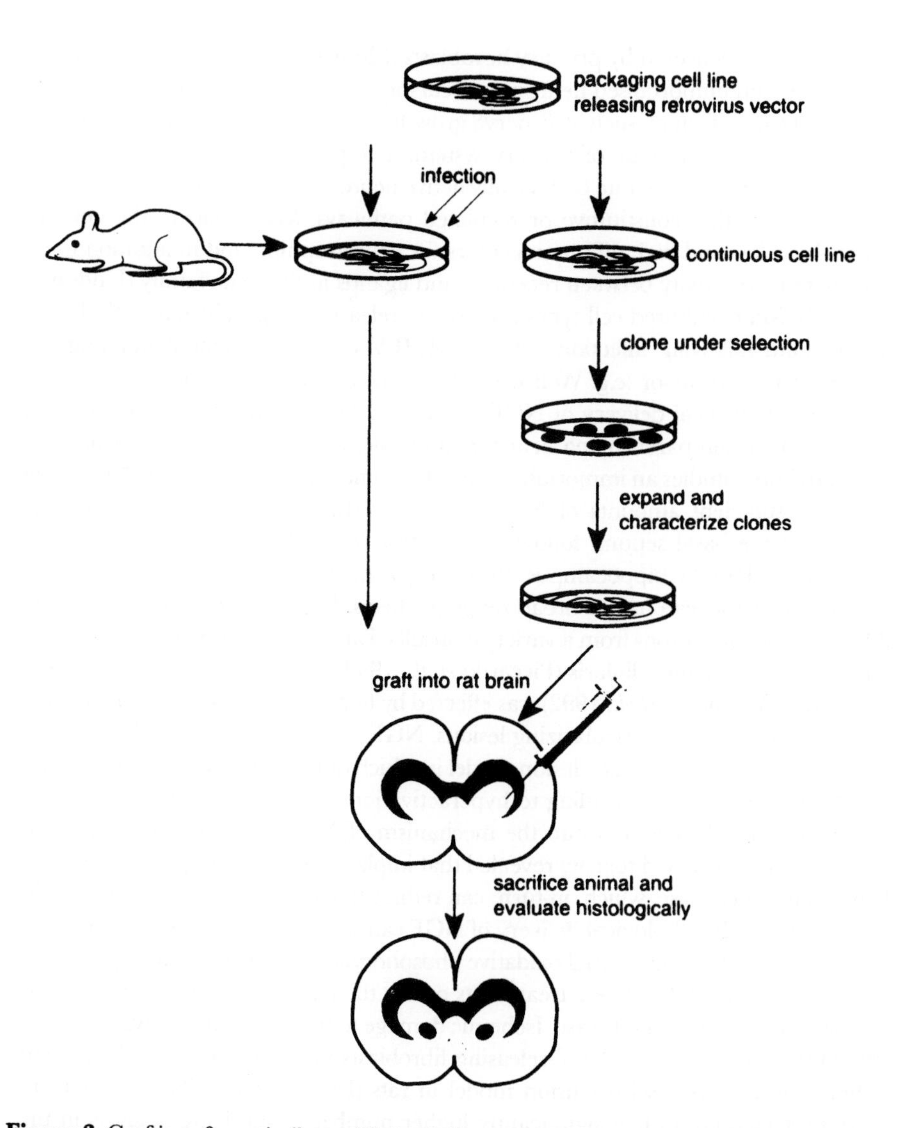

Figure 3 Grafting of genetically modified cells into rodents. Conditioned medium from packaging cells is used as a source of retrovirus vector. Transgenes can be delivered to cultured primary cells by serial infection with the vector to maximize the percentage of the population transfected, or by single infection of continuous cells and selection of clones representing stable transfected lines. Cells are grafted into rodent brain by stereotactic surgery and the effects of these grafts are assessed by histochemical analyses. Modified and reprinted with permission from Fisher and Gage (1993).

somal enzymes delivered by primary fibroblasts (Moullier *et al.*, 1993a; Sena-Esteves *et al.*, 1994) and immortalised neuronal precursors (Snyder *et al.*, 1995).

Neurotrophic factors, such as β-nerve growth factor (NGF), have proved particularly amenable to this type of delivery system. The precursor polypeptide requires simple processing to generate biologically active neurotrophin, which can be released from cells by either constitutive or regulated pathways. Many cells in the nervous system bear low and high affinity receptors for neurotrophins, and in addition there is some cross-reactivity between receptors and ligands in this gene family (Lindsay *et al.*, 1994). Many cultured cell types are able to release biologically active NGF into the medium following infection with an MMLV retroviral vector containing the cDNA for the precursor (e.g. Wolf *et al.*, 1988). In typical rodent models of neuroprotection, biological delivery of NGF is achieved by grafting genetically modified cells into the brain parenchyma near the future site of injury about a week prior to injury. In initial studies an immortalized fibroblast line transfected with the NGF gene produced sufficient amounts of NGF to prevent the degeneration of cholinergic neurons in the basal septum following a fimbria fornix lesion, which disrupts the supply of NGF to the hippocampus (Rosenberg *et al.*, 1988).

Trophic factors can have a broad range of effects. For example, NGF can protect different types of neurons from a variety of insults. Partial protection of neurons in the nucleus basalis magnocellularis (Piccardo *et al.*, 1992) and recovery of cortical function in rats (Maysinger *et al.*, 1992) was effected by fibroblasts producing NGF grafted at the site of cortical devascularizing lesions. NGF-secreting fibroblasts also reduced neuronal loss in an excitotoxic lesion model in which quinolinic acid was injected into the striatum, presumably leading to hyperactivation of glutamate receptors (Frim *et al.*, 1993a). An investigation into the mechanism of NGF protection against glutamate-receptor-mediated toxicity revealed that implantation of NGF-producing fibroblasts induced catalase, which, in turn, can reduce free radicals that are toxic to cells (Frim *et al.*, 1994b). Biological delivery of NGF can also prevent striatal degeneration after blockade of mitochondrial oxidative phosphorylation induced by nitropropionic acid (Frim *et al.*, 1993c). These treatments mimic the type of striatal neurodegeneration seen in Huntington's disease. Ischemic damage to the hippocampus was also prevented by implantation of NGF-releasing fibroblasts prior to induction of transient ischemia in a four-vessel occlusion model in rats (Pechan *et al.*, 1995). Animals in NGF-protected group had significantly higher number of surviving neurons in the grafted right CA1 and CA2 regions, as compared to the contralateral control hemispheres (Figure 4).

Retrovirally mediated expression of growth factors can also change the phenotype of the modified cell or provide sustaining functions to co-grafted cells. For example, infection of rat pheochromocytoma PC12 cells with the NGF vectors resulted in autocrine differentiation, as assessed by extensive neurite formation, which occurred within hours after infection and was maintained for weeks in culture (Short *et al.*, 1990a). Genetically modified PC12 cells, expressing NGF, implanted into the striatal parenchyma of a rodent model of Parkinson's disease, gradually lost their catecholaminergic phenotype, as assessed by immunocytochemistry for tyrosine hydroxylase (TH), but maintained a

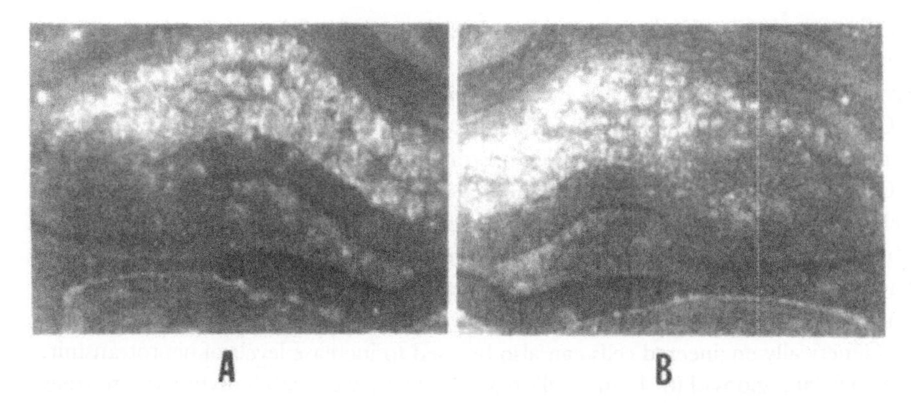

Figure 4 Effect of implanted genetically modified NGF-secreting fibroblasts in the post-ischemic hippocampus. NGF-secreting *Rat*I cells were grafted near the right hippocampus of adult rats 1 week prior to a vascular occlusion model of ischemia and histological analysis was carried out 1 week later (Pechan *et al.*, 1995). This photograph shows a microscopic image of hematoxylin–eosin staining in dark field. Eosin-stained (light) areas represent damaged tissue. (A) Left contralateral non-protected part of hippocampus; (B) right ipsilateral part of hippocampus, protected by NGF-secreting fibroblasts. Note the reduction of eosin-stained tissue compared to the contralateral side. Magnification ×65.

neuronal morphology (Cunningham *et al.*, 1991a). Use of genetically altered, NGF-secreting primary astrocytes enhanced survival of co-grafted adrenal chromaffin cells in a dopamine-denervated striatum of the rat (Cunningham *et al.*, 1991b).

A number of retrovirus vectors have been generated for delivery of other neurotrophic factors, including basic fibroblast growth factor (bFGF), brain-derived neurotrophic factor (BDNF), neurotrophin-3 (NT3) and ciliary neurotrophic factor (CNTF) (Frim *et al.*, 1993d; Fisher, 1995). There is some specificity for the type of neurotrophin that will protect defined neuronal populations, for example in striatal injury caused by glutamate toxicity bFGF was less effective than NGF, and BDNF was not effective (Frim *et al.*, 1993d). On the other hand BDNF is more potent than NGF in promoting the survival of other neurons, including retinal ganglion cells (RGCs) in culture (Castillo *et al.*, 1994) and *in vivo* following lesion of the optic tract (Mey and Thanos, 1993). Rat fibroblasts engineered to secrete human BDNF also protected nigral dopaminergic neurons when implanted near the substantia nigra prior to striatal infusion of the neurotoxin, 1-methyl-4-phenylpyridinium (MPP+) (Frim *et al.*, 1994a).

Grafts of fibroblasts genetically modified to secrete trophic factors also merit study as potential tools for promoting neurite regeneration after injury (Breakefield, 1994). Following the implantation of fibroblasts producing NGF into the striatum, cholinergic axons arising from the nucleus basalis grow toward and penetrate these grafts in close association with reactive astrocytic processes (Kawaja and Gage, 1991). When these fibroblasts are embedded in a collagen matrix, regnerating cholinergic axons from axotomized septal neurons traverse the graft and appear to reform synaptic connections with neurons on the other side, in the hippocampus (Kawaja *et al.*, 1992).

NGF-producing grafts also resulted in robust in-growth of damaged neurites in the spinal cord, as compared to bFGF-producing grafts, which did not promote neurite growth (Tuszynski *et al.*, 1994). Electron microscopy revealed that neurites within NGF-secreting grafts were enveloped in glial cell processes and that axons frequently became myelinated. These results demonstrate that sensory neurites maintain robust NGF responsiveness into adulthood and that sprouting neurites can follow glial channels and become myelinated in the adult spinal cord. Grafting of NGF-producing fibroblasts into the brain reduced behavioral deficits in rats with lesions of the nucleus basalis, probably through the partial restoration of the lesioned projection from the nucleus basalis magnocellularis to neocortex (Dekker *et al.*, 1994).

Genetically engineered cells can also be used to increase levels of neurotransmitters in focal regions of the brain. Cells have been endowed with biosynthetic enzymes for neurotransmitter synthesis, including TH for catecholamines, choline acetyl transferase (ChAT) for acetylcholine, and glutamic acid decarboxylase (GAD) for γ-aminobutyric acid (GABA). Primary fibroblasts and other cells transfected with a transgene for TH have the capacity to deliver L-dopa locally to the striatum in quantities sufficient to compensate partially for the loss of striatal dopaminergic input caused by lesions of the substantia nigra (e.g. Horellou *et al.*, 1990a,b; Fisher *et al.*, 1991; Uchida *et al.*, 1992). Administration of biopterin, the precursor for the TH co-factor tetrahydrobiopterin, can increase synthesis of dopa further. In some studies the gene for L-amino acid decarboxylase (AADC), which converts L-dopa to dopamine, has also been transfected into grafted cells (Kang *et al.*, 1993). Retrovirus-mediated gene transfer of the cDNA for ChAT allows cells to synthesize and release acetylcholine (Xi *et al.*, 1993; Fisher *et al.*, 1993). Fibroblasts modified to express ChAT were implanted into the hippocampus of adult rats and microdialysis was performed 7–10 days after grafting to verify the ability of the cells to produce acetylcholine and to increase production in response to exogenous choline (Fisher *et al.*, 1993). When implanted into the neocortex of rats these cells can improve deficits in spatial learning and memory paradigms in rats (Winkler *et al.*, 1995). Retrovirus vectors bearing the GAD gene have been used to confer GABA synthesis on cells in culture which can be used for grafting (Ruppert *et al.*, 1993).

Ex vivo therapies have also been evaluated for genetic deficiency states. Since the first description of a lysosomal disorder, enzyme replacement strategies have been considered (Baudhuin *et al.*, 1964). Essentially all cells can secrete lysosomal enzymes and these enzymes can be taken up by other cells through the mannose-6-phosphate and other receptor-mediated pathways (von Figura and Hasilik, 1986). Co-culturing normal cells with mutant cells results in diminution of the lysosomal storage accumulations in the affected cells (Neufeld and Muenzer, 1989). Over-expression of a gene for a lysosomal enzyme in genetically modified cells promotes the 'trafficking of the enzyme out of the cell', since the abundantly produced enzyme forms aggregates which are unable to enter the lysosome and are routed out through a constitutive secretory pathway (Ioannou *et al.*, 1992). Genetically altered cells which express the gene for a lysosomal enzyme can be transplanted into an organ to serve as a 'biological pump' for constant delivery of that enzyme. The transplantation of cells avoids having to repeatedly access the organ and avoids the potentially toxic effects of direct vector administration. Previous studies in an animal model of lysosomal storage, the murine model of mucopolysaccharidosis type VII (Sly syndrome), have demonstrated correction of lysosomal storage in the liver and spleen following intraperitoneal implantation of genetically modified fibroblasts (Moullier *et al.*, 1993a). Gene delivery to hematopoietic stem cells can reduce lysosomal stores to some extent in peripheral nerves and in the CNS (Ohashi *et al.*, 1992; Yeager *et al.*, 1984). In our studies we have used a Portuguese water dog which transmits an autosomal recessive form of β-galactosidase deficiency, genetically and phenotypically similar to G_{M1} gangliosidosis. Affected animals provide a model for the neurological features of lysosomal storage diseases, as the neurological features are much more prominent than the systemic problems and the animals die from neurological complications long before any significant problems with other organs develop (Alroy *et al.*, 1989). When fibroblast cells excreting large quantities of β-galactosidase are grafted into the brains of affected animals, neuronal cells proximate to the transplant show a marked diminution of the lysosomal storage products (W. Rosenberg, J. Alroy and E.M. Kaye, unpublished studies). These studies suggest that transplantation of genetically modified cells into brain may be a potential treatment for the central nervous system manifestations in the lysosomal storage diseases (see also Snyder *et al.*, 1995).

10.5 Brain tumor therapy

Gene transfer techniques are also being explored as a possible therapeutic option for cancer and may be particularly applicable to malignant, primary tumors of the brain, such as glioblastomas (for review see Chiocca *et al.*, 1994). These tumors generally fail to respond to chemotherapy and radiation therapy; and since tumor cells are migratory, it is virtually impossible to completely remove them surgically. These tumors usually induce death by local expansion within the brain, and systemic metastases are very rare. Retrovirus-mediated gene transfer is a particularly applicable technique for brain tumors because of the selectivity of retroviruses for proliferating cells. In the adult brain, essentially the only proliferating cells are glioma cells, when present, as well as endothelial cells associated with tumor-induced neovascularization and astrocytes involved in the process of reactive astrocytosis. Thus retroviral-mediated transfer of a tumor killing gene, which acts in a cell-intrinsic or division-dependent manner, will act specifically on dividing neoplastic cells. When compared to other vectors, like adenovirus and herpes vectors, retrovirus vectors are more selective for transgene delivery to tumor cells in the brain and produce less necrosis and inflammation (Boviatsis *et al.*, 1994; for further details of this approach using adenoviral vectors see Chapter 4).

Retrovirus-mediated therapy for tumor cells has been limited to date by a fairly low efficiency of gene delivery. Delivery is most efficient when a packaging cell line is grafted surgically within the tumor mass, as compared to direct injection of retrovirus vectors. Short *et al.* (1990b) found that direct intratumoral injection of a retroviral

Table 1 Genes for tumor therapy

Drug enhancing
Viral thymidine kinase (tk)/ganciclovir
Cytosine deaminase/5-fluorouracil
Xanthine guanine phosphoribosyltransferase/6-thioxanthine
Cytochrome P450 B1/cyclophosphamide
Immune enhancing
Antigenic presentation: MHC
Cytokines: GM-CSF, IL-2, IL-4
Antisence: IGF1 and 2
Immunotoxins
Suppress neovascularization
Dominant negative FIK-1/VEGF receptor
Inhibit growth and promote apoptosis
Tumor suppressor genes: Rb, p53
Apoptotic genes: ICE

vector bearing the *Escherichia coli lacZ* marker gene introduced the transgene into only 0.1% of tumor cells, while intratumoral grafting of a retroviral packaging line releasing the vector increased the efficiency of gene transfer to approximately 10%. Efficiencies of 50–70% have been reported in glioma tumors in rodents under conditions where immune rejection of the packaging cells is compromised (Ram *et al.*, 1993; Tamiya *et al.*, 1995). But, even under ideal conditions, it is impossible with this strategy to infect all tumor cells in the primary mass (as many are in G_0 during the infection period) or to infect cells that have migrated away from the primary mass.

While attempts are being made to extend the length and range of retrovirus-mediated gene delivery *in vivo*, at present the best way to spread the tumoricidal effect is by using a gene that generates a toxic product that can be transferred to other tumor cells, or a cytokine that enhances specific immune responses to the tumor. Several approaches used so far fulfill these criteria (Table 1; Wei *et al.*, 1995a). One approach involves transfer of genes encoding enzymes that, when expressed in the tumor cell, can transform a prodrug into an active analog that is selectively toxic to dividing cells. The first example of this approach to killing glioma cells was expression of the thymidine kinase enzyme (tk) of herpes simplex type 1 virus (HSV) under the control of the MMLV LTR in a retroviral vector, which was used to infect C6 rat glioma cells in culture (Ezzeddine *et al.*, 1991). Transfected tumor cells were injected into nude mice and subsequently inhibited from forming tumors by systemic treatment with ganciclovir, a prodrug converted by the HSV-tk enzyme into a toxic phosphorylated derivative which causes disruption of DNA synthesis and cell death.

In order to achieve *in vivo* transfer of the HSV-tk transgene, a retrovirus packaging line *psi*2STK, with and without wild-type retrovirus, was co-grafted with C6 glioma cells subcutaneously into nude mice (Figure 5; Takamiya *et al.*, 1993). Retrovirus-

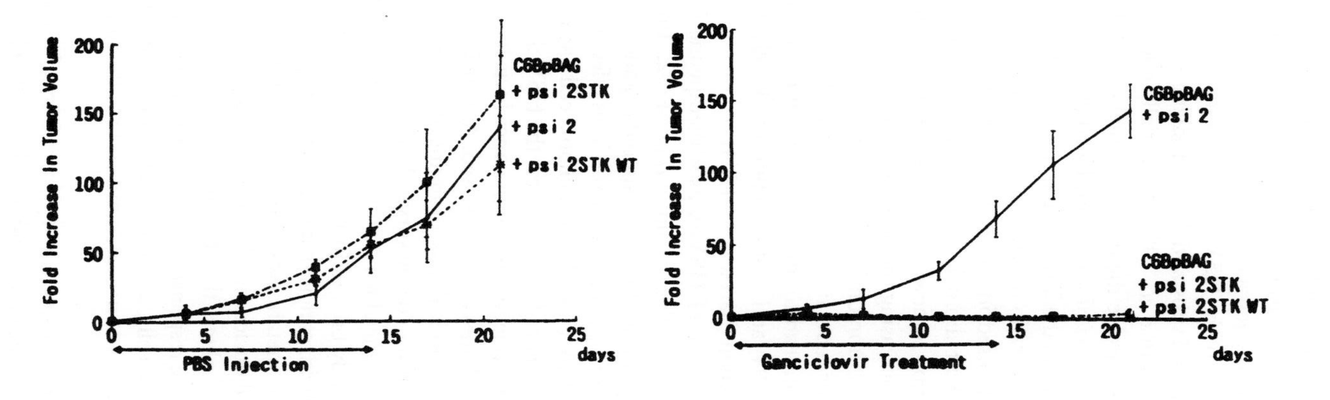

Figure 5 Gene therapy of tumor cells in mice using HSV-tk gene and ganciclovir. Mixtures of 10^6 recipient C6BpBAG glioma cells and 10^6 donor cells: (•) retrovirus packaging line, psi 2; (■) packaging line generating retrovirus vector with HSV-TK gene, psi 2STK; or (*) an HSV-tk vector-producing packaging cell line infected with wild-type MMLV, psi 2STKWT, were injected subcutaneously into nude mice (20 mice in each group). When tumors reached 1 cm in diameter, the animals were treated with phosphate-buffered saline (PBS, left) or 25 mg kg^{-1} day^{-1} of ganciclovir (right) for 14 days. Tumor growth was calculated as the fold increase in volume compared to that at the time treatment started. Vertical bars indicate the standard error of the mean. Reprinted with permission from Takamiya *et al.* (1993).

mediated gene transfer was allowed to occur over 3 days before beginning a 14 day treatment with ganciclovir. In this model, inhibition of tumor growth was greater when infectious wild-type MMLV retrovirus was included, as compared to when it was not. The wild-type virus presumably acts as a helper to generate the vector on site in the tumor and thus expand delivery of the HSV-tk gene to more tumor cells. Other groups, using a similar paradigm, found substantial tumor killing even without wild-type retrovirus (Culver *et al.*, 1992; Ram *et al.*, 1993). Potentially relevant differences between the two sets of experiments involve the type of animal model (subcutaneous flank of nude mice vs rat brain), the titers of vector produced by the retrovirus packaging cell line used (which were lower in the Takamiya *et al.* study), the promoter used for HSV-tk (SV40 vs CMV), and the ratio of packaging cells to tumor cells (1:1 vs 10:1). In addition the former study used immune incompetent animals, while the latter study used immune competent animals *albeit* treated with dexamethasone (which compromises immune function somewhat) and an 'antigenic' tumor line, 9L. In particular, it would appear that increasing the number of vector particles available to the tumor cells and recruiting the immune system of the host can overcome the advantages of using wild-type helper retrovirus. Subsequent studies have shown that transgene expression by 9L cells can lead to immune rejection even without ganciclovir treatment (Tapscott *et al.*, 1994) and that rejection of 9L cells in this model depends primarily on immune rejection (Barba *et al.*, 1994).

An important component in the use of HSV-tk as a therapeutic gene was the demonstration by Moolten (1990) and others (Culver *et al.*, 1992; Takamiya *et al.*, 1992; Freeman *et al.*, 1993) that HSV-tk-positive cells exposed to ganciclovir are lethal to HSV-tk-negative cells as a result of a 'bystander' effect. This effect is illustrated by the following set of experiments (Figure 6). C6BAG, a glioma cell line expressing the *lacZ* gene, was co-cultured with C6VIK cells (that express the HSV-tk transgene) in a ratio of 1:2 for 7 days. Cells were then incubated with ganciclovir for 14 days, after which the number of *lacZ*-positive colonies can be assessed. At ganciclovir concentrations above 2 μM, C6BAG colony survival was dramatically decreased when co-cultured with C6VIK, whereas C6BAG colony survival in the presence of parental C6BU1 cells was not affected by ganciclovir at doses as high as 500 μM. This bystander effect did not appear to rely on a diffusible product, because conditioned medium from C6VIK cells treated with ganciclovir did not promote killing of C6BAG cells. The transfer of ganciclovir metabolites through direct cell contacts and incorporation into the recipient cell nucleus has been verified by Li Bi *et al.* (1993). In addition to transfer of toxic metabolites, the 'bystander' effect *in vivo* apparently involves enhancement of the immune response to the tumor through foreign antigens introduced by grafting of packaging cells, as well as ganciclovir-mediated killing of dividing endothelial cells in the process of neovascularization of the tumor bed (Ram *et al.*, 1994b).

A more recent example of retrovirus-mediated metabolism of a prodrug is the demonstration that intratumoral grafting of packaging cells, which release a retrovirus vector bearing the gene for cytochrome P450 2B1 and express this enzyme, followed by intrathecal administration of cyclophosphamide, led to regression of intracerebral

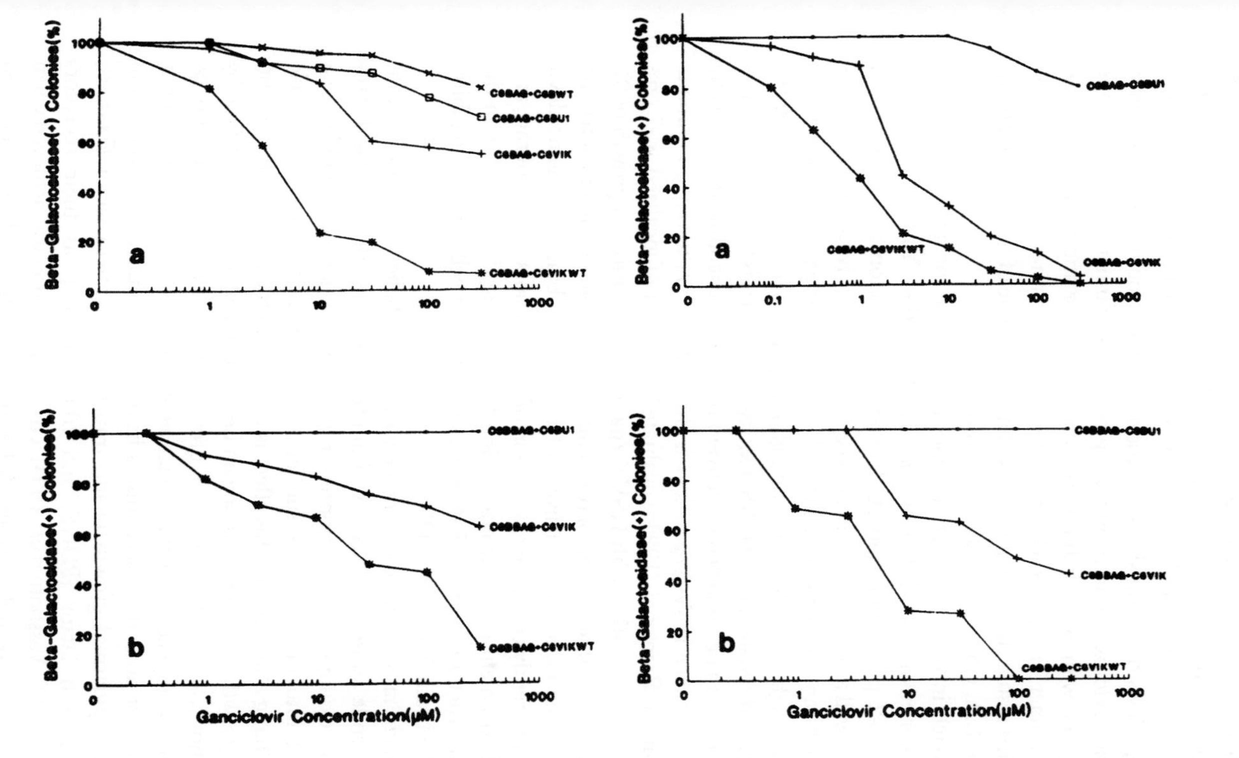

Figure 6 Ganciclovir sensitivity of C6 glioma cells co-cultured with other C6-derived lines. Left panels: (a) Seven days after C6BAG glioma cells (recipients; BAG=expressing *lacZ* gene) were plated C6BU1 (*lacZ*-negative), C6BWT (*lacZ*-negative infected with wild-type MMLV), C6BVIK (*lacZ*-negative expressing HSV-tk gene), or C6BVIKWT (*lacZ*-negative, expressing HSV-tk gene, infected with wild-type MMLV) cells (all=donors) were added at a ratio of 1:2 (recipient:donor). Ganciclovir treatment was begun 3 days later and continued for 9–12 days. β-Galactosidase-positive colonies were scored, and expressed as a percentage of those seen for parallel cultures without ganciclovir. (b) This analysis was carried out as in (a) except C6BBAG cells which lack endogenous tk were used instead of C6BAG. Right panels: same as left panels except recipient and donor cells were plated simultaneously. Reprinted with permission from Takamiya *et al.* (1992).

rat C6 gliomas (Figure 7;. Wei *et al.*, 1994). Cytochrome P450 2B1 is normally expressed only in the liver. It can convert the anticancer drug, cyclophosphamide, into two active metabolites: phosphoramide mustard, which acts as a DNA-alkylating agent, and acrolein, a protein cross-linking agent. These metabolites have restricted entry to the brain due to the blood–brain barrier. Expression of this enzyme in tumor cells in the brain, allows direct on-site generation of these toxic metabolites. These metabolites transfer cytotoxicity to neighboring cells in a manner analogous to the 'bystander' effect seen with HSV-tk, but in this case the donor and recipient cells do not need to be in contact for transfer of metabolites and DNA damage caused by phosphoramide mustard results in death of cells dividing at the time of or after drug treatment (M. Wei and E.A. Chiocca, personal communication).

Another approach involves transferring genes that encode for immune response stimulators, like interleukin-2 (IL-2; Fearon *et al.*, 1990) and IL-4 (Yu *et al.*, 1993). The observation that IL-2 production and IL-2 receptor expression are decreased in glioma patients, as well as the correlation between the degree of lymphocytic infiltration of brain tumors and the length of survival of patients, suggested that immune response modifiers might be useful in treating glioma patients (Rozman *et al.*, 1991). One way of reducing the systemic and CNS toxicity of cytokines is by selective expression in tumors, which can be mediated by delivery of the gene encoding them directly to tumor cells *in vivo*. Since HSV-tk/ganciclovir therapy appears to rely on immune rejection of the tumor, Ram *et al.* (1994a) evaluated the effect of localized IL-2 production alone or in combination with HSV-tk/ganciclovir therapy on growth of 9L gliomas. While IL-2 caused regression of over half of the peripheral tumors, it had no effect on tumors in the brain. Gene delivery of HSV-tk combined with ganciclovir treatment led to apparently complete eradication of brain tumors in 50% of rats, and there was no enhancement of tumor eradication by including the IL-2 gene in the HSV-tk vector construct. In contrast, IL-4 expression by tumor cells appears to inhibit their growth in the brain (Wei *et al.*, 1995b).

A third approach has been to introduce genes that metabolically suppress tumor growth or that encode antisense RNA which inhibits the translation of growth promoting factors. For examples, tumor suppresser genes, such as *p53* or retinoblastoma genes, introduced into tumor cells result in a resumption of normal cell cycle characteristics (Chang *et al.*, 1992). This effect is intrinsic to the transfected cells and slows down their growth. This suppression can be overcome at a subsequent time by additional mutations acquired during cell growth. Antisense RNA can also be designed to inhibit expression of growth factors which promote autocrine tumor growth. Several major autocrine mechanisms have been identified in human brain tumors, including epidermal growth factor, transforming growth factor (TGF)-α, platelet-derived growth factor (PDGF) and bFGF, as well as their receptors (Hermansson *et al.*, 1992; Vassbotn *et al.*, 1994). Growth inhibition was achieved when A172 glioma cells were treated with an 18 bp antisense oligonucleotide designed against the translation initiation site of the *c-sis* proto-oncogene encoding the β-chain of PDGF (Nitta and Satto, 1992). Clearly other growth factors can also have a role in tumor propagation. Trojan *et al.* (1993) transfected C6 glioma cells with an episome-

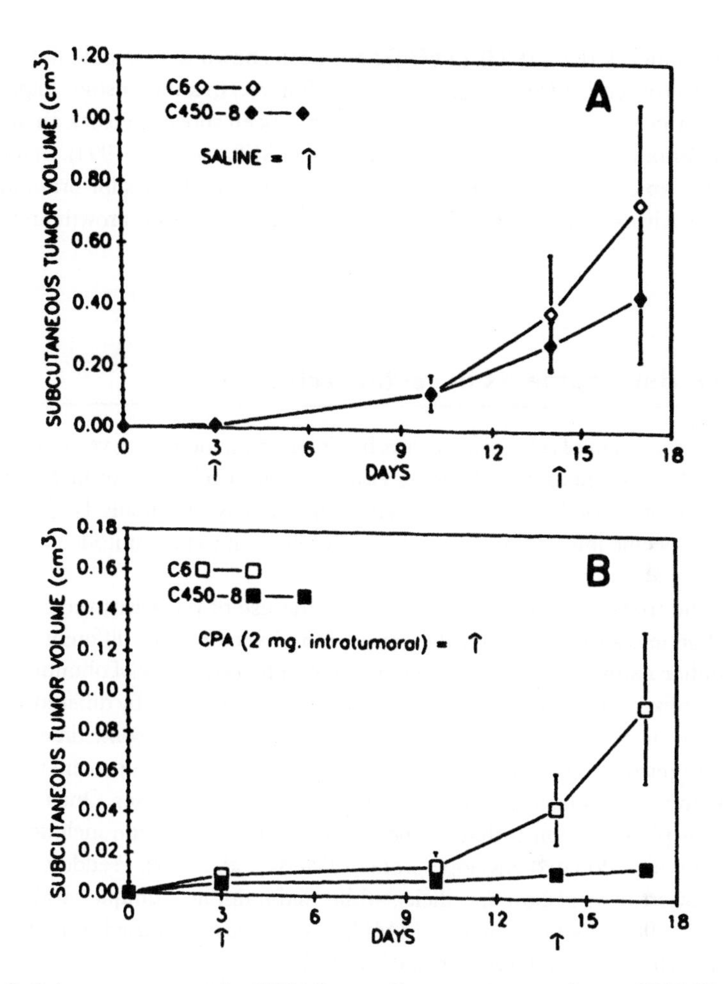

Figure 7 Subcutaneous growth of C6 glioma cells expressing cytochrome P450 B1 in nude mice with and without cyclophosphamide treatment. One million C6 or C450-8 (transfected with cytochrome P450 B1 gene) cells were injected subcutaneously into the flanks of nude mice. Tumors (5 animals per group) were injected on days 3 and 14 with saline (A) or cyclophosphamide (B). The average tumor volume ± SD is shown for each group. Note the Y-axis scale difference between A and B. Reprinted with permission from Wei *et al.* (1994).

based transgene cassette encoding antisense RNA for insulin-like growth factor 1 (IGF-1). The transfected cells showed reduced expression of IGF-1 mRNA and protein, were growth inhibited in culture, and were no longer tumorigenic when implanted into rats. Inhibition of IGF-1 also enhanced the immune response to the tumor cells, possibly by changing their antigenic profile. Thus cells inoculated subcutaneously on one side of the rat suppressed growth of C6 glioma cells inoculated intracerebrally or subcutaneously on the other side. Another group has shown increased antigenicity of glioblastoma cell lines incubated with antisense oligonucleotides to

TGF-β2 (Jachimczak *et al.*, 1993). (TFG-β2 has been shown to be a potent immunosuppressive factor secreted by malignant human gliomas.) In another study expression of a dominant negative mutant receptor protein has been shown to block angiogenesis of brain tumors and hence to restrict their growth (Millauer *et al.*, 1994). In the next few years we can expect a 'landslide' of new therapeutic modalities for brain tumors based on retrovirus-mediated delivery of genes designed to inhibit growth and to kill tumor cells.

10.6 New developments in vector technology

Work on improvement of retrovirus vectors has focused on increasing vector titers and particle stability, as well as on enhancing virus entry into cells and/or increasing the cell selectivity of entry. Some studies suggest that if may eventually be possible to mediate site-specific integration of retrovirus sequences into post-mitotic and mitotic cells.

One of the major technological advances in the use of retrovirus vectors for the purpose of gene transfer has been the generation of vectors with different envelope proteins, either naturally occurring or engineered. The possibility of obtaining retroviruses with mixed phenotypes, through infection of the same cell by different viruses, from the same or different taxonomic groups, has been known for some time (Zavada, 1972). These retroviruses, in which virus particles have at least one protein that is not encoded in the encapsulated genome, are called pseudotyped viruses. One interesting aspect of this phenomenon is that the host range of the new virion includes that of both viruses which donated envelope proteins (Weiss *et al.*, 1977). Pseudotyping thus allows alteration of viral host specificity, while maintaining a unique viral genome (Wilson *et al.*, 1989). Pseudotyping of MMLV vectors has been used to increase the stability of particles and to target specific host cells.

Retroviruses recognize a discrete number of different cell surface proteins as receptors for the envelope protein, which is responsible for the fusion of the virus particles with the cell membrane (Dubay *et al.*, 1992; Doms *et al.*, 1993; Weiss, 1993). Generation of retrovirus particles bearing the vesicular stomatitis virus G glycoprotein (VSV-G) has increased stability and provided a much wider host range, as compared to those with the MMLV glycoprotein (Emi *et al.*, 1991). In contrast to most retroviral envelope proteins that interact with species-specific cell surface proteins, VSV-G recognizes a phospholipid component common to all cell membranes (Mastomarino *et al.*, 1987). The ubiquitous distribution of this 'receptor' combined with a highly fusogenic activity allow VSV-G-pseudotyped virus to efficiently infect a broad range of different host cell types (Yee *et al.*, 1994), including species which are poorly infected by current vectors, such as Zebrafish (Hopkins, 1993). These pseudotyped viruses can also be concentrated 100- to 300-fold by ultracentrifugation with 94–100% recovery of infectious particles, thus achieving titers in the range of 10^9 infectious units ml^{-1} (Burns *et al.*, 1993; Yee *et al.*, 1994). This combination of high titers with more efficient infection of cells should allow VSV-G-pseudotyped retroviral vectors to be used for direct *in vivo* gene therapy (Kay *et al.*, 1992). VSV-G has also been used in synthetic vector systems for the production of highly fusogenic virosomes (Hug and Sleight, 1994).

Targeting viral vectors to specific cell types has been achieved by alteration of the envelope protein. The design of envelope proteins to achieve cell-restricted specificity has been approached in three different ways: chemical modification, use of antibody ligands and creation of recombinant envelope proteins. Chemical modification of the surface of ectropic MMLV particles with lactose produces artificial asialoglycoproteins that can target particles to asialoglycoprotein receptors present on human hepatocytes (Neda *et al.*, 1991). Construction of bispecific streptavidin–biotin–streptavidin bridges using streptavidin-bound antibodies specific for both the virus particle and cell surface epitopes (e.g. MHC class I and II molecules, and EGF and insulin receptors) has yielded a low efficiency of infection of target cells (Roux *et al.*, 1989; Etienne-Julan *et al.*, 1992).

The latest development in the tissue-specific targeting of retroviral vectors involves creation of chimeric envelope proteins containing sequences encoding protein domains that interact with cell-specific receptors. Several molecules have been designed. Single chain antigen-binding proteins (designated scFv; Bird *et al.*, 1988) have been introduced into the C-terminal domain of the envelope protein, which after processing by viral protease becomes the transmembrane protein that anchors the surface protein (SU) to the membrane, thus replacing part or all of the SU domain (Russel *et al.*, 1993; Chu *et al.*, 1994). Some cell specificity was achieved in this way, but it was not clear whether virus penetration was mediated by the chimeric protein. There have also been attempts to design fusion molecules with smaller foreign peptide sequences inserted within the SU domain. One of these chimeras was constructed by introducing the polypeptide hormone erythropoietin into the MMLV ecotropic envelope protein. The virus particles containing both chimeric and wild-type envelope proteins were able to infect human cell lines of erythroid lineage (Kasahara *et al.*, 1994) as efficiently as murine lines, achieving both cell targeting and altered species specificity. The wild-type ecotropic envelope included in these virion particles may account for its apparent intact ability to mediate membrane fusion and thereby infection. In the same line of research another study made use of the integrin receptors to cross the avian/mammalian barrier. After computer modeling, four small regions in the host range determinant domain of the avian leukosis virus (ALV) envelope protein were identified as the optimal sites for insertion of a 16 amino acid peptide containing the consensus sequence (arginine–glycine–asparagine), which is found in extracellular matrix proteins recognized by integrins (Yamada, 1991; Hynes, 1992). Pure pseudotyped virus particles, without any wild-type envelope protein, however, had a low efficiency of infection of human and rat cell lines (Valsesia-Wittmann *et al.*, 1994).

All these studies show the potential to direct viral infection to certain cells. These studies are of major significance for brain tumor gene therapy since some types of receptor are over-expressed on tumor cells (see Section 10.5). Chimeric envelope protein-bearing ligands for such receptors could also be inserted into replication competent vectors, derived from either the ALV or ecotropic murine virus (Hughes *et al.*, 1987; Stuhlmann *et al.*, 1989; Petropoulos and Hughes, 1991), bearing 'suicide' gene(s). In this way infection, replication and eventual destruction could be restricted to dividing tumor cells. In a very interesting study that addresses the tissue-specific targeting issue in a somewhat different way, a transgenic mouse was generated in which the viral receptor for a subtype of ALV was expressed under the control of a muscle-specific promoter. Although the promoter had some basal activity in other tissues, infection with avian-specific retroviruses was shown to occur primarily in muscle cells (Federspiel *et al.*, 1994). This study demonstrates the potential of introducing chimeric envelope proteins into replication-competent virus vectors for restricted uses, such as gene therapy for brain tumors.

Several vector strategies have developed around treatment of AIDS, including use of HIV-based pseudotyped vectors (Young *et al.*, 1990; Joshi *et al.*, 1992; Mosca *et al.*, 1992) and novel packaging cell lines (Buchschacher and Panganiban, 1992; Carroll *et al.*, 1994). Onco-retroviruses, such as MMLV, depend on cell division to integrate into the genome and thus to produce viral progency (Humphries and Temin, 1974; Chen and Temin, 1982; Miller *et al.*, 1990). The critical event for the integration is mitosis, and not DNA replication *per se* (Palmer *et al.*, 1990; Lewis and Emerman, 1994; Muller and Varmus, 1994). This requirement is thought to be related to the size of the MMLV pre-integration complex, which is apparently too large to be imported through the nuclear pores, and can gain access only when the nuclear membrane disassembles during mitosis. It is also possible that the MMLV pre-integration complex does not enter the nucleus because it lacks the proper nuclear localization signal (NLS) needed to bind to chaperone proteins which mediate ATP-dependent nuclear import (Garcia-Bustos *et al.*, 1991; Forbes, 1992). In contrast, HIV-1, which is a lentivirus, can infect and propagate in non-dividing cells, such as $CD4^+$ T lymphocytes (helper T cells) and macrophages (Lewis *et al.*, 1992; Lewis and Emerman, 1994). The function of HIV-1 which allow integration into the genome of non-dividing cells maps to the matrix protein, and more specifically to a sequence that resembles the SV40 large T antigen NLS (Bukrinsky *et al.*, 1993; Schwedler *et al.*, 1994).

The most relevant aspect of HIV infection for the purpose of gene transfer to the nervous system is that neural tissue damage in this disease results not only from immunological aspects of infection, but also from direct infection of cells in the CNS (Navia *et al.*, 1986; Wiley *et al.*, 1986; Michaels *et al.*, 1988). Brain tissue can harbor a large reservoir of HIV-1 provirus and unintegrated DNA sequences (Shaw *et al.*, 1985; Pang *et al.*, 1990; Vazeux *et al.*, 1992). Immunohistochemistry and *in situ* hybridization analysis of HIV-1 p24 (capsid protein) and mRNAs indicates that viral proteins and RNA are localized primarily in macrophages, microglia and multinucleated giant cells, and not in neurons, astrocytes or oligodendrocytes (Koenig *et al.*, 1986; Wiley *et al.*, 1986; Michaels *et al.*, 1988). However, there is some evidence that the percentage of HIV-1-infected astrocytes is actually very high and that the inability to detect viral mRNAs by *in situ* hybridization may be a result of transcriptional silencing of the HIV-1 provirus by glial specific factors, since when astrocytes become reactive, HIV message can be detected by *in situ* hybridization (Epstein and Gendelman, 1993). The ability of HIV-1 to infect non-dividing cells in the CNS suggests the possibility of pseudotyping MMLV and HIV-1 viruses for gene delivery. Attempts to create a pseudotyped MMLV by introducing analogous matrix and capsid proteins from HIV-1 did not result, however, in a functional virus (Deminie and Emerman, 1993). Another unsuccessful attempt involved replacing only the matrix protein, which appears to be responsible for nuclear targeting of HIV-1 pre-integration complexes (Bukrinsky *et al.*, 1993; Deminie and Emerman, 1994). A recent study also suggests that cellular transcription factors coupled to the pre-integration complex may be able to direct site-specific integration of retrovirus sequences (Kirchner *et al.*, 1995). These early attempts at incorporating HIV components into retrovirus vectors suggest that with more knowledge about the integration mechanism of HIV-1 and factors that regulate this event, it may be possible to generate a hybrid vector based on MMLV and HIV which could achieve site-specific integration of transgenes in non-mitotic cells of the nervous system.

References

Adam, M.A., Ramesh, N., Miller, A.D. & Osborne, W.R. (1991) Internal initiation of translation in retroviral vectors carrying picornavirus 5′ nontranslated regions. *J. Virol.* **65**, 4985–4990.

Adra, C.N., Boer, P.H. & McBurney, M.W. (1987) Cloning and expression of the mouse pgk-1 gene and the nucleotide sequence of its promoter. *Gene* **60**, 65–74.

Almazan, G. & McKay, R. (1992) An oligodendrocyte precursor cell line from rat optic nerve. *Brain Res.* **579**, 234–245.

Alroy, J., Warren, C.D., Raghavan, S.S. & Kolodny, E.H. (1989) Animal models for lysosomal storage diseases: their past and future contribution. *Hum. Pathol.* **20**, 823–826.

Anton, R., Kordower, J.H., Maidment, N.T., Manaster, J.S., Kane, D.J., Rabizadeh, S., Schueller, S.B., Yang, J., Rabizadeh, S., Edwards, R.H. *et al.* (1994) Neural-targeted gene therapy for rodent and primate hemiparkinsonism. *Exp. Neurol.* **127**, 207–218.

Barba, D., Hardin, J., Sadelair, M. & Gage, F.H. (1994) Development of anti-tumor immunity following thymidine kinase-mediated killing of experimental brain tumors. *Proc. Natl. Acad. Sci. USA* **91**, 4348–4352.

Baudhuin, P., Hers, H.G. & Loeb, H. (1964) An electron microscopic and biochemical diagnosis of Type II glycogenosis. *Lab. Invest.* **13**, 1139.

Bird, R.E., Hardman, K.D., Jacobson, J.W. Johnson, K., Kaufman, B.M., Lee, S.M., Lee, T., Pope, S.H., Riordan, G.S. & Whitlow, M. (1988), Single-chain antigen-binding proteins. *Science* **242**, 423–426. [Published erratum appears in *Science* **244**, 409].

Boviatsis, E.J., Chase, M., Wei, M.X., Tamiya, T., Hurford, R.K., Kowall, N.W., Hurford, R.K., Kowall, N.W., Tepper, R.I., Breakefield, X.O. & Chiocca, E.A. (1994) Gene transfer into experimental brain tumors mediated by adenovirus, herpes simplex virus, and retrovirus vectors. *Hum. Gene Ther.* **5**, 183–191.

Breakefield, X.O. (1994) Gene therapy for spinal cord injury? *ISRT Res. Digest* **6**, 2.

Breakefield, X.O. & Geller, A.I. (1987) Gene transfer into the nervous system. *Mol. Neurobiol. Rev.* **1**, 339–371.

Bredesen, D.E., Hisanaga, K. & Sharp, F.R. (1990) Neural transplantation using temperature-sensitive immortalized neural cells: A preliminary report. *Am. Neurol. Assoc.* **205**.

Brook, G.A., Lawrence, J.M., Shah, B. & Raisman, G. (1994) Extrusion transplantation of

Schwann cells into the adult rat thalamus induces directional host axon growth. *Exp. Neurol.* **126**, 31–43.

Buchschacher, G.L.J. & Panganiban, A.T. (1992) Human immunodeficiency virus vectors for inducible expression of foreign genes. *J. Virol.* **66**, 2731–2739.

Buetti, E. & Kuhnel, B. (1986) Distinct sequence elements involved in the glucocorticoid regulation of the mouse mammary tumor virus promoter identified by linker scanning mutagenesis. *J. Mol. Biol.* **190**, 379–389.

Bukrinsky, M.I., Haggerty, S., Dempsey, M.P., Sharova, N., Adzhubel, A., Spitz, L., Lewis, P., Goldfarb, D., Emerman, M. & Stevenson, M. (1993) A nuclear localization signal within HIV-1 matrix protein that governs infection of non-dividing cells. *Nature* **365**, 666–669.

Bunge, M.B., Wood, P.M., Tynan, L.B., Bates, M.L. & Sanes, J.R. (1989) Perineurium originates from fibroblasts: demonstration *in vitro* with a retroviral marker. *Science* **243**, 229–231.

Burns, J.C., Friedmann, T., Driever, W., Burrascano, M. & Yee, J.-K. (1993) Vesicular stomatitis virus G glycoprotein pseudotyped retroviral vectors: concentration to very high titer and efficient gene transfer into mammalian and nonmammalian cells. *Proc. Natl. Acad. Sci. USA* **90**, 8033–8037.

Carroll, R., Lin, J.T., Dacquel, E.J., Mosca, J.D., Burke, D.S. & St. Louis, D.C. (1994) A human immunodeficiency virus type 1 (HIV-1)-based retroviral vector system utilizing stable HIV-1 packaging cell lines. *J. Virol.* **68**, 6047–6051.

Castillo, B., del Cerro, M., Breakefield, X.O., Frim, D.M., Barnstable, C.J., Dean D.O. & Bohn, M.C. (1994) Retinal ganglion cell survival is promoted by genetically modified astrocytes designed to secrete brain-derived neurotrophic factor (BDNF). *Brain Res.* **647**, 30–36.

Cepko, C. (1988) Retrovirus vectors and their applications in neurobiology. *Neuron* **1**, 345–353.

Cepko, C.L., Roberts, B.E. & Mulligan, R.C. (1984) Construction and applications of a highly transmissible murine retrovirus shuttle vector. *Cell* **37**, 1053–1062.

Chang, T., Yee, J.K., Yeargin, T., Friedman, T. & Haas, M. (1992) Suppression of acute lymphoblastic leukemia by the human *p53* gene. *Cancer Res.* **52**, 222–226.

Chen, I.S.Y. & Temin, H. (1982) Establishment of infection by spleen necrosis virus: inhibition in stationary cells and the role of secondary infection. *J. Virol.* **41**, 183–191.

Chiocca, E.A., Andersen, J., Takamiya, Y., Martuza, R. & Breakefield, X.O. (1994) Virus-mediated genetic treatment of rodent gliomas. In *Gene Therapeutics* (ed. Wolff, J.A.), pp. 245–262. Boston, Birkhauser.

Chu, T.-H., Martinez, I., Sheay, W. & Dornburg, R. (1994) Cell targetting with retroviral vector particles containing antibody–envelope fusion proteins. *Gene Therapy* **1**, 292–299.

Culver, K.W., Ram, X., Waebridge, S., Ishii, H., Oldfield, E.H. & Blaese, R.M. (1992) *In vivo* gene transfer with retroviral vector-producer cells for treatment of experimental brain tumors. *Science* **256**, 1550–1552.

Cunningham, L.A., Short, M.P., Vilkind, U., Breakefield, X.O. & Bohn, M.C. (1991a) Survival and differentiation within the adult mouse striatum of grafted rat pheochromocytoma cells (PC12) genetically modified to express recombinant beta-NGF. *Exp. Neurol.* **112**, 174–182.

Cunningham, L.A., Hansen, J.T., Short, M.P. & Bohn, M.C. (1991b) The use of genetically altered astrocytes to provide nerve growth factor to adrenal chromaffin cells grafted into the striatum. *Brain Res.* **561**, 192–202.

Cunningham, L.A., Short, M.P., Breakefield, X.O. & Bohn, M.C. (1994) Nerve growth factor released by transgenic astrocytes enhances the function of adrenal chromaffin cell grafts in a rat model of Parkinson's disease. *Brain Res.* **658**, 219–231.

Dai, Y., Roman, M., Naviaux, R.K. & Verma, I.M. (1992) Gene therapy via primary myoblasts: long-term expression of factor IX protein following transplantation *in vivo*. *Proc. Natl. Acad. Sci. USA* **89**, 10892–10895.

Danos, O. & Mulligan, R.C. (1988) Safe and efficient generation of recombinant retroviruses with amphotropic and ecotropic host ranges. *Proc. Natl. Acad. Sci. USA* **85**, 6460–6464.

Dekker, A.J., Winkler, J., Ray, J., Thal, L.J. & Gage, F.H. (1994) Grafting of nerve growth factor-producing fibroblasts reduces behavioural deficits in rats and lesions of the nucleus basalis magnocellularis. *Neuroscience* **60**, 299–309.

Deminie, C.A. & Emerman, M. (1993) Incorporation of Human Immunodeficiency Virus type 1 Gag proteins into murine leukemia virus virions. *J. Virol.* **67**, 6499–6506.

Deminie, C.A. & Emerman, M. (1994) Functional exchange of an oncoretrovirus and lentivirus matrix protein. *J. Virol.* **68**, 4442–4449.

Doering, L.C. (1994) Nervous system modification by transplants and gene transfer. *BioEssays* **16**, 825–831.

Doms, R.W., Lamb, R.A., Rose, J.K. & Helenius, A. (1993) Folding and assembly of viral membrane proteins. *Virology* **193**, 545–562.

Dubay, J.W., Roberts, S.J., Brody, B. & Hunter, E. (1992) Mutations in the leucine zipper of the human immunodeficiency virus type 1 transmembrane glycoprotein affect fusion and infectivity. *J. Virol.* **66**, 4748–4756.

Eccleston, P.A., Mirsky, R. & Jessen, K.R. (1991) Spontaneous immortalization of Schwann cells in culture: short-term cultured Schwann cells secrete growth inhibitory activity. *Development* **112**, 33–42.

Emi, N., Friedmann, T. & Yee, J.-K. (1991) Pseudotype formation of murine leukemia virus with the G protein of vesicular stomatitis virus. *J. Virol.* **65**, 1202–1207.

Epstein, L.G. & Gendelman, H.E. (1993) Human immunodeficiency virus type 1 infection of the nervous system: pathogenic mechanisms. *Ann. Neurol.* **33**, 429–436.

Etienne-Julan, M., Roux, P., Carillo, S., Jeanteur, P. & Piechaczyk, M. (1992) The efficiency of cell targeting by recombinant retroviruses depends on the nature of the receptor and the composition of the artificial cell-virus linker. *J. Gen. Virol.* **74**, 3251–3255.

Ezzeddine, Z.D., Martuza, R.L., Platika, D., Short, M.P., Malick, A., Choi, B. & Breakefield, X.O. (1991) Selective killing of glioma cells in culture and *in vivo* by retrovirus transfer of the herpes simplex virus thymidine kinase gene. *New Biol.* **3**, 608–614.

Fearon, E.R., Pardoll, D.M., Itaya, T., Golumbek, P., Levitsy, H.I., Simons, J.W., Karasuyama, H., Vogelstein, B. & Frost, P. (1990) Interleukin-2 production by tumor cells bypasses T helper function in the generation of an antitumor response. *Cell* **60**, 397–403.

Federoff, H.J., Geschwind, M.D., Geller, A.I. & Kessler, J.A. (1992) Expression of nerve growth factor *in vivo* from a defective herpes simplex virus 1 vector prevents effects of axotomy on sympathetic ganglia. *Proc. Natl. Acad. Sci. USA* **89**, 1636–1640.

Federspiel, M.J., Bates, P., Young, J.A., Varmus, H.E. & Hughes, S.H. (1994) A system for tissue-specific gene targeting: transgenic mice susceptible to subgroup A avian leukosis virus-based retroviral vectors. *Proc. Natl. Acad. Sci. USA* **91**, 11241–11245.

Fisher, L.J. (1995) Engineered cells: a promising therapeutic approach for neural disease. *Rest. Neurol. Neurosci.* (in press).

Fisher, L.J. & Gage, F.H. (1993) Grafting in the mammalian central nervous system. *Physiol. Rev.* **73**, 583–616.

Fisher, L.J., Jinnah, H.A., Kale, L.C., Higgins, G.A. & Gage, F.H. (1991) Survival and function of intrastriatally grafted primary fibroblasts genetically modified to produce L-dopa. *Neuron* **6**, 371–380.

Fisher, L.J., Schinstine, M., Salvaterra, P., Dekker, A.J., Thal, L. & Gage, F.H. (1993) *In vivo* production and release of acetylcholine from primary fibroblasts genetically modified to express choline acetyltransferase. *J. Neurochem.* **61**, 1323–1332.

Forbes, D.J. (1992) Structure and function of the nuclear pore complex. *Ann. Rev. Cell Biol.* **8**, 495–527.

Frederiksen, K., Jat, P.S., Valtz, N., Levy, D. & McKay, R. (1988) Immortalization of precursor cells from the mammalian CNS. *Neuron* **1**, 439–448.

Freeman, S.M., Abboud, C.N., Whartenby, K.A., Packman, C.H., Koeplin, D.S. Moolten, F.L. & Abraham, G.N. (1993) The 'bystander effect': tumor regression when a fraction of the tumor mass is genetically modified. *Cancer Res.* **53**, 5274–5283.

Frim, D.M., Yee, W.M. & Isacson, O. (1993a) NGF reduces striatal excitotoxic neuronal loss without affecting concurrent neuronal stress. *NeuroReport* **4**, 655–658.

Frim, D.M., Short, M.P., Rosenberg, W.S., Simpson, J., Breakefield, X.O. & Isacson, O. (1993b) Local protective effects of nerve growth factor-secreting fibroblasts against excitotoxic lesions in the rat striatum. *J. Neurosurg.* **78**, 267–273.

Frim, D.M., Simpson, J., Uhler, T.A., Short, M.P., Bossi, S.R., Breakefield, X.O. & Isacson, O. (1993c) Striatal degeneration induced by mitochondrial blockade is prevented by biologically delivered NGF. *J. Neurosci. Res.* **35**, 452–458.

Frim, D.M., Uhler, T.A., Short, M.P., Ezzeddine, Z.D., Klagsbrun, M., Breakefield, X.O. & Isacson, O. (1993d) Effects of biologically delivered NGF, BDNF and bFGF on striatal excitotoxic lesions. *NeuroReport* **4**, 367–370.

Frim, D.M., Uhler, T.A., Galpern, W.R., Beal, M.F., Breakefield, X.O. & Isacson, O. (1994a) Implanted fibroblasts genetically engineered to produce brain-derived neurotrophic factor prevent 1-methyl-4-phenylpyridinium toxicity to dopaminergic neurons in the rat. *Proc. Natl. Acad. Sci. USA* **91**, 5104–5108.

Frim, D.M., Wullner, U., Beal, M.F. & Isacson, O. (1994b) Implanted NGF-producing fibroblasts induce catalase and modify ATP levels but do not affect glutamate receptor binding or NMDA receptor expression in the rat striatum. *Exp. Neurol.* **128**, 172–180.

Gage, F.H., Wolff, J.A., Rosenberg, M.B., Xu, L., Yee, J.K., Shults, C. & Friedmann, T. (1987) Grafting genetically modified cells to the brain: possibilities for the future. *Neuroscience* **23**, 795–807.

Galileo, D.S., Majors, J., Horwitz, A.F. & Sanes, J.R. (1992) Retrovirally introduced antisense integrin RNA inhibits neuroblast migration *in vivo*. *Neuron* **9**, 1117–1131.

Garcia-Bustos, J., Heitman, J. & Hall, M.N. (1991) Nuclear protein localization. *Biochim. Biophys. Acta* **1071**, 83–101.

Ghattas, I.R., Sanes, J.R. & Majors, J.E. (1991) The encephalomyocarditis virus internal ribosome entry site allows efficient coexpression of two genes from a recombinant provirus in cultured cells and in embryos. *Mol. Cell Biol.* **11**, 5848–5859.

Gilboa, E., Eglitis, M.A., Kantoff, P.W. & Anderson, W.F. (1986) Transfer and expression of cloned genes using retroviral vectors. *Biotechnology* **4**, 504–512.

Goodman, M.N., Silver, J. & Jacobberger, J.W. (1993) Establishment and neurite outgrowth properties of neonatal and adult rat olfactory bulb glial cell lines. *Brain Res.* **619**, 199–213.

Gossen, M. & Bujard, H. (1992) Tight control of gene expression in mammalian cells by tetracycline-responsive promoters. *Proc. Natl. Acad. Sci. USA* **89**, 5547–5551.

Hannan, G.N., Lehnert, S.A., MacAvoy, E.S., Jennings, P.A. & Molloy, P.L. (1993) An engineered PGK promoter and *lac* operator-repressor system for the regulation of gene expression in mammalian cells. *Gene* **130**, 233–239.

Hantzopoulos, P.A., Sullenger, B.A., Ungers, G. & Gilboa, E. (1989) Improved gene expression upon transfer of the adenosine deaminase minigene outside the transcriptional unit of a retrovirus vector. *Proc. Natl. Acad. Sci USA* **86**, 3519–3523.

Hawley, R.G., Sabourin, L.A. & Hawley, T.S. (1989) An improved retroviral vector for gene transfer into undifferentiated cells. *Nucl. Acids Res.* **17**, 4001.

Hawley, R.G., Lieu, F.H.L., Fong, A.Z.C. & Hawley, T.S. (1994) Versatile retroviral vectors for potential use in gene therapy. *Gene Ther.* **1**, 136–138.

Hermansson, M., Funa, K., Hartman, M., Claesson-Welsh, L., Heldin, C.-H., Westermark, G. & Nister, M. (1992) Platelet-derived growth factor and its receptors in human glioma tissue: expression of messenger RNA and protein suggests the presence of autocrine and paracrine loops. *Cancer Res.* **52**, 3213–3219.

Hoffman, D., Breakefield, X.O., Short, M.P. & Aebischer, P. (1993) Transplantation of polymer encapsulated cell line genetically engineered to release NGF. *Exp. Neurol.* **122**, 100–106.

Hopkins, N. (1993) High titers of retrovirus (vesicular stomatitis virus) pseudotypes, at last. *Proc. Natl. Acad. Sci. USA* **90**, 8759–8760.

Horellou, P., Brundin, P., Kalen, P., Mallet, J. & Bjorklund, A. (1990a) *In vivo* release of dopa and dopamine from genetically engineered cells grafted to the denervated rat striatum. *Neuron* **5**, 393–402.

Horellou, P., Marlier, L., Privat, A. & Mallet, J. (1990b) Behavioural effect of engineered cells that synthesize L-dopa or dopamine after grafting into the rat neostriatum. *Eur. J. Neurosci.* **2**, 116–119.

Hug, P. & Sleight, R.G. (1994) Fusogenic virosomes prepared by partitioning of vesicular stomatitis virus G protein into preformed vesicles. *J. Biol. Chem.* **269**, 4050–4056.

Hughes, S.H., Greenhouse, J.J., Petropoulos, C.J. & Sutrave, P. (1987) Adaptor plasmids simplify the insertion of foreign DNA into helper-independent retroviral vectors. *J. Virol.* **61**, 3004–3012.

Humphries, E.H. & Temin, H.M. (1974) Requirement for cell division for initiation of transcription of Rous sarcoma virus RNA. *J. Virol.* **14**, 531–546.

Hynes, R.O. (1992) Integrins: Versatility, modulation, and signaling in cell adhesion. *Cell* **69**, 11–25.

Ioannou, Y.A., Bishop, D.F. & Desnick, R.F. (1992) Overexpression of human alpha-galactosidase A results in its intracellular aggregation, crystalization in lysosmes, and selective secretion. *J. Cell. Biol.* **119**, 1137–1150.

Isacson, O. (1993) On neuronal health. *Trends Neurosci.* **16**, 306–308.

Jachimczak, P., Bogdahn, U., Schenider, J., Behl, C., Meixensberger, J., Apfel, R., Dorries, R., Schlingensiepen, K.H. & Brysch, W. (1993) The effect of transforming growth factor-beta 2-specific phosphorothioate-antisense oligodeoxynucleotides in reversing cellular immunosuppression in malignant glioma. *J. Neurosurg.* **78**, 944–951.

Jiao, S., Gurevich, V. & Wolff, J.A. (1993) Long-term correction of rat model of Parkinson's disease by gene therapy. *Nature* **362**, 450–453.

Joshi, S., Bangimaier, R. & Liem, S.E. (1992) Development of HIV-1 like vector particles for CD4+ cell gene therapy. *Int. Conf. AIDS*, **8(3)**: 24.

Kang, U.J., Fisher, L.J., Joh, T.H., O'Malley, K.L. & Gage, F.H. (1993) Regulation of dopamine production by genetically modified primary fibroblasts. *J. Neurosci.* **13**, 5203–5211.

Kasahara, N., Dozy, A.M. & Kan, Y.W. (1994) Tissue-specific targeting of retroviral vectors through ligand-receptor interactions. *Science* **266**, 1373–1376.

Kawaja, M.D. & Gage, F.H. (1991) Reactive astrocytes are substrates for the growth of adult CNS axons in the presence of elevated levels of nerve growth factor. *Neuron* **7**, 1019–1030.

Kawaja, M.D., Rosenberg, M.B., Yoshida, K. & Gage, F.H. (1992) Somatic gene transfer of nerve growth factor promotes the survival of axotomized septal neurons and the regeneration of their axons in adult rats. *J. Neurosci.* **12**, 2849–2864.

Kay, M.A., Baley, P., Rothenberg, S., Leland, F., Fleming, L., Ponder, K.P., Liu, T., Finegold, M., Darlinton, G. & Pokorny, W. *et al.* (1992) Expression of human alpha 1-antitrypsin in dogs after autologous transplantation of retroviral transduced hepatocytes. *Proc. Natl. Acad. Sci. USA* **89**, 89–93.

Kirchner, J., Connolly, C.M. & Sandmeyer, B. (1995) Requirement of RNA polymerase III transcription factors for *in vitro* position-specific integration of a retrovirus like element. *Science* **267**, 1488–1490.

Koenig, S., Gendelman, H.E., Orenstein, J.M. *et al.* (1986) Detection of AIDS virus in macrophages in brain tissue from AIDS patients with encephalopathy. *Science* **233**, 1089–1093.

Kromer, L.F. & Cornbrooks, C.J. (1985) Transplants of Schwann cell cultures promote axonal regeneration in the adult mammalian brain. *Proc. Natl. Acad. Sci. USA* **82**, 6330–6334.

Krushel, L.A., Johnston, J.G., Fishell, G., Tibshirani, R. & van der Kooy, D. (1993) Spatially localized neuronal cell lineages in the developing mammalian forebrain. *Neuroscience* **53**, 1035–1047.

Lampson, L.A., Lampson, M.A. & Dunne, A.D. (1993) Exploiting the *lacZ* reporter gene for

quantitative analysis of disseminated tumor growth with the brain. *Cancer Res.* **53**, 176–180.
Langford, L.A. & Owens, G.C. (1990) Resolution of the pathway taken by implanted Schwann cells to a spinal cord lesion by prior infection with a retrovirus encoding beta-galactosidase. *Acta Neuropathol.* **80**, 514–520.
Langford, L.A., Porter, S. & Bunge, R.P. (1988) Immortalized rat Schwann cells produce tumors *in vivo*. *J. Neurocytol.* **17**, 521–529.
Lewis, P.F. & Emerman, M. (1994) Passage through mitosis is required for oncoretroviruses but not for the human immunodeficiency virus. *J. Virol.* **68**, 510–516.
Lewis, P.F., Hensel, M. & Emerman, M. (1992) Human immunodeficiency virus infection of cells arrested in the cell cycle. *EMBO J.* **11**, 3053–3058.
Li Bi, W., Parysek, L.M., Warnick, R. & Stambrook, P.J. (1993) *In vivo* evidence that metabolic cooperation is responsible for the bystander effect observed with HSV tk retroviral gene therapy. *Human Gene Ther.* **4**, 725–731.
Lieber, A., Sandig, V., Sommer, W., Bahring, S. & Strauss, M. (1993) Stable high-level gene expression in mammalian cells by T7 phage RNA. *Methods Enzymol.* **217**, 47–66.
Lindsay, R.M., Wiegand, S.J., Altar, C.A. & DiStefano, P.S. (1994) Neurotrophic factors: from molecule to man. *TINS* **17**, 182–190.
Luskin, M.B. & McDermott, K. (1994) Divergent lineages for ologodendrocytes and astrocytes originating in the neonatal forebrain subventricular zone. *Glia* **11**, 211–226.
Mann, R., Mulligan, R.C. & Baltimore, D. (1983) Construction of a retrovirus packaging mutant and its use to produce helper-free defective retrovirus. *Cell* **33**, 153–159.
Markovitz, D., Goff, S. & Bank, A. (1988) Construction and use of a safe and efficient amphotropic packaging cell line. *Virology* **167**, 400–406.
Mastromarino, P., Conti, C., Goldoni, P., Hauttecoeur, B. & Orsi, N. (1987) Characterization of membrane components of the erythrocyte involved in vesicular stomatitis virus attachment and fusion at acidic pH. *J. Gen. Virol.* **68**, 2359–2369.
Maysinger, D., Piccardo, P., Goiny, M. & Cuello, A.C. (1992) Grafting of genetically modified cells: effects of acetylcholine release *in vivo*. *Neurochem. Int.* **21**, 543–458.
Maysinger, D., Piccardo, P., Filipovic-Grcic, J. & Cuello, A.C. (1993) Microencapsulation of genetically engineered fibroblasts secreting nerve growth factor. *Neurochem. Int.* **23**, 123–129.
Mey, J. & Thanos, S. (1993) Intravitreal injections of neurotrophic factors support the survival of axotomized retinal ganglion cells in adult rats *in vivo*. *Brain Res.* **602**, 304–317.
Michaels, J., Sharer, L.R. & Epstein, L.G. (1988) Human immunodeficiency virus type 1 (HIV-1) infection of the nervous system: a review. *Immunodefic. Rev.* **1**, 71–104.
Millauer, B., Shawver, L.K., Plate, K.H., Risau, W. & Ullrich, A. (1994) Glioblastoma growth inhibited *in vivo* by dominant-negative Flk-1 mutant. *Nature* **367**, 576–579.
Miller, A.D. (1990) Retrovirus packaging cells. *Hum. Gene Ther.* **1**, 5–14.
Miller, D.G., Adam, M.A. & Miller, A.D. (1990) Gene transfer by retrovirus vectors occurs only in cells that are actively replicating at the time of infection. *Mol. Cell. Biol.* **10**, 4239–4242.
Moolten, F.L. (1990) Mosaicism induced by gene insertion as a means of improving chemotherapeutic selectivity. *Crit. Rev. Immunol.* **10**, 203–233.
Morshead, C.M., Reynolds, B.A., Craig, C.G., McBurney, M.W., Staines, W.A., Morassutti, D., Weiss, S. & van der Kooy, D. (1994) Neural stem cells in the adult mammalian forebrain: a relatively quiescent subpopulation of subependymal cells. *Neuron* **13**, 1071–1082.
Mosca, J., Kim, J., Perera, P, Yu, Z., Ritchey, D., D'Arcy, L., Kaushal, I., Xu, J., Vahey, M., Louis, D.S. *et al.* (1992) Determining if an HIV based retroviral vector system will target HIV-infected cells. *Int. Conf. AIDS*, **8(2)**: A80 (Abstract No. PoA 2467).
Moullier, P., Bohl, D., Heard, J.-M. & Danos, O. (1993a) Correction of lysosomal storage in the liver and spleen of MPS VII mice by implantation of genetically modified skin fibroblasts. *Nature Genet.* **4**, 154–159.

Moullier, P., Marechal, V., Danos, O. & Heard, J.M. (1993b) Continuous systemic secretion of lysosomal enzyme by genetically modified skin fibroblasts. *Transplantation* **56**, 427–432.

Mroz, P.J. & Moolten, F.L. (1993) Retrovirally transduced *Escherichia coli gpt* genes combine selectability with chemosensitivity capable of mediating tumor eradication. *Hum. Gene Ther.* **4**, 589–595.

Muller, H.P. & Varmus, H.E. (1994) DNA bending creates favored sites for retroviral integration: an explanation for preferred insertion sites in nucleosomes. *EMBO J.* **13**, 4704–4714.

Mulligan, R.C. (1993) The basic science of gene therapy. *Science* **260**, 926–931.

Mulligan, R.C. & Berg, P. (1981) Selection for animal cells that express the *Escherichia coli* gene coding for xanthineguanine phosphoribosyltransferase. *Proc. Natl. Acad. Sci. USA* **78**, 2072–2076.

Navia, B.A., Cho, E.-S., Petito, C.K. & Price, R.W. (1986) The AIDS dementia complex: II. Neuropathology. *Ann. Neurol.* **19**, 525–535.

Neda, H., Wu, C.H. & Wu, G.Y. (1991) Chemical modification of an ecotropic murine keukemia virus results in redirection of its target cell specificity. *J. Biol. Chem.* **266**, 14143.

Neufeld, E.F. & Muenzer, J. (1989) The mucopolysaccharidoses. In: *The Metabolic Basis of Inherited Disease* (eds Scriver, C.J. & Beadet, A.L.), pp. 1565–1587. New York, McGraw-Hill.

Nitta, T. & Sato, K. (1992) Inhibition of *c-sis* protein synthesis and cell growth with antisense oligonucleotides in human glioma cells. *No Shinkei Geka* **20**, 857–863.

Ohashi, T., Boggs, S., Robbins, P., Bahnson, A., Patrene, K., Wei, F.S., Wei, J.F., Li, J., Lucht, L. & Fei, Y. *et al.* (1992) Efficient transfer and sustained high expression of the human glucocerebrosidase gene in mice and their functional macrophages following transplantation of bone marrow transduced by a retroviral vector. *Proc. Natl. Acad. Sci. USA* **89**, 11332–11336.

Okoye, G.S., Freed, W.J. & Geller, H.M. (1994) Short-term immunosuppression enhances the survival of intracerebral grafts of A7-immortalized glial cells. *Exp. Neurol.* **128**, 191–201.

Onifer, S.M., Whittemore, S.R. & Holets, V.R. (1993) Variable morphological differentiation of a raphe-derived neuronal cell line following transplantation into the adult rat CNS. *Exp. Neurol.* **122**, 130–142.

Owens, G.C. & Boyd, C.J. (1991) Expressing antisence PO RNA in Schwann cells perturbs myelination. *Development* **112**, 639–649.

Owens, G.C. & Bunge, R.P. (1991) Schwann cells infected with a recombinant retrovirus expressing myelin-associated glycoprotein antisence RNA do not form myelin. *Neuron* **7**, 565–575.

Palmer, T.D., Rosman, G.J., Osborne, W.R. & Miller, A.D. (1990) Genetically modified skin fibroblasts persist long after transplantation but gradually inactivate introduced genes. *Proc. Natl. Acad. Sci. USA* **88**, 1330–1334.

Pang, S., Koyanagi, Y., Miles, S., Wiley, C., Vinters, H.V. & Chen, I.S.Y. (1990) High levels of unintegrated HIV-1 DNA in brain tissue of AIDS dementia patients. *Nature* **343**, 85–89.

Pear, W.S., Nolan, G.P., Scott, M.L. & Baltimore, D. (1993) Production of high-titer helper-free retroviruses by transient transfection. *Proc. Natl. Acad. Sci. USA* **90**, 8392–8396.

Pechan, P.A., Yoshida, T., Panahian, N., Moskowitz, M.A. & Breakefield, X.O. (1995) Genetically modified fibroblasts producing NGF protect hippocampal neurons after ischemia in the rat. *NeuroReport* **6**, 669–672.

Petropoulos, C.J. & Hughes, S.H. (1991) Replication-competent retrovirus vectors for the transfer and expression of gene cassettes in avian cells. *J. Virol* **65**, 3728–3737.

Piccardo, P., Maysinger, D. & Cuello, A.C. (1992) Recovery of nucleus basalis cholinergic neurons by grafting NGF secretor fibroblasts. *NeuroReport* **3**, 353–356.

Ram, Z., Culver, K.W., Walbridge, S., Blaese, R.M. & Oldfield, E.H. (1993) *In situ* retroviral-mediated gene transfer for treatment of brain tumors in rats. *Cancer Res.* **53**, 83–88.

Ram, Z., Walbridge, S., Heiss, J.D., Culver, K.W., Blaese, R.M. & Oldfield, E.H. (1994a) *In vivo* transfer of the human interleukin-2 gene: negative tumoricidal results in experimental brain tumors. *J. Neurosurg.* **80**, 535–540.

Ram, Z., Walbridge, S., Shawker, T., Culber, K., Blaese, R.M. & Oldfield, E.H. (1994b) The effect of thymidine kinase transduction and ganciclovir therapy on tumor vasculature and growth of 9L gliomas in rats. *J. Neurosurg.* **81**, 256–260.

Rosenberg, M.B., Friedmann, T., Robertson, R.C., Tuszynski, M., Wolff, J.A., Breakefield, X.O. & Gage, F.H. (1988) Grafting genetically modified cells to the damaged brain: restorative effects of NGF expression. *Science* **242**, 1575–1578.

Roux, P., Jeanteur, P. & Piechaczyk, M. (1989) A versatile and potentially general approach to the targeting of specific cell types by retroviruses: application to the infection of human cells by means of major histocompatibility complex class I and class II antigens by mouse ecotropic murine leukemia virus-derived viruses. *Proc. Natl. Acad. Sci. USA* **86**, 9079–9083.

Rozman, T., Elliot, L. & Brooks, W. (1991) Modulation of T cell function by gliomas. *Immunol. Today* **12**, 370–374.

Ruppert, C., Sandrasagra, A., Anton, B., Evans, C. Schweitzer, E.S. & Tobin, A.J. (1993) Rat-1 fibroblasts engineered with GAD65 and GAD67 cDNAs in retroviral vectors produce and release GABA. *J. Neurochem.* **61**, 768–771.

Russell, S.J., Hawkins, R.E. & Winter, G. (1993) Retroviral vectors displaying functional antibody fragments. *Nucleic Acids Res.* **21**, 1081–1085.

Ryder, E.G. & Cepko, C.L. (1994) Migration patterns of clonally related granule cells and their progenitors in the developing chick cerebellum. *Neuron* **12**, 1011–1028.

Sanes, J.R., Rubenstein, J.L.R. & Nicolas, J.-F. (1986) Use of a recombinant retrovirus to study post-implantation cell lineage in mouse embryos. *EMBO J.* **5**, 3133–3142.

Scharf, J.M., Boviatsis, E.J., Fleet, C., Breakefield, X.O. & Chiocca, E.A. (1994) Genetically modified rat 9L gliosarcoma cells facilitate detection of infiltrating tumor cells in a rat model of brain neoplasms. *Transgenics* **1**, 219–224.

Scharfmann, R., Axelrod, J.H. & Verma, I.M. (1991) Long-term *in vivo* expression of retrovirus-mediated gene transfer in mouse fibroblast implants. *Proc. Natl. Acad. Sci. USA* **88**, 4626–4630.

Schwedler, U.V., Kornbluth, R.S. & Trono, D. (1994) The nuclear localization signal of the matrix protein of human immunodeficiency virus type 1 allows the establishment of infection in macrophages and quiescent T lymphocytes. *Proc. Natl. Acad. Sci. USA* **91**, 6992–6996.

Sena-Esteves, M., Sims, K.B., Breakefield, X.O. & Kaye, E.M. (1994) Construction of retroviral vectors containing the human β-galactosidase gene for use in animal models of G_{M1} gangliosidosis. *Soc. Neurosci.* Abst. 8.9.

Shaw, G.M., Harper, M.E., Hahn, B.E. *et al.* (1985) HTLV-III infection in brains of children and adults with AIDS encephalopathy. *Science* **1**, 177–182.

Shih, C.-C., Stoye, J.P. & Coffin, J.M. (1988) Highly preferred targets for retrovirus integration. *Cell* **53**, 531–537.

Shimohama, S., Fisher, L.J. & Gage, F.H. (1993) Intracerebral grafting of genetically modified cells. Applications to a rat model of Parkinson's disease. *Adv. Neurol.* **60**, 744–748.

Short, M.P., Rosenberg, M.B., Ezzedine, Z.D., Gage, F.H., Friedmann, T. & Breakefield, X.O. (1990a) Autocrine differentiation of PC12 cells mediated by retroviral vectors. *Devel. Neurosci.* **12**, 34–45.

Short, M.P., Choi, B., Lee, Malick, A., Breakefield, X.O. & Martuza, R.L. (1990b) Gene delivery to glioma cells in rat brain by grafting of a retrovirus packaging cell line. *J. Neurosci. Res.* **27**, 427–439.

Snyder, E.Y., Deitcher, D.L., Walsh, C., Arnold-Aldea, S., Hartwieg, E.A. & Cepko, C.L. (1992) Multipotent neural cell lines can engraft and participate in development of mouse cerebellum. *Cell* **68**, 33–51.

Snyder, E.Y., Taylor, R.M. & Wolfe, J.H. (1995) Neural progenitor cell engraftment corrects lysosomal storage throughout the MPS VII mouse brain. *Nature* **374**, 367–370.

Stocker, K.M., Brown, A.M. & Ciment, G. (1993) Gene transfer of lacZ into avian neural tube and neural creast cells by retroviral infection of grafted embryonic tissues. *J. Neurosci. Res.* **34**, 135–145.

Stuhlmann, H., Jaenisch, R. & Mulligan, R.C. (1989) Construction and properties of replication-competent murine retroviral vectors encoding methotrexate resistance. *Mol. Cell. Biol.* **9**, 100–108.

Sullenger, B.A., Lee, T.C., Smith, C.A., Ungers, G.E. & Gilboa, E. (1990) Expression of chimeric tRNA-driven antisense transcripts renders NIH 3T3 cells highly resistant to Moloney Murine Leukemia virus replication. *Mol. Cell. Biol.* **10**, 6512–6523.

Takamiya, Y., Short, M.P., Ezzeddine, Z.D., Moolten, F.L., Breakefield, X.O. & Martuza, R.L. (1992) Gene therapy of malignant brain tumors: a rat glioma line bearing the herpes simplex virus type 1-thymidine kinase gene and wild type retrovirus kills other tumor cells. *J. Neurosci. Res.* **33**, 493–503.

Takamiya, Y., Short, M.P., Moolten, R.L., Fleet, C., Mineta, T., Breakefield, X.O. & Martuza, R.L. (1993) An experimental model of retrovirus gene therapy for malignant brain tumors. *J. Neurosurg.* **79**, 104–110.

Tamiya, T., Wei, M.X., Chase, M., Breakefield, X.O. & Chiocca, E.A. (1995) Transgene inheritance and retroviral infection contribute to the efficiency of gene expression in solid tumors inoculated with retrovirus vector producer cells. *Gene Ther.* (in press).

Tapscott, S.J., Miller, A.D., Olson, J.M., Berger, M.S., Groudine, M. & Spence, A.M. (1994) Gene therapy of rat 9L gliosarcoma tumors by transduction with selectable genes does not require drug selection. *Proc. Natl. Acad. Sci. USA* **91**, 8185–8189.

Trojan, J., Johnson, T.R., Rudin, S.D., Ilan, J., Tykocinski, M.L. & Ilan, J. (1993) Treatment and prevention of rat glioblastoma by immunogenic C6 cells expressing antisense insulin-like growth factor I RNA. *Science* **259**, 94–97.

Turner, D.L., Snyder, E.Y. & Cepko, C.L. (1990) Lineage-independent determination of cell type in the embryonic mouse retina. *Neuron* **4**, 833–845.

Tuszynski, M.H., Peterson, D.A., Ray, J., Baird, A., Nakahara, Y. & Gage, F.H. (1994) Fibroblasts genetically modified to produce nerve growth factor induce robust neuritic ingrowth after grafting to the spinal cord. *Exp. Neurol.* **126**, 1–14.

Uchida, K., Tsuzaki, N., Nagatsu, T. & Kohsaka, S. (1992) Tetrahydrobiopterin-dependent functional recovery in 6-hydroxydopamine-treated rats by intracerebral grafting of fibroblasts transfected with tyrosine hydroxylase cDNA. *Devel. Neurosci.* **14**, 173–180.

Valsesia-Wittmann, S., Drynda, A., Deleage, G., Aumailley, M. & Heard, J.M. (1994) Modifications in the binding domain of avian retrovirus envelope protein to redirect the host range of retroviral vectors. *J. Virol.* **68**, 4609–4619.

Vassbotn, F.S., Ostman, A., Langeland, N., Holmsen, H., Westermark, B., Heldin, C.H. & Nister, M.J. (1994) Activated growth factor autocrine pathway drives the transformed phenotype of a human glioblastoma cell line. *Cell. Physiol.* **158**, 381–389.

Vazeux, R., Lacroix-Ciaudo, C., Blanche, S., Cumont, M.C., Henin, D., Gray, F., Boccon-Gibod, L. & Tardieu, M. (1992) Low levels of human immunodeficiency virus in the brain tissue of children with severe acquired immunodeficiency syndrome encephalopathy. *Am. J. Pathol.* **140**, 137–144.

von Figura, K. & Hasilik, A. (1986) Lysosomal enzymes and their receptors. *Annu. Rev. Biochem.* **55**, 167–193.

Wei, M.X., Tamiya, T., Chase, M., Boviatsis, E.J., Chang, T.K.H., Kowall, N.W., Hochberg, F.H., Waxman, D.J., Breakefield, X.O. & Chiocca, E.A. (1994) Experimental tumor therapy in mice using the cyclophosphamide-activating cytochrome P450 2B1 gene. *Hum. Gene Ther.* **5**, 969–978.

Wei, M.X., Tamiya, T., Breakefield, X.O. & Chiocca, E.A. (1995a) Virus vector-mediated transfer of drug-sensitivity genes for the treatment of experimental brain tumor tumors. In *Viral Vectors: Tools for Study and Genetic Manipulation of the Nervous System* (eds Kaplitt, M.G. & Loewy, A.D.) Orlando, FL, Academic Press, pp. 239–257.

Wei, M.X., Tamiya, T., Hurford, R.K.Jr., Boviatsis, E.J., Tepper, R.I. & Chiocca, E.A. (1995b) Enhancement of interleukin 4-mediated tumor regression in athymic mice by *in situ* retroviral gene transfer. *Human Gene Ther.* (in press).

Weiss, R.A. (1993) Cellular receptors and viral glycoproteins involved in retrovirus entry. In: *The Retroviridae*, (J.A. Levy, ed) Plenum Press, New York, pp. 1–108.

Weiss, R.A., Boettiger, D. & Murphy, H.M. (1977) Pseudotypes of avian sarcoma viruses with the envelope properties of vesicular stomatitis virus. *Virology* **76**, 808–825.

Wiley, C.A., Schrier, R.D., Nelson, J.A., Lampert, P.W. & Oldstone, M.B.A. (1986) Cellular localization of human immunodeficiency virus infection within the brains of acquired immune deficiency syndrome patients. *Proc. Natl. Acad. Sci. USA* **83**, 7089–7093.

Wilson, C., Reitz, M.S., Okayama, H. & Eiden, M.V. (1989) Formation of infectious hybrid virions with gibbon ape leukemia virus and human T-cell leukemia virus retroviral envelope glycoproteins and the *gag* and *pol* proteins of Moloney murine leukemia virus. *J. Virol.* **63**, 2374–2378.

Wilson, J.M., Danos, O., Grossman, M., Raulet, D.H. & Mulligan, R.C. (1990) Expression of human adenosine deaminase in mice reconstituted with retrovirus-transduced hematopoietic stem cells. *Proc. Natl. Acad. Sci. USA* **87**, 439–443.

Winkler, J., Suhr, S.T., Gage, F.H., Thal, L.J. & Fisher, L.J. (1995) Essential role of neocortical acetylocholine in spatial memory. *Nature* **375**, 484–487.

Wojcik, B.E., Nothias, F., Lazar, M., Jouin, H., Nicolas, J.F. & Peschanski, M. (1993) Catecholaminergic neurons result from intracerebral implantation of embryonal carcinoma cells. *Proc. Natl. Acad. Sci. USA* **90**, 1305–1309.

Wolf, D., Richter-Landsberg, C., Short, M.P., Cepko, C., Breakefield, X.O. (1988) Retrovirus-mediated gene transfer of beta-nerve growth factor into mouse pituitary line AtT-20. *Mol. Biol. Med.* **5**, 43–59.

Wolff, J.A., Fisher, L.J., Xu, L., Jinnah, H.A., Langlais, P.J., Iuvone, P.M., O'Malley, K.L., Rosenberg, M.B., Shimohama, S., Friedman, T. *et al.* (1989) Grafting fibroblasts genetically modified to produce L-dopa in a rat model of Parkinson disease. *Proc. Natl. Acad. Sci. USA* **86**, 9011–9014.

Xi, X.G., Horellou, P., Leroy, C. & Mallet, J. (1993) Retroviral-mediated gene transfer of the porcine choline acetyltransferase: a model to study the synthesis and secretion of acetylcholine in mammalian cells. *Neurochem. Int.* **22**, 511–516.

Xu, X.M., Guenard, U., Kleitman, N. & Bunge, M.B. (1994) Axonal regeneration into Schwann cell-seeded guidance channels grafted into transected adult rat spinal cord. *J. Comp. Neurol.* **351**, 145–160.

Yamada, K.M. (1991) Adhesive recognition sequences. *J. Biol. Chem.* **266**, 12809–12812.

Yeager, A.M., Brennan, S., Tiffany, C., Moser, H.W. & Santos, G.W. (1984) Prolonged survival and remyelination after hematopoietic cell transplantation in the twitcher mouse. *Science* **225**, 1052–1054.

Yee, J.K., Moores, J.C., Jolly, D.J., Wolff, J.A., Respess, J.G. & Friedmann, T. (1987) Gene expression from transcriptionally disabled retroviral vectors. *Proc. Natl. Acad. Sci. USA* **84**, 5197–5201.

Yee, J.-K., Miyanohara, A., LaPorte, P., Bouic, K., Burns, J.C. & Friedmann, T. (1994) A general method for the generation of high-titer, pantropic retroviral vectors: Highly efficient infection of primary hepatocytes. *Proc. Natl. Acad. Sci. USA* **91**, 9564–9568.

Young, J.A.T., Bates, P., Willert, K. & Varmus, H.E. (1990) Efficient incorporation of human CD4 protein into avian leukosis virus particles. *Science* **250**, 1421–1423.

Yu, J.S., Wei, M.X., Chiocca, E.A., Martuza, R.L. & Tepper, R.I. (1993) Treatment of glioma by engineered interleukin 4-secreting cells. *Cancer Res.* **53**, 3125–3128.

Zavada, J. (1972) Pseudotypes of vesicular stomatitis virus with the coat of murine leukaemia and of avian myeloblastosis viruses. *J. General Virol.* **125**, 183–191.

Zhou, H.F., Lee, L.H. & Lund, R.D. (1990) Timing and patterns of astrocyte migration from xenogenic transplants of the cortex and corpus callosum. *J. Comp. Neurol.* **292**, 320–330.

CHAPTER 11

TRANSPLANTATION OF GENETICALLY MODIFIED NON-NEURONAL CELLS IN THE CENTRAL NERVOUS SYSTEM

Marie-Claude Senut,[*†] Steven T. Suhr[*] and Fred H. Gage[*]

[*]*Department of Neurosciences, University of California, San Diego, La Jolla, CA 92093-0627, USA, and* [†]*INSERM U161, 2, rue d'Alésia, 75014 Paris, France*

Table of Contents

11.1 Introduction

Gene therapy is developing as a potent new method for treating both systemic and neurological disease by introducing novel genetic material into the host organism (Gage *et al.*, 1987, 1990; Kawaja *et al.*, 1992a; Mulligan, 1993; Wolff, 1993; Suhr and Gage, 1993, 1994; Friedmann, 1994; Fisher and Ray, 1994; Senut *et al.*, 1995). This therapeutic approach is also both 'energy' and cost efficient as the host organism itself provides the means of production of the therapeutic factor. The delivery of biologically active compounds to highly localized regions of the body is of particular importance to the nervous system which is 'screened' from the general circulation by the blood–brain barrier. This barrier limits the systemic, or blood-borne, delivery of proteins and other macromolecules to the central nervous system (CNS). Consequently, to have an effect on tissues of the CNS, the majority of proteins must also be delivered within the CNS.

As described in other chapters, a variety of viral and non-viral agents have been developed to transfer therapeutic genes into existing neural cells *in situ*. An alternative to this '*in vivo*' gene therapy is the '*ex vivo*' approach, whereby cells are genetically modified in culture and then later transplanted into the CNS resulting in intraparenchymal

Genetic Manipulation of the Nervous System
ISBN 0-12-437165-5

delivery of neuroactive substances to the CNS in chronic high doses. Since non-tumorigenic cells of neural origin are frequently difficult to maintain and transfect *in vitro*, researchers have explored non-neuronal cell types for potential utility in providing therapeutic proteins to the CNS. For many neurologic diseases or disease models examined to date, transplanted non-neuronal cells have proved effective in producing palliative responses via the secretion of cell survival or neuroactive compounds to surrounding neural tissue (see references in Suhr and Gage, 1993; Senut *et al.*, 1995).

In addition to functioning as a 'biological pump', a list of criteria should be satisfied by potential candidate cell types for implantation to the CNS. Although this list may be different according to the chosen cell type and/or targeted disease, it should include the following common concerns.

(1) *Autologous or non-immunogenic.* Barring immunosuppression of the target, cells selected for implantation should be immunocompatible with the host organism.
(2) *Readily obtained, propagated and genetically modified.* Candidate cells should be easily obtained from individuals and they must be amenable to growth, maintenance and transfection in culture. For example, since retroviral vectors require cell division and DNA synthesis for the integration of foreign transgenes into the host genome, donor cells must be capable of continued cell division in culture.
(3) *Survival post-implantation.* Donor cells must survive the grafting procedure, and more importantly, survive *in situ* long enough to yield significant therapeutic value.
(4) *Non-pathogenic in the target location.* The most common pathogenic trait of cells for potential use in gene therapy is tumor formation, although individual cell types may produce locally cytotoxic or detrimental effects within the target region.

Many cell types used in the development of potential *ex vivo* gene therapeutic approaches to neurological dysfunction violate one or more of the above criteria. Although a degree of latitude may be allowed in the development of gene therapies, clinical application will require far more exacting considerations. Several different cell types have, in recent years, been put forward as candidates for *ex vivo* gene therapy. Each has its particular drawbacks and benefits for future application to human gene therapy. Below, many cell types, a wide range of therapeutic applications (excepting application to cancer, described in another chapter) and their attributes are discussed.

11.2 Non-neuronal cells as delivery systems for gene products into the CNS

The choice of cells for transplantation into the CNS depends on a variety of parameters that have been outlined above. Two main classes of non-neuronal donor cell – 1) immortalized lines and 2) primary cells, satisfy some or all of the criteria described. Both classes of cell types have been used for gene transfer and subsequent intracerebral grafting in model systems of human neurological diseases.

11.2.1 Cell lines

Established cell lines were used in the early *ex vivo* somatic gene therapy experiments because of the advantages offered by their use (Tables 1 and 2). Cell lines are easily obtained through sources such as the American Tissue Type Collection for a nominal fee, they survive well, and they can be easily maintained in culture. Because they are mitotically active in culture, they provide a large population of cells that may be readily genetically modified by any number of methods (Ray and Gage, 1992) and from which engineered clonal lines can be readily established. Most importantly, they survive after implantation into the CNS.

The use of genetically modified cell lines to deliver neuroactive substances (trophic factors or transmitters) to the mammalian CNS has been primarily assessed in animal models reproducing deficits observed in human diseases. Neuronal degeneration and neurotransmitter loss are common features of various neurological disorders, such as Alzheimer's and Parkinson's diseases, and animal models have been established which mimic the pathology observed in humans. One example is memory loss in Alzheimer's disease, which is thought to correlate, in part, with the loss of cholinergic neurons (Bartus *et al.*, 1982). Fimbria–fornix-lesioned animals, in which a loss of forebrain cholinergic neurons occurs, have been commonly used to assess the potential protective effects of trophic factors such as nerve growth factor (NGF) upon cholinergic neurons (Hefti, 1986; Williams *et al.*, 1986; Gage *et al.*, 1986; Kromer, 1987). Different fibroblast cell lines (rat 208F, Rat 1, Fisher rat 3T3 and mouse NIH 3T3) have been transduced to express a variety of constitutively secreted gene products, including NGF (Rosenberg *et al.*, 1988; Horellou *et al.*, 1989, 1990; Wolff *et al.*, 1989; Shimohama *et al.*, 1989; Ernfors *et al.*, 1989; Strömberg *et al.*, 1990; Piccardo *et al.*, 1992; Maysinger *et al.*, 1992; Arenas and Persson, 1994). Rosenberg *et al.* (1988) demonstrated that as early as 2 weeks after grafting of NGF-producing 208F fibroblasts in the fimbria–fornix-lesioned rat brain, 92% of septal neurons immunoreactive for choline acetyltransferase (ChAT, the synthesizing enzyme of acetylcholine) survived on the side ipsilateral to the lesion. In contrast, grafts of control 208F fibroblasts had no effect upon cell survival. In addition to the rescue of cholinergic septal neurons, a sprouting of cholinergic fibers in response to the NGF-producing grafts was consistently observed. Strömberg *et al.* (1990) also reported a 90% survival of septal cells positive for acetylcholinesterase (AChE, the degradative enzyme of acetylcholine) 4–6 weeks after grafting of collagen-embedded NGF-producing 3T3 cells in unilaterally fimbria–fornix-lesioned rats.

Other studies have demonstrated increased survival of cholinergic neurons following lesion of the basalocortical pathway by implantation of NGF-producing fibroblast cell lines. NGF-producing 3T3 cells were grafted using both collagen embedding and suspension techniques into the cholinergically denervated frontal and parietal cortices of rats 3 weeks after the excitotoxic lesion of the nucleus basalis magnocellularis (nBM; Ernfors *et al.*, 1989). Grafted cells survived up to 7 weeks within the brain, forming a dense network. Additionally, the density of AChE-positive fibers around the grafts was increased more than 2-fold from the levels seen around control non-NGF-producing

Table 1 Advantages and disadvantages of the various cell types

Cell type	Advantages	Disadvantages
Fibroblast cell lines (R208F, Rat 1, NIH 3T3, Rat 3T3, BHK)	Large numbers of cells available Grow actively in culture Can be easily transfected High levels of transgene expression *in vitro* Survive transplantation Long-term expression of the transgene following transplantation	Tumorigenic Cells require encapsulation before grafting to prevent tumorigenecity Heterologous Require further characterization
Others (RIN, C6)	Large numbers of cells available Grow actively in culture Can be easily transfected Survive transplantation	Tumorigenic Heterologous Require further characterization
Fibroblasts	Can be readily obtained from biopsies Autologous Can be maintained in culture Can be easily transfected Survive well after transplantation	Although cells survive, their number decreases with time Long-term expression of the transgenes is problematic

Table 1 *contd*

Cell type	Advantages	Disadvantages
Myoblasts	Can be readily obtained from biopsies Can be maintained in culture Can be transfected Survive transplantation for long periods of time Long-term expression after grafting	Non autologous, since only fetal cells can be used Not responsive to every method of transfection Require further characterization
Astrocytes	Can be maintained in culture Can be transfected Survive brain transplantation Provide good substrate for neurite outgrowth	Difficult to get autologous cells Long-term expression of the transgenes is problematic Dispersement of the cells post-implantation
Schwann cells	Can be maintained in culture Can be transfected (1%, 5%) Survive brain transplantation Provide good substrate for neurite outgrowth and also myelinate	Difficult to get autologous cells, since only neonatal cells are used Transfection requires chemically induced cell proliferation Require further characterization

Table 2 Transplantation of genetically modified non-neuronal cell lines in the intact and lesioned CNS

Cell type	Method of gene transfer	Genes of interest	Type of graft	Model	Disease	Reference
NIH 3T3	Retroviral	TH	Suspension	Mesostriatal pathway lesion	Parkinson	Horellou *et al.* (1990)
NIH 3T3	Ca Phos	NGF	Mixed suspension	nBM lesion	Alzheimer	Enfors *et al.* (1989)
NIH 3T3	Retroviral	NGF	Collagen	Fimbria–fornix lesion	Alzheimer	Strömberg *et al.* (1990)
Rat 3T3	Retroviral	NGF, NT3, NT4	Suspension	Locus coeruleus lesion	Alzheimer	Arenas and Persson (1994)
Rat 1	Retroviral	NGF	Suspension	Intact brain	—	Frim *et al.* (1994)
Rat 1	Retroviral	NGF	Suspension	Striatal lesion	Huntington	Frim *et al.* (1993a)
Rat 1	Retroviral	BDNF	Suspension	Substantia nigra lesion	Parkinson	Frim *et al.* (1994b)
Rat 1	Retroviral	NGF	Suspension	Cortical lesion	Alzheimer	Piccardo *et al.* (1992)
Rat 1	Retroviral	NGF	Gelfoam	Cortical lesion	Alzheimer	Maysinger *et al.* (1992)
Rat 1	Retroviral	NGF, BDNF, bFGF	Suspension	Striatal lesion	Huntington	Frim *et al.* (1993b)
Rat 1	Retroviral	NGF	Suspension	Striatal lesion	Huntington	Schumacher *et al.* (1991)
R208F	Retroviral	*lacZ*	Suspension	Intact brain	—	Shimohama *et al.* (1989)
R208F	Retroviral	NGF	Encapsulation	Fimbria–fornix lesion	Alzheimer	Hoffman *et al.* (1993)
R208F	Retroviral	TH	Suspension	Nigrostriatal pathway lesion	Parkinson	Wolff *et al.* (1989)
R208F	Retroviral	NGF	Suspension	Fimbria–fornix lesion	Alzheimer	Rosenberg *et al.* (1988)
BHK	Ca Phos	NGF	Encapsulation	Aged intact brain	Alzheimer	Kordower *et al.* (1994)
BHK	Ca Phos	NGF	Encapsulation	Fimbria–fornix lesion	Alzheimer	Winn *et al.* (1994)
BHK	Ca Phos	NGF	Encapsulation	Fornix lesion	Alzheimer	Emerich *et al.* (1994)
Rat C6	Retroviral	*lacZ*	Suspension	Intact brain	—	Shimohama *et al.* (1989)
RIN	Retroviral	TH	Suspension	Mesostriatal pathway lesion	Parkinson	Horellou *et al.* (1990)

3T3 cell grafts. These results suggest that grafts of NGF-producing 3T3 cells were able to enhance cholinergic activity within the denervated cortex. Additional studies have assessed the effects of grafting NGF-producing cells in the devascularizing model of nBM degeneration (Piccardo *et al.*, 1992; Maysinger *et al.*, 1992). Recombinant retroviruses were used to transduce NGF cDNA into the Rat 1 fibroblast line. NGF-producing Rat 1 cells were then placed in a piece of gelfoam that was gently positioned over the exposed cortex. In both studies, NGF-producing cells reduced the atrophy of ChAT- and NGF receptor-immunoreactive neurons within the nBM and maintained neurite density at the level of unlesioned controls. Correspondingly, ChAT activity in the nBM of NGF-treated rats remained within the range normally seen in the intact nBM. In addition to the complete rescue of the cholinergic environment within the nBM, the NGF-producing fibroblasts enhanced ChAT activity approximately 38% above control levels in the cortex (Piccardo *et al.*, 1992) and improved KCl-stimulated release of ACh from cortical terminals (Maysinger *et al.*, 1992). The potent effect of the NGF grafts on the host brain at doses that were over 40-fold lower than those used for intraventricular infusions suggests that the discrete application of NGF to lesioned areas of the brain is an effective technique for preventing cholinergic neurodegenerative changes.

The effects of transplanting NGF-producing cells in rat lesion models of Huntington's disease were assessed in several studies (Schumacher *et al.*, 1991; Frim *et al.*, 1993a,b). Suspensions of control and NGF-producing Rat 1 cells were prepared and injected into the corpus callosum. One week after grafting and 11 days before sacrifice, rats received an excitotoxic lesion of the striatum. Animals with NGF-producing grafts ipsilateral to the lesioned side exhibited an 80% reduction in the maximum lesion cross-sectional area. In contrast, when NGF-producing cells were implanted on the side contralateral to the lesion, a reduction of only 5% of the lesion cross-sectional area was measured (Frim *et al.*, 1993a,b). These findings indicate a distance-dependent protective effect of NGF.

In addition to an examination of the therapeutic potential of NGF within the adult brain, other studies have used the intracerebral grafting paradigm to address more fundamental questions of neurotrophin action. In an attempt to further determine the cellular mechanisms through which NGF exerts its neuroprotective effects, NGF-producing Rat 1 fibroblasts were implanted into the corpus callosum of adult intact rats (Frim *et al.*, 1994a). At 1 week post-grafting, NGF grafts induced an increase in catalase synthesis associated with a decrease in striatal ATP, whereas no changes were observed in glutamate receptor binding or expression of NMDA receptors, suggesting that the protective properties of NGF were mediated through an active enzymatic synthesis.

Besides NGF, the role of other neurotrophins, such as brain-derived neurotrophic factor (BDNF) and neurotrophin 3 (NT3), has also been assessed in the damaged rat brain using grafting of engineered cell lines. The effects of grafting NT3-producing Fischer rat 3T3 fibroblasts into the rat brain prior to a neurotoxic lesion rostral to the locus coeruleus were recently reported. Pre-implantation of NT3-producing cells salvaged 80% of the noradrenergic locus coeruleus (LC) neurons on the side ipsilateral to the lesion, whereas no specific survival effects were observed in animals that received control grafts (Arenas and Persson, 1994). Although no survival effects were described in brains grafted following LC lesion, this study may reveal novel properties of NT3 in the adult brain.

Grafts of BDNF-secreting Rat 1 cells in the dorsal tegmentum of the mesencephalon were shown to increase the survival of dopaminergic nigral neurons in the MPTP-lesioned rat brain (Frim *et al.*, 1994b). BDNF- and fibroblast growth factor (FGF)-producing Rat 1 cells were also grafted near the excitotoxically lesioned rat striatum (Frim *et al.*, 1993b). Whereas FGF-Rat 1 grafts reduced the lesion area by 30%, BDNF-producing Rat 1 cells were found to have no protective effect.

In addition to the strategy of increasing neuronal survival in the injured brain, another option is to restore lost cerebral function through neurotransmitter replacement. This strategy has been evaluated using implants of engineered cell lines in animal models of Parkinson's disease, in which the loss of dopaminergic nigro-striatal neurons results in bradykinesia proportional to the loss of dopamine. Fibroblast cell lines 208F (Wolff *et al.*, 1989) and NIH 3T3 (Horellou *et al.*, 1989, 1990) have been genetically modified to produce tyrosine hydroxylase (TH), the enzyme converting tyrosine to L-dopa, the CNS precursor of dopamine. These cells were found to produce and release L-dopa *in vitro*, whereas no detectable levels could be measured from control cells (Wolff *et al.*, 1989; Horellou *et al.*, 1989, 1990). TH-producing cells were grafted into the caudate of rats previously depleted of their dopaminergic innervation by unilateral 6-OHDA injections in the medial forebrain bundle. When stimulated with apomorphine or amphetamine, 6-OHDA-lesioned rats exhibited a behavioral abnormality characterized by rotations towards the direction of the lesion (Ungerstedt, 1971a,b). Grafts of TH-producing 208F cells survived implantation into the rostral caudate and reduced rotational asymmetry by 30% at 2 weeks post-grafting (Wolff *et al.*, 1989). Similar results were observed using grafts of TH-producing 3T3 cells: 3T3-grafted animals displayed a 50% average decrease in apomorphine-induced rotation 1 week after grafting. In addition, 3T3 cells secreted high amounts of dopa that were shown to be decarboxylated to dopamine in the surrounding host striatal tissue (Horellou *et al.*, 1990).

Besides cell lines of fibroblastic origin, other non-neuronal cell lines have also been used to deliver genes of interest into the CNS of lesioned animals. The rat pancreatic endocrine cell line RIN has also been genetically modified to produce TH. TH-producing RIN cells grafted in the striatum of 6-OHDA-lesioned rats survived implantation, expressed TH immunoreactivity and secreted high levels of L-dopa into the surrounding striatum. In addition, grafted TH-positive RIN cells induced a significant decrease in apomorphine-induced rotations 1 week after grafting in some cases (Horellou *et al.*, 1990). The *lacZ* gene was introduced in the rat glioma cell line C6 by retroviral infection (Shimohama *et al.*, 1989). One week after grafting in the rat caudate and hippocampus, genetically modified cells expressed β-galactosidase.

The studies cited above clearly demonstrate that cell lines genetically modified to produce transgenes *in vitro* continue to produce the gene product *in vivo* in a bioactive form. Despite the fact that encouraging results were observed in most of the *in vivo* studies, genetically modified cell lines, including those discussed above, exhibit invasive growth and result in the formation of tumors within 2–4 weeks post-implantation (Horellou *et al.*, 1990). The ultimate fate of these tumors is either to continue in unchecked growth resulting finally in a lethal brain tumor, or more frequently, to ultimately be completely destroyed by the host immune system. The use of cell lines raises an important problem of immunohistoincompatibility: because of the immunological mismatch between the engineered cells and the implanted host brain, grafted rats have to be treated with immunosuppressive drugs such as cyclosporin to allow graft survival (Ernfors *et al.*, 1989). The consequent uncontrolled tumor growth is the crucial issue that must be overcome before established cell lines can be widely used in intracerebral transplantation.

Several recent studies offer a possible solution by demonstrating that transplantation of encapsulated genetically modified cell lines could be an alternative and efficient strategy to circumvent the previously described problems of tumorization and rejection (Chang *et al.*, 1993). The transplantation of encapsulated cell lines prevents the cells from leaving the designed capsule and thereby avoids risk of *in vivo* tumorization. In addition, by protecting the implanted cells from the host immune system, it allows grafting of heterologous cells without immunosuppressive treatment. Polymer fibers and other biocompatible substrates have been used as devices for cellular encapsulation (Maysinger *et al.*, 1992; Hoffman *et al.*, 1993; Emerich *et al.*, 1994; Winn *et al.*, 1994; Kordower *et al.*, 1994). For example, 208F cells genetically modified to produce NGF were encapsulated in a hollow fiber-based capsule and grafted in the ventricle of fimbria–fornix-lesioned rats (Hoffman *et al.*, 1993). Two weeks after implantation, grafted animals did not show any reaction to the implanted capsules. Additionally, encapsulated cells remained viable and secreted NGF at levels high enough to prevent cholinergic septal neurons from degenerating. In another study, baby hamster kidney (BHK) fibroblast cells were genetically modified to produce NGF and were grafted in polymer capsules into the brain of fimbria–fornix-lesioned rats (Winn *et al.*, 1994). The levels of NGF produced by recovered capsules were 5 ng capsule^{-1} day^{-1} 3 weeks after grafting. Grafted cells survived up to 6 months *in vivo* and were able to prevent the degeneration of lesioned cholinergic septal neurons. Rescue and sprouting of degenerating cholinergic basal forebrain neurons were also recently reported following implantation of NGF-secreting BHK cells in the brain of adult (Emerich *et al.*, 1994) and aged (Kordower *et al.*, 1994) monkeys with fornix lesions.

Although it offers many advantages, grafting encapsulated engineered cell lines still presents several risks. One possibility is the eventual rupture of a capsule leading to intracerebral release and growth of the grafted cells. Another risk that cannot be excluded is the progressive blockage of material/nutrient movement due to obstruction of the pores of the polymer. Finally, the effects on mitotically active cells of long-term spatial confinement have yet to be fully evaluated. These potential problems need to be cautiously evaluated before considering the use of genetically modified cell lines for clinical trials.

11.2.2 Primary cells

Studies using engineered cell lines emphasize the need to choose donor cell candidates that have limited growth and good survival in the brain after transplantation. A more progressive strategy for somatic gene therapy in neurological diseases would be to use autografts, implying that the host also functions as the cell donor. This strategy minimizes or negates possible immune rejection and increases graft survival in the syngenic host brain. These unique advantages may be found in the use of primary cells, i.e. cells with conditional proliferative capacity that are taken directly from host tissue. To this end, a number of primary cell types have been evaluated and proven effective as transgene carriers because of their accessibility, their ease of culturing and genetic modification, and most importantly, their ability to deliver therapeutic compounds *in vivo*.

11.2.2.1 Primary fibroblasts

There are several advantages to the use of primary fibroblasts as donor cells for gene transfer into the brain (Tables 1 and 3). First, fibroblasts can be easily obtained from small skin biopsies that can be grown and readily maintained in culture. Second, primary fibroblasts are efficiently transfected by physical, chemical or retroviral techniques (Ray and Gage, 1992; Suhr and Gage, 1993; Senut *et al.*, 1995). The efficiency of retroviral infection and long-term gene expression is similar to that obtained with fibroblast cell lines. Finally, primary fibroblasts survive in a contact-inhibited non-proliferative state within the rodent and non-human primate brain for extended periods of times (up to 18 months in the rat; unpublished data). As shown using electron microscopy, grafts of primary fibroblasts progressively incorporate to the host brain, as assessed by their vascularization and astrocytic infiltration (Kawaja and Gage, 1991; Kawaja *et al.*, 1991). Genetically modified fibroblasts have been inserted into the mammalian brain primarily in two ways: cell suspensions or cells embedded in collagen 'plugs'. Cell suspension is the preferred method into most sites, whereas collagen or gelfoam plugs were generally chosen to be grafted into larger spaces such as wound cavities. The latter method offers two advantages: (1) collagen matrix grafts may allow the formation of bridges between the neurons and their targets, thereby facilitating the reconstitution of the lesioned pathways; (2) primary cells survive well in the collagen embedding.

Intracerebral implantation of genetically modified primary fibroblasts has been used in the development of delivery strategies for neurological disease, but also as a tool in the elucidation of the functional role of various substances in the brain (Table 3). In our laboratory, we have used genetically modified skin fibroblasts to study the role of various neurotrophins on the functioning of the brain. The implantation of NGF-producing primary fibroblasts in the rat striatum was shown to induce a robust sprouting from the host cholinergic neurons that lasted up to 3 weeks post-grafting, suggesting the continuous release of NGF from the grafted cells (Kawaja and Gage, 1991). A sprouting response from the host axonal processes was also observed when

Table 3 Transplantation of genetically modified non-neuronal primary cells in the intact and lesioned CNS

Cell type	Method of gene transfer	Genes of interest	Type of graft	Model	Disease	Reference
Fibroblast	Retroviral	ChAT, *lacZ*	Suspension	Intact brain	—	Fisher *et al.* (1993)
Fibroblast	Retroviral	NGF	Suspension	Intact brain	—	Kawaja *et al.* (1991b)
Fibroblast	Retroviral	NGF, *lacZ*	Suspension	Intact brain	—	Tuszynski *et al.* (1994b)
Fibroblast	Retroviral	NT3, *lacZ*	Suspension	Intact brain, spinal cord	—	Senut *et al.* (1993)
Fibroblast	Retroviral	bFGF, NGF, *lacZ*	Suspension	Intact spinal cord	—	Tuszynski *et al.* (1994a)
Fibroblast	Retroviral	NGF, *lacZ*	Suspension	Intact brain	—	Tuszynski *et al.* (1994b)
Fibroblast	Retroviral	NGF, *lacZ*	Suspension	Aged intact brain	Alzheimer	Chen and Gage (1995)
Fibroblast	Retroviral	NGF, *lacZ*	Suspension	nBM lesion	Alzheimer	Dekker *et al.* (1994)
Fibroblast	Retroviral	NGF	Collagen	Fimbria–fornix lesion	Alzheimer	Kawaja *et al.* (1992b)
Fibroblast	Retroviral	BDNF, NGF, *lacZ*	Suspension	Fimbria–fornix lesion Substantia nigra lesion	Alzheimer Parkinson	Lucidi-Philippi *et al.* (1995)
Fibroblast	Retroviral	NGF, *lacZ*	Mixed suspension	Nigrostriatal pathway lesion	Parkinson	Niijima *et al.* (1995)
Fibroblast	Retroviral	NGF, *lacZ*, bFGF	Mixed suspension	Nigrostriatal pathway lesion	Parkinson	Chalmers *et al.* (1995)
Fibroblast	Retroviral	TH, *lacZ*	Suspension	Nigrostriatal pathway lesion	Parkinson	Fisher *et al.* (1991)
Fibroblast	Retroviral	NGF, *lacZ*	Suspension	Fornix lesion	Alzheimer	Roberts *et al.* (1994)
Fibroblast	Retroviral	bFGF	Mixed suspension	Substantia nigra lesion	Parkinson	Takayama *et al.* (1995)
Astrocyte	Retroviral	Luciferase	Suspension	Intact brain	—	Gage *et al.* (1987)
Astrocyte	Retroviral	NGF	Mixed suspension	Nigrostriatal pathway lesion	Parkinson	Cunningham *et al.* (1991)
Astrocyte	Transgenic	*lacZ*	Suspension	Lesioned brain	Injury	Mucke and Rockenstein (1993)
Schwann cells	Retroviral	*lacZ*	Suspension	Spinal cord lesion	Injury	Langford and Owen (1990)
Myoblast	Lipofection	*lacZ*, Luciferase	Suspension	Intact brain	—	Jiao *et al.* (1992b)
Myoblast	Ca Phos	TH	Suspension	Substantia nigra lesion	Parkinson	Jiao *et al.* (1993)

NGF-producing fibroblasts were grafted into the intact rat spinal cord (Tuszynski *et al.*, 1994a). Monkey and human primary fibroblasts have also been successfully transduced to produce human NGF. When monkey NGF-producing primary fibroblasts were grafted into various brain areas of the intact adult monkey brain, they were robustly penetrated by host cholinergic fibers up to 6 months post-grafting, whereas control grafts exhibited little if any penetration (Tuszynski *et al.*, 1994b). In a co-graft study, NGF-producing fibroblasts were also shown to increase the survival of adrenal chromaffin cells grafted in the adult rat brain and to induce their transdifferentiation as assessed by the expression of neuronal features (Niijima *et al.*, 1995).

The role of neurotrophins other than NGF has also been assessed in the rat brain using somatic gene transfer. Six weeks after grafting, the transplantation of BDNF-producing fibroblasts in the intact substantia nigra elicited sprouting from host axonal processes immunoreactive for TH. In contrast, when these cells were grafted into the striatum or the midbrain of 6-OHDA-lesioned rats, BDNF-producing fibroblasts did not elicit any behavioral or histological effects (Lucidi-Phillipi *et al.*, 1995). Another study analyzing the effects of transplantation of fibroblasts genetically modified to produce human NT3 (hNT3) in various areas of the rat CNS revealed the existence of a regional difference in the responsiveness of the host axons to local hNT3 over-expression. Whereas host axonal processes showed similar penetration into grafts of hNT3-expressing and control cell grafts located in various regions of the brain, hNT3-producing cells grafted in the spinal cord induced an NT3-dependent specific sprouting as early as 2 weeks post-transplantation (Senut *et al.*, 1993, and unpublished data).

The transplantation of engineered primary fibroblasts has also been shown to be effective in animal models of degenerative diseases. The implantation of NGF-producing fibroblasts in the brains of rats with aspirative lesions of the fimbria–fornix prevented cholinergic septal neurons from dying but also promoted reinnervation of the denervated target (hippocampus). Four to 8 weeks after grafting NGF-producing primary fibroblasts in fimbria–fornix-lesioned rats, a survival of 65% (collagen matrix) and 75% (cell suspension) of septal neurons immunoreactive for the low-affinity receptor for NGF was observed by Kawaja *et al.* (1992b). Control grafts supported a survival of only 44% of cells. In these studies using primary fibroblasts as a delivery vehicle for NGF, the survival of cholinergic septal neurons was less than that observed using fibroblast cell lines. The discrepancy in the result observed between cell lines and primary fibroblasts may be in part explained by differences in the number of cells grafted, or the proliferative vs quiescent nature of the two cell types and consequent effects on transcriptional levels.

Grafts of NGF-producing primary fibroblasts induced sprouting from the host axonal processes towards the NGF-producing grafts and resulted in the rescue of cholinergic septal neurons as efficiently as grafted NGF-producing cell lines. In the case of collagen plug transplants, cholinergic fibers were observed growing toward the grafts and reinnervating the dentate gyrus. Kawaja *et al.* (1992b) reported the establishment of a new topographically organized input to the deafferented hippocampus. Using retrograde labeling techniques, these authors demonstrated that the axons growing in the deafferented hippocampus were issued from septal neurons located on the side ipsilateral to the lesion. Electron microscopy revealed synaptic contacts between AChE-positive terminals and dendritic profiles of dentate granular neurons similar to the synaptic organization normally observed in the hippocampus. This demonstrates that NGF-producing primary fibroblasts were not only capable of preventing cholinergic septal neurons from dying but also promoted reinnervation of the denervated target. These results have been reproduced in non-human primate fornix lesion models by demonstration of the rescue of 68% of septal cholinergic neurons in animals that received NGF-producing primary fibroblasts compared to 27% in control animals (Roberts *et al.*, 1994). The protective effects of NGF were also shown to be dependent on the size and location of the NGF-producing grafts.

Another lesion model in which *in vivo* effects of grafting primary fibroblasts have been studied is the basalo-cortical model. It consists of experimentally induced cholinergic degeneration within the nBM following excitotoxic or devascularizing cortical lesions. In a recent study, the effects of grafting NGF-producing fibroblasts upon neurons of the nBM were analyzed in rats with bilateral ibotenic acid lesions (Dekker *et al.*, 1994). Two weeks after excitotoxic lesions of the nBM, lesioned rats received either control or NGF-producing grafts. Six weeks after implantation, lesioned, impaired animals receiving grafts of NGF-producing cells were found to have reduced behavioral impairment relative to controls, as assessed by Morris water maze tasks. In addition, a 25% size increase in the remaining nBM neurons immunoreactive for NGF receptor was observed. Grafts of genetically modified fibroblasts have also been used to assess the role of NGF in aging. NGF-producing cells were grafted in the nBM of behaviorally impaired aged rats (Chen and Gage, 1995). Grafts survived up to 6 weeks post-implantation and continued *in vivo* to express NGF mRNA. In animals that received NGF grafts, the size and the number of basal forebrain cells immunoreactive for NGF receptor were increased. Furthermore, as compared to control grafts, NGF grafts were able to improve age-related memory loss.

As mentioned previously, besides delivering surviving factors into the lesioned brain, another strategy aimed at restoring function through neurotransmitter replacement has been developed. For instance, the implantation of cells engineered to produce acetylcholine (ACh) into damaged areas of the brain may partially replace the missing neurotransmitter and facilitate the restoration of neuronal transmission. ChAT, which synthesizes ACh, has also been introduced into rat primary fibroblasts. Drosophila (d) ChAT-transduced fibroblasts were produced and characterized *in vitro* as producing roughly 700 nmol ACh h^{-1} mg^{-1} protein. Fisher *et al.* (1993) demonstrated that these dCHAT fibroblasts also produced ACh both *in vitro* and *in vivo* after implantation into the rat hippocampus. Microdialysis studies revealed a 4-fold increase in the intrahippocampal level of ACh 7–10 days after grafting in animals receiving ACh-producing grafts as compared to control grafts. Studies with longer time courses will indicate whether this approach may be applicable to ACh replacement for neurodegenerative disorders such as Alzheimer's disease.

Genetically modified primary fibroblasts have also been grafted as a palliative strategy in animal models of Parkinson's disease. Rat primary fibroblasts infected with a retroviral vector containing a copy of human TH produced *in vitro* L-dopa, the precursor of dopamine. When grafted into the striatum of 6-OHDA-lesioned animals, grafts of TH-producing fibroblasts survived more than 10 weeks and elicited a significant decrease (65%) in drug-induced rotation 2 weeks after implantation (Fisher *et al.*, 1991). In contrast, animals implanted with control fibroblasts did not show any significant difference in the rotational behavior. The degree of behavioral improvement decreased significantly between 4 and 8 weeks following implantation, an observation which may be explained by a diminution of the number of grafted fibroblasts and/or by a progressive decrease in transgene expression by the grafted cells. Clinical use of fetal dopaminergic neuronal grafts for patients with Parkinson's disease has been limited because of poor graft survival. Since FGF has a survival effect on dopaminergic neurons, a recent study examined the effects of co-grafting FGF-producing fibroblasts with fetal dopaminergic neurons in the brain of unilaterally 6-OHDA-lesioned rats. Results showed that FGF-producing fibroblasts increased the survival of TH-positive grafted neurons and induced outgrowth of TH-positive fibers. Furthermore, animals that received co-grafts of FGF-producing cells and fetal dopaminergic neurons demonstrated a significant reduction in the drug-induced rotational behavior as compared to controls. Interestingly, better survival of grafted dopaminergic neurons was noticed when the latter were co-grafted with fibroblasts genetically modified to produce, but not release, FGF (Takayama *et al.*, 1995). In a recent study, bFGF-producing fibroblasts were co-grafted with neonatal chromaffin cells in the striatum of 6-OHDA-lesioned rats (Chalmers *et al.*, 1995). Cells producing bFGF induced the transdifferentiation of grafted chromaffin cells as assessed by neurite outgrowth and expression of neuronal proteins. In addition, distally located NGF grafts increased the survival of transdifferentiated cells and induced growth of neurites.

Primary fibroblasts have numerous advantages over the fibroblast cell lines described in the previous section. Problems of immune rejection are minimal or non-existent with primary cells if they are derived from the same graft recipient. In addition, in the grafted state, primary fibroblasts have proven repeatedly to be post-mitotic and non-tumorigenic. Progress in prolonging cell survival and maintaining therapeutic levels of gene expression will make the primary fibroblast an excellent delivery vehicle for foreign genes.

11.2.2.2 Primary myoblasts

Primary muscle myoblasts are a second primary cell type justifiably receiving considerable attention within recent years as a potential donor cell type (Tables 1 and 3). Like fibroblasts, primary myoblasts may be proliferated in culture (Blau and Webster, 1981) but are contact-inhibited and post-mitotic when reimplanted either intramuscularly (Partridge *et al.*, 1989) or into the rodent brain (Jiao and Wolff, 1992). Fetal myoblasts differentiate both *in vitro* and *in vivo* and undergo cell fusion to form the myotubes characteristic of striated muscle (Blau and Webster, 1981). Individual myoblasts transfected or infected at the single-cell stage can also continue to express transgenes subsequent to the differentiation–fusion event (Parmacek *et al.*, 1990). More importantly, the primary benefit of grafted genetically modified myoblasts/tubes is that they may maintain their original level of transgene expression even after prolonged quiescence post-implantation (Barr and Leiden, 1991; Dhawan *et al.*, 1991). This finding differs from most reports of prolonged transgene expression in post-mitotic fibroblasts, which, as described in the previous section, decreases within weeks of implantation. Primary myoblasts harvested from 1–3 day old rats, and transfected with a plasmid expressing either luciferase or β-galactosidase have been grafted into the brain of adult Lewis rats. Implanted cells were observed to survive for >6 months and, although there was a precipitous drop in overall enzymatic activity within the first 2 weeks following grafting, luciferase and β-galactosidase were detectable and stable up to 2 months post-implantation (Jiao *et al.*, 1992). A further demonstration of the utility of myoblasts in the amelioration of nervous system dysfunction is suggested in a study that describes the implantation of myoblasts genetically modified to produce TH into the denervated striatum of adult rats. For the 6 month duration of the study, a 75% decrease in apomorphine-induced rotation was observed (Jiao *et al.*, 1993). No significant reduction in the ameliorating effect was observed over this same time.

Myoblasts, like fibroblasts, are post-mitotic and do not form tumors following implantation to the CNS. Evidence to date suggests that myoblasts may be superior to other primary cell types in their ability to maintain prolonged transgene expression. Whether this prolonged expression results from the multinucleate nature of the myotube, the 'sequestration' of pools of plasmid within the myoblast cytoplasm, or some other mechanism, remains unknown. However, numerous questions persist about the utility of myoblasts in general application to human *ex vivo* gene therapy. Of serious concern is whether or not an adult human individual can function as an autologous source of proliferative and transducible primary myoblasts. It is also not known if such cells would survive for prolonged periods of time in a heterotypic environment. Further research will better define the role of muscle-derived cells in the development of *ex vivo* gene therapies for the CNS.

11.2.2.3 Glial cells

One final non-neuronal cell type that is being examined as a vehicle for gene transfer to the CNS is cells of glial lineage (Tables 1 and 3). Glial cells are normally CNS derived and are therefore predicted to integrate and survive well when engrafted back into their native environment. Grafted astrocytes would also have therapeutic value beyond the immediate function of the transgene that they carry, performing a variety of supportive functions such as myelination within the host.

For glial-mediated gene transfer to the brain, most research has focused on the astrocyte as a transgene carrier. Astrocytes may be cultured from the adult brain and, as a consequence, can be used as autologous grafts. In 1987, Gage *et al.* proposed and demonstrated that astrocytes from 1 day old newborn rats could be explanted, infected with recombinant luciferase retrovirus, selected and reimplanted into the adult brain where they survived well. More recently, Mucke and Rockenstein (1993) used transgenic astrocytes to demonstrate that grafted astrocytes migrated and functioned similarly to their endogenous counterparts. Cultured astrocytes from the brains of mice transgenic for the *lacZ* gene under the control of the GFAP promoter and grafted into the thalamus or striatum of non-transgenics were monitored for migration and GFAP expression by β-galactosidase histochemistry. The grafted cells migrated to specific regions of the brain and displayed increased β-galactosidase expression as a result of local neuronal injury. Astrocytes genetically modified to over-express neurotrophic factors such as NGF (Cunningham *et al.*, 1991) and BDNF (Castillo *et al.*, 1994) have also both been reported. In the study by Cunningham *et al.* (1991), NGF-producing primary cortical astrocytes from newborn rat brains were co-grafted with adrenal chromaffin cells intrastriatally into 6-OHDA-lesioned rats. Adrenal chromaffin cells have been examined clinically as a non-neuronal cell type with potential use as a source of dopamine in a replacement therapy strategy for Parkinson's disease (Madrazo *et al.*, 1987). This therapy has proven largely ineffective in human clinical trials; poor survival of the implanted chromaffin cells in the heterotypic environment of the host striatum may be responsible (Goetz *et al.*, 1989). NGF has been found to provide trophic support for adrenal chromaffin cells in *in vitro* (Aloe and Levi-Montalcini, 1979) and *in vivo* grafting experiments (Strömberg *et al.*, 1985). An approximately 4-fold increase was found in survival of adrenal chromaffin cells co-grafted with NGF-producing astrocytes compared to non-genetically altered controls. Further studies will determine if this increase in chromaffin cell fitness proportionally increases the effectiveness of these cells in supplementing depleted dopamine.

Another glial cell type with potential use in the delivery of therapeutic transgenes is the Schwann cell. Like astrocytes, Schwann cells may also be grown and maintained in culture, and also survive well after reintroduction into the host CNS (Montgomery and Robson, 1993; Brook *et al.*, 1994). Schwann cells are also amenable to infection with recombinant retroviruses. In one report, Schwann cells were infected with a construct producing an antisense myelin-associated glycoprotein (MAG) mRNA that was found to suppress the ability of the infected cells to myelinate axons, supporting the notion that MAG is an important mechanistic component of myelination (Owens and Bunge, 1991). The same group has also produced Schwann cells that over-produce tyrosine hydroxylase as a method of increasing dopamine production in these cells. The L-dopa-secreting cells were found to function indistinguishably from native Schwann cells as regards myelination of neurons *in vitro* (Owens *et al.*, 1991).

Schwann cells have also been used in the development of gene therapy for spinal cord injury. In a preliminary study, genetically modified Schwann cells expressing β-galactosidase were grafted into the lesioned rat spinal cord. Grafted Schwann cells were found both at the site of implantation and also in the region of the demyelinating lesion (Langford and Owens, 1990). This finding suggests that genetically modified Schwann cells may be efficacious in targeting therapeutic factors to areas of spinal cord damage.

11.3 Conclusion

There is no single cell type that satisfies all of the necessary criteria for a universal, non-neuronal donor cell. Two of the main problems that need to be addressed – prolonged gene expression and cell survival/immunogenicity – are general issues that confront the entire field of gene transfer therapy.

Regardless of the method used to transfer genes into the CNS, prolonged transgene expression is not satisfactory. There are several favored explanations for the loss of transgene expression observed in gene transfer experiments. One possibility that has been advanced for direct viral infection strategies, such as the *in vivo* use of recombinant adenovirus, is that cells expressing viral proteins are slowly targeted for destruction by the host immune system (Yang *et al.*, 1994). If foreign protein products expressed intracellularly are eventually displayed and targeted by the immune system, this poses an extremely difficult problem to overcome. In such a scenario, cells that express the greatest amount of protein would be selectively eliminated the quickest, leaving behind only cells with very low or no transgene expression. A second possibility is suppression of transgene expression within the genetically modified cell (Fisher *et al.*, 1991; Scharfmann *et al.*, 1991; Palmer *et al.*, 1991; Schinstine and Gage, 1993). This suppression may occur at either the level of mRNA or protein and may result from inhibition by host-derived or endogenous donor cell factors. In either case, this problem of attenuated transgene expression in the CNS will need to be overcome before long-term gene therapies are practical in the clinical setting.

Significant progress has been made in developing strategies and methods for transplanting genetically modified cells to the brain. In testing basic principles of cell survival and gene expression the limits and promises of this approach to CNS therapy have been revealed. Much more basic information is required to optimize the present approaches. However, with this new information about the cellular and molecular biology of gene transfer to the brain, still more creative innovations will be discovered which will add to the success of this expanding basic and applied field of study.

Acknowledgements

We thank Mary Lynn Gage for her help in the preparation of the manuscript. Supported by NIH grants AG10435, NS28121, AG06088, AG00216, TW04813, NATO grant 58C91FR, the Metropolitan Life Foundation and the Alliance for Aging Research.

References

Aloe, L. & Levi-Montalcini, R. (1979) Nerve growth factor-induced transformation of immature chromaffin cells *in vivo* in sympathetic neurons: effect of antiserum to nerve growth factor. *Proc. Natl. Acad. Sci. USA* **76**, 1246–1250.

Arenas, E. & Persson, H. (1994) Neurotrophin-3 prevents the death of adult central noradrenergic neurons. *Nature* **367**, 368–371.

Barr, E. & Leiden, J.M. (1991) Systemic delivery of recombinant proteins by genetically modified myoblasts. *Science* **254**, 1507–1509.

Bartus, R., Dean, R.L., Beer, C. & Lipa, A.S. (1982) The cholinergic hypothesis of geriatric memory dysfunction. *Science* **217**, 408–417.

Blau, H.M. & Webster, C. (1981) Isolation and characterization of human muscle cells. *Proc. Natl. Acad. Sci. USA* **78**, 5623–5627.

Brook, G.A., Lawrence, J.M., Shah, B. & Raisman, G. (1994) Extrusion transplantation of Schwann cells into the adult rat thalamus induces directional host axon growth. *Exp. Neurol.* **126**, 31–43.

Castillo, B. Jr., del Cerro, M., Breakefield, X.O., Frim, D.M., Barnstable, C.J., Dean, D.O. & Bohn, M.C. (1994) Retinal ganglion cell survival is promoted by genetically modified astrocytes designed to secrete brain-derived neurotrophic factor (BDNF). *Brain Res.* **647**, 30–36.

Chalmers, G.R., Fisher, L.J., Niijima, K., Patterson, P.H. & Gage, F.H. (1995) Adrenal chromaffin cells transdifferentiate in response to basic fibroblast growth factor and show direct outgrowth to a nerve growth factor source *in vivo*. *Exp. Neurol.* (in press).

Chang, P.L., Shen, N. & Westcott, A.J. (1993) Delivery of recombinant gene products with microencapsulated cells *in vivo*. *Hum. Gene Ther.* **4**, 433–440.

Chen, K.S. & Gage, F.H. (1995) Somatic gene transfer of NGF to the aged brain: behavioral and morphological responses. *J. Neurosci.* (in press).

Cunningham, L.A., Hansen, J.T., Short, M.P. & Bohn, M.C. (1991) The use of genetically altered astrocytes to provide nerve growth factor to adrenal chromaffin cells grafted into the striatum. *Brain Res.* **561**, 192–202.

Dekker, A.J., Winkler, J., Ray, J., Thal, L.J. & Gage, F.H. (1994) Grafting of nerve growth factor-producing fibroblasts reduces behavioral deficits in rats with lesions of the nucleus basalis magnocellularis. *Neuroscience* **60**, 299–309.

Dhawan, J., Pan, L.C., Pavlath, G.K., Travis, M.A., Lanctot, A.M. & Blau, H.M. (1991) Systemic delivery of human growth hormone by injection of genetically engineered myoblasts. *Science* **254**, 1509–1512.

Emerich D.F., Winn, S.R., Harper, J., Hammang, J.P., Baetge, E.E. & Kordower, J.H. (1994) Implants of polymer-encapsulated human NGF-secreting cells in the nonhuman primate: rescue and sprouting of degenerating cholinergic basal forebrain neurons. *J. Comp. Neurol.* **349**, 148–164.

Ernfors, P., Ebendal, T., Olson, L., Mouton, P., Strömberg, I. & Persson, H. (1989) A cell line producing recombinant nerve growth factor evokes growth responses in intrinsic and grafted central cholinergic neurons. *Proc. Natl. Acad. Sci. USA* **86**, 4756–4760.

Fisher, L.J. & Ray, J. (1994) *In vivo* and *ex vivo* gene transfer to the brain. *Curr. Opin. Neurobiol.* **4**, 735–741.

Fisher, L.J., Jinnah, H.A., Kale, L.C., Higgins, G.A. & Gage, F.H. (1991) Survival and function of intrastriatally grafted primary fibroblasts genetically modified to produce L-Dopa. *Neuron* **6**, 371–380.

Fisher, L.J., Schinstine, M., Salvaterra, P., Dekker, A.J., Thal, L. & Gage, F.H. (1993) *In vivo* production and release of acetylcholine from primary fibroblasts genetically modified to express choline acetyltransferase. *J. Neurochem.* **61**, 1323–1333.

Friedmann, T. (1994) Gene therapy for neurological disorders. *Trends Genet.* **10**, 210–214.

Frim, D.M., Short, M.P., Rosenberg, W.S., Simspson, J., Breakefield, X.O. & Isacson, O. (1993a) Local protective effects of nerve growth factor-secreting fibroblasts against excitotoxic lesions in the rat striatum. *J. Neurosurg.* **78**, 267–273.

Frim, D.M., Uhler, T.A., Short, M.P., Ezzedine, Z.D., Klagsbrun, M., Breakefield, X.O. & Isacson, O. (1993b) Effects of biologically delivered NGF, BDNF, and bFGF on striatal excitotoxic lesions. *NeuroReport* **4**, 367–370.

Frim, D.M., Wüllner, U., Beal, F.M. & Isacson, O. (1994a) Implanted NGF-producing fibroblasts induce catalase and modify ATP levels but do not affect glutamate receptor binding or NMDA receptor expression in the rat striatum. *Exp. Neurol.* **128**, 172–180.

Frim, D.M., Uhler, T.A., Galpern, W.R., Beal, M.F., Breakefield, X.O. & Isacson, O. (1994b) Implanted fibroblasts genetically engineered to produce brain-derived neurotrophic factor prevent 1-methyl-4-phenylpyridinium toxicity to dopaminergic neurons in the rat. *Proc. Natl. Acad. Sci. USA* **91**, 5104–5108.

Gage, F.H., Wictorin, K., Fisher, W., Williams, L.R., Varon, S. & Björklund, A. (1986) Retrograde cell changes in medial septum and diagonal band following fimbria–fornix transection: quantitative temporal analysis. *Neuroscience* **19**, 241–255.

Gage, F.H., Wolff, J.A., Rosenberg, M.B., Xu, L., Yee, J.-K., Shults, C. & Friedmann, T. (1987) Grafting genetically modified cells to the brain: possibilities for the future. *Neuroscience* **23**, 795–807.

Gage, F.H., Rosenberg, M.B., Tuszynski, M.H., Yoshida, K., Armstrong, D.M., Hayes, R.C. & Friedmann, T. (1990) Gene therapy in the CNS: intracerebral grafting of genetically modified cells. *Prog. Brain Res.* **86**, 205–216.

Goetz, C.G., Olanow, C.W., Koller, W.C., Penn, R.D., Cahill, D., Morantz, R., Stebbins, G., Tanner, C.M., Klawans, H.L., Shannon, K.M., Comella, C.L., Witt, T., Cox, C., Waxman, M. & Gauger, L. (1989) Multicenter study of autologous adrenal medullary transplantation to the corpus striatum in patients with advanced Parkinson's disease. *N. Engl. J. Med.* **320**, 337–341.

Hefti, F. (1986) Nerve growth factor promotes survival of septal cholinergic neurons after fimbrial transections. *J. Neurosci.* **6**, 2155–2162.

Hoffman, D., Breakefield, X.O., Short, M.P. & Aebischer, P. (1993) Transplantation of a polymer-encapsulated cell line genetically engineered to release NGF. *Exp. Neurol.* **122**, 100–106.

Horellou, P., Guibert, B., Leviel, V. & Mallet, J. (1989) Retroviral transfer of a human tyrosine hydroxylase cDNA in various cell lines: regulated release of dopamine in mouse anterior pituitary AtT20 cells. *Proc. Natl. Acad. Sci. USA* **86**, 7233–7237.

Horellou, P., Brundin, P., Kalen, P., Mallet, J. & Björklund, A. (1990) *In vivo* release of DOPA and dopamine from genetically engineered cells grafted to the denervated rat striatum. *Neuron* **5**, 393–402.

Jiao, S. & Wolff, J.A. (1992) Long-term survival of autologous muscle grafts in rat brain. *Neurosci. Lett.* **137**, 207–210.

Jiao, S., Schultz, E. & Wolff, J.A. (1992) Intracerebral transplants of primary muscle cells: a potential 'platform' for transgene expression in the brain. *Brain Res.* **575**, 143–147.

Jiao, S., Gurevich, V. & Wolff, J.A. (1993) Long-term correction of rat model of Parkinson's disease by gene therapy. *Nature.* **362**, 450–453.

Kawaja, M.D. & Gage, F.H. (1991) Reactive astrocytes are substrates for the growth of adult CNS axons in the presence of elevated levels of nerve growth factor. *Neuron* **7**, 1019–1030.

Kawaja, M.D., Fagan, A.M., Firestein, B.L. & Gage, F.H. (1991) Intracerebral grafting of cultured autologous skin fibroblasts into the rat striatum: an assessement of grafts size and ultrastructure. *J. Comp. Neurol.* **307**, 695–706.

Kawaja, M.D., Fisher, L.J., Schinstine, M., Jinnah, H.A., Ray, J., Chen, L.S. & Gage, F.H. (1992a) Grafting genetically modified cells within the rat central nervous system: methodological considerations. In *Neural Transplantation: A Practical Approach* (eds Dunnett, S.B. & Björklund, A.), pp. 21–25. New York, Oxford University Press.

Kawaja, M.D., Rosenberg, M.B., Yoshida, K. & Gage, F.H. (1992b) Somatic gene transfer of nerve growth factor promotes the survival of axotomized septal neurons and the regeneration of their axons in adult rats. *J. Neurosci.* **12**, 2849–2864.

Kordower, J.H., Winn, S.R., Liu, Y.-T., Mufson, E.J., Sladek, J.R., Hammang, J.P., Baetge, E.E. & Emerich, D.F. (1994) The aged monkey basal forebrain: rescue and sprouting of axotomized basal forebrain neurons after grafts of encapsulated cells secreting human nerve growth factor. *Proc. Natl. Acad. Sci. USA* **91**, 10898–10902.

Kromer, L.F. (1987) Nerve growth factor treatment after brain injury prevents neuronal death. *Science* **235**, 214–216.

Langford, L.A. & Owens, G.C. (1990) Resolution of the pathway taken by implanted Schwann cells to a spinal cord lesion by prior infection with a retrovirus encoding β-galactosidase. *Acta Neuropathol.* **80**, 514–520.

Lucidi-Phillipi, C.A., Gage, F.H., Shults, C.W., Jones, K.R., Reichardt, L.F. & Kang, U.J. (1995) Brain-derived neurotrophic factor-transduced fibroblasts: production of BDNF and effects of grafting to the adult rat brain. *J. Comp. Neurol.* **351**, 1–16.

Madrazo, I., Drucker-Colin, R., Diaz, V., Martinez-Mata, J., Torres, C. & Becerril, J.J. (1987) Open microsurgical autograft of adrenal medulla to the right caudate nucleus in two patients with intractable Parkinson's disease. *New Engl. J. Med.* **316**, 831–834.

Maysinger, D., Piccardo, P., Goiny, M. & Cuello, C. (1992) Grafting of genetically modified cells: effects of acetylcholine release *in vivo*. *Neurochem. Int.* **21**, 543–548.

Montgomery, C.T. & Robson, J.A. (1993) Implants of cultured Schwann cells support axonal growth in the central nervous system of adult rats. *Exp. Neurol.* **122**, 107–124.

Mucke, L. & Rockenstein, E.M. (1993) Prolonged delivery of transgene products to specific brain regions by migratory astrocyte grafts. *Transgene* **1**, 3–9.

Mulligan, R.C. (1993) The basic science of gene therapy. *Science* **260**, 926–931.

Niijima, K., Chalmers, G.R., Peterson, D.A., Fisher, L.J., Patterson, P.H. & Gage, F.H. (1995) Enhanced survival and neuronal differentiation of adrenal chromaffin cells cografted into the striatum with NGF-producing fibroblasts. *J. Neurosci.* **15**, 1180–1194.

Owens, G.C. & Bunge, R.P. (1991) Schwann cells infected with a recombinant retrovirus expressing myelin-associated glycoprotein antisence RNA do not form myelin. *Neuron* **7**, 565–575.

Owens, G.C., Johnson, R., Bunge, R.P. & O'Malley K.L. (1991) L-3,4-Dihydroxyphenylalanine synthesis by genetically modified Schwann cells. *J. Neurochemistry* **56**, 1030–1036.

Palmer, T.D., Rosman, G.J., Osborne, W.R.A. & Miller, D. (1991) Genetically modified skin fibroblasts persist long after transplantation but gradually inactivate introduced genes. *Proc. Natl. Acad. Sci. USA* **88**, 1330–1334.

Parmacek, M.S., Bengur, A.R., Vora, A.J. & Leiden, J.M. (1990) The structure and regulation of expression of the murine fast skeletal troponin C gene. Identification of a developmentally regulated, muscle-specific transcriptional enhancer. *J. Biol. Chem.* **265**, 15970–15976.

Partridge, T.A., Morgan, J.E., Coulton, G.R., Hoffmann, E.P. & Kunkel, L.M. (1989) Conversion of mdx myofibres from dystrophic-negative to-positive by injection of normal myoblasts. *Nature* **337**, 176–179.

Piccardo, P., Maysinger, D. & Cuello, A.C. (1992) Recovery of nucleus basalis cholinergic neurons by grafting NGF secretor fibroblasts. *NeuroReport*, **3** 353–356.

Ray, J. & Gage, F.H. (1992) Gene transfer into established and primary fibroblast cell lines: comparison of transfection methods and promoters. *Biotechniques* **13**, 598–603.

Roberts, J., Senut, M.-C., Eberling, J., Gage, F.H. & Tuszynski, M.H. (1994) Intraparenchymal NGF delivery to the adult primate brain by gene transfer prevents cholinergic neuronal degeneration in a distance- and size-dependent manner. *Soc. Neurosci. Abstr.* 368–15.

Rosenberg, M.B., Friedmann, T., Robertson, R.C., Tuszynski, M., Wolff, J.A., Breakefield, X.O. & Gage, F.H. (1988) Grafting genetically modified cells to the damaged brain: restorative effects of NGF expression. *Science* **242**, 1575–1578.

Scharfmann, R., Axelrod, J.H. & Verma, I.M. (1991) Long-term *in vivo* expression of retrovirus-mediated gene transfer in mouse fibroblast implants. *Proc. Natl. Acad. Sci. USA* **88**, 4626–4630.

Schinstine, M. & Gage, F.H. (1993) Factors affecting proviral expression in primary cells grafted into the CNS. In *Molecular and Cellular Approaches to the Treatment of Neurological Diseases* (ed. Waxman, S.G.), pp. 311–323. New York, Raven Press.

Schumacher, J.M., Short, M.P., Hyman, B.T., Breakefield, X.O. & Isacson, O. (1991) Intracerebral implantation of nerve growth factor-producing fibroblasts protects striatum against neurotoxic levels of excitatory amino acids. *Neuroscience* **45**, 561–570.

Senut, M.-C., Liou, N.H., Suhr, S.T., Raymon, H.K., Jones, K.R., Reichardt, L.F. & Gage, F.H. (1993) Expression of hNT3 in rat fibroblasts: implications for intracerebral grafting. *Soc. Neurosc. Abstr.*, 709–713.

Senut, M.-C., Fisher, L.J., Ray, J. & Gage, F.H. (1995) Somatic gene therapy in the brain. In *Methods in Cell Transplantation* (ed Landers, R.G.), pp. 200–214. Boca Raton, FL, CRC Press.

Shimohama, S., Rosenberg, M.B., Fagan, A.M., Wolff, J.A., Short, M.P., Breakefield, X.O., Friedmann, T. & Gage, F.H. (1989) Grafting genetically modified cells into the rat brain: characteristics of *E. coli* beta-galactosidase as a reporter gene. *Mol. Brain Res.* **4**, 271–278.

Strömberg, I., Herrera-Marschitz, M., Ungerstedt, U., Ebendal, T. & Olson, L. (1985) Chronic implants of chromaffin tissue into the dopamine-denervated striatum. Effects of NGF on graft survival, fiber growth and rotational behavior. *Exp. Brain Res.* **60**, 335–349.

Strömberg, I., Wetmore, C.J., Ebendal, T., Ernfors, P., Persson, H. & Olson, L. (1990) Rescue of basal forebrain cholinergic neurons after implantation of genetically modified cells producing recombinant NGF. *J. Neurosci. Res.* **25**, 405–411.

Suhr, S.T. & Gage, F.H. (1993) Gene therapy for neurologic disease. *Arch. Neurol.* **50**, 1252–1268.

Suhr, S.T. & Gage, F.H. (1994) Gene transfer therapy for diseases of the nervous system. In *Neurodegenerative Diseases.* (eds Jolles, G. & Stutzmann, J.M.), pp. 105–212. London, Academic Press.

Takayama, H., Ray, J., Raymon, H.K., Baird, A., Hogg, J., Fisher, L.J. & Gage, F.H. (1995) Basic fibroblast growth factor increases dopaminergic graft survival and function in a rat model of Parkinson's disease. *Nature Med.* **1**, 53–58.

Tuszynski, M.H., Peterson, D.A., Ray, J., Baird, A., Nakahara, Y. & Gage, F.H. (1994a) Fibroblasts genetically modified to produce nerve growth factor induce robust neuritic ingrowth after grafting to the spinal cord. *Exp. Neurol.* **126**, 1–14.

Tuszynski, M.H., Senut, M.-C., Ray, J., Roberts, J., U, H.-S. & Gage, F.H. (1994b) Somatic gene transfer to the adult primate central nervous system: *in vitro* and *in vivo* characterization of cells genetically modified to secrete nerve growth factor. *Neurobiol. Dis.* **1**, 67–78.

Ungerstedt, U. (1971a) Postsynaptic supersensitivity after 6-hydroxydopamine induced degeneration of the nigro-striatal dopamine system. *Acta Physiol. Scand.* **367**, 69–93.

Ungerstedt, U. (1971b) Striatal dopamine release after amphetamine or nerve degeneration revealed by rotational behavior. *Acta Physiol. Scand.* **367** (Suppl), 49–68.

Williams, L.R., Varon, S., Peterson, G.F., Wictorin, K., Björklund, A., Gage, F.H. (1986) Continuous infusion of nerve growth factor prevents basal forebrain neuronal death after fimbria fornix transection. *Proc. Natl. Acad. Sci. USA* **83**, 9231–9235.

Winn, S.R., Hammang, J.P., Emerich, D.F., Lee, A., Palmiter, R.D. & Baetge, E.E. (1994) Polymer-encapsulated cells genetically modified to secrete human nerve growth factor promote the survival of axotomized septal cholinergic neurons. *Proc. Natl. Acad. Sci. USA* **91**, 2324–2328.

Wolff, J.A. (1993) Postnatal gene transfer into the central nervous system. *Curr. Opin. Neurobiol.* **3**, P743–748.

Wolff, J.A., Fisher, L.J., Xu, L., Jinnah, H.A., Langlais, P.J., Iuvone, P.M., O'Malley, K.L., Rosenberg, M.B., Shimohama, S., Friedmann, T. & Gage, F.H. (1989) Grafting fibroblasts genetically modified to produce L-dopa in a rat model of Parkinson's disease. *Proc. Natl. Acad. Sci. USA* **86**, 9011–9014.

Yang, Y., Nunes, F.A., Berensci, K., Gönczöl, E., Engelhardt, J.F. & Wilson, J.M. (1994) Inactivation of E2a in recombinant adenoviruses improves the prospect for gene therapy in cystic fibrosis. *Nature Genet.* **7**, 362–369.

CHAPTER 12

GENETIC ANIMAL MODELS FOR ALZHEIMER'S DISEASE

Rachael L. Neve,[*] Frederick M. Boyce[†] and Mary Lou Oster-Granite[‡]

[*]Department of Genetics, Harvard Medical School and McLean Hospital, Belmont MA 02178, [†]Department of Neurology, Harvard Medical School and Massachusetts General Hospital, Charlestown, MA 02129, and [‡]Division of Biomedical Sciences, University of California, Riverside, CA 92521, USA

Table of Contents

12.1 Introduction

The central feature of Alzheimer's disease is a remarkably specific loss of cognitive functions in affected individuals. This loss is presumably due to the progressive neurodegeneration occurring in the disease, the classic pathological hallmarks of which are the deposition of amyloid in plaques and in the cerebrovasculature, and the appearance of neurofibrillary tangles in a population of neurons. During the last decade, additional phenotypic abnormalities that are equally characteristic of the disease have been described. For example, over-production of lysosomal hydrolases both intracellularly and in plaques (Cataldo and Nixon, 1990), and accumulations of lipofuscin

Genetic Manipulation of the Nervous System
ISBN 0-12-437165-5

(Mann and Yates, 1974; Benowitz *et al.*, 1989) are seen throughout the Alzheimer's disease brain. In addition, synaptic loss in defined regions of the cortex correlates strongly with loss of cognitive function in Alzheimer's disease (Terry *et al.*, 1991).

The molecular events that link these distinct pathological entities remain obscure. Furthermore, it is not clear which are causes and which consequences of the death of neurons in Alzheimer's disease. It has been assumed by the scientific community that any information we can acquire about molecules comprising the pathological markers of Alzheimer's disease will necessarily yield insights into the mechanisms by which nerve cells degenerate in the disease. In particular, the study of β-amyloid (also known as Aβ and A4) and its derivation from the β-amyloid protein precursor (APP) should reveal at least some aspects of the etiology of Alzheimer's disease.

One of the most compelling pieces of evidence that links Alzheimer's disease neurodegeneration and pathology with APP and/or its Aβ-containing derivatives is the early finding that the APP gene is on chromosome 21, for virtually all individuals trisomic for this chromosome will show AD-like neurodegeneration by the age of 40. More recently, the disclosure that specific point mutations in the APP gene cause some forms of familial Alzheimer's disease (reviewed in Mullan and Crawford, 1993) has contributed to the increased interest in the role of APP in the disease. Our laboratory has focused on a specific aspect of APP and its connection with the neuronal destruction of Alzheimer's disease. Most of this work emerged from our initial finding that the carboxyterminal 100 amino acids of the amyloid precursor protein (APP-C100; previously termed AB1 or APP-C104), a normal intermediate in the lysosomal APP processing pathway, is neurotoxic (Yankner *et al.*, 1989). Work from other groups subsequently revealed that APP-C100, or more simply C100, was itself amyloidogenic (Wolf *et al.*, 1990; Maruyama *et al.*, 1990). The neurotoxicity of APP-C100 has since been confirmed by other laboratories (Martin *et al.*, 1991; Fukuchi *et al.*, 1993).

These data formed the crux of our hypothesis that C100 or a similar Aβ-containing fragment of APP may be crucial to the development of amyloid accumulations and neurodegeneration in Alzheimer's disease. To test this hypothesis, we designed and generated two *in vivo* models for the action of APP-C100, one based on transplantation of C100-producing neuronal cells into mouse brains (Neve *et al.*, 1992), the second based on transgenic mice expressing C100 in the brain (Kammesheidt *et al.*, 1992). Mice in both animal models displayed neuropathology that resembles features of Alzheimer's disease neuropathology, lending strength to our hypothesis that APP-C100, or an APP fragment very much like it, may be the perpetrator of neurodegeneration in Alzheimer's disease. We present below data describing the animal models which implicate APP-C100 in both amyloidogenic and neurodegenerative processes. We also discuss the significance of these findings within the context of additional published animal models for Alzheimer's disease.

12.2 APP-C100 is toxic specifically to neurons

We discovered previously (Yankner *et al.*, 1989) that PC12 cells transfected with a retroviral recombinant expressing the carboxyterminal 100 amino acids of APP die when they are induced to acquire a neuronal phenotype with nerve growth factor (NGF), while transfectants expressing exogenous APP-695 or Aβ differentiate normally without cell loss when exposed to NGF. Further, conditioned medium from the C100-transfected cells, but not from control cells transfected with recombinant APP-695 or with vector alone, is toxic to neurons but not to non-neuronal cells in primary rat hippocampal cultures. The fact that the neurotoxicity can be removed from the medium by immunoabsorption with an antibody to APP-C100 (Yankner *et al.*, 1989) suggests that APP-C100 is secreted by the transfected cells and is the primary agent of the neurotoxicity.

12.3 Transplantation of APP-C100-transfected PC12 cells into the brains of newborn mice results in neuropathology

To determine whether APP-C100 causes Alzheimer's disease-like neuropathology *in vivo*, we transplanted PC12 cells transfected with the APP-C100 retroviral recombinant (Yankner *et al.*, 1989), or with the retroviral vector alone, into the hippocampal–cortical region of postnatal day (PD) 1–2 or PD6 mice (Neve *et al.*, 1988). Because we had previously shown (Yankner *et al.*, 1989) that glia protected neurons from C100 toxicity in culture, we elected to carry out the transplantation studies in newborn mice, in which glia have not reached full maturity. We observed clusters of grafted PC12 cells that had clearly proliferated post-transplantation in the hippocampi of both the experimental (C100) and control animals (DO vector only) sacrificed 20 days following transplantation. These clusters had largely disappeared by 2 months post-transplantation, leaving only scars to mark the locations of the transplants. At 4 months post-transplantation, experimental animals exhibited statistically significant cortical atrophy relative to controls; the atrophy was most prominent in regions of the cortex proximal to the site of the transplant. Interestingly, this atrophy was not evident at the earlier age of 2 months, suggesting that the deleterious consequences of the C100 persisted even after the cells making the exogenous C100 had disappeared.

The Alz-50 antibody was originally reported to be selectively immunoreactive with a neuronal antigen in the brains of Alzheimer's disease patients (Wolozin *et al.*, 1986) and was subsequently shown to recognize a modification of the microtubule-associated protein τ that precedes its incorporation into neurofibrillary tangles (Ueda *et al.*, 1990; Lee *et al.*, 1991). Examination of the mouse brains with Alz-50 revealed that at 4 months post-transplantation, some of the mice that had been transplanted with APP-C100-transfected cells stained positively with this antibody in the somatodendritic domain of neurons in the cortex surrounding the transplants, and in dystrophic-appearing fibers in the same region. The staining resembled the Alz-50 immunoreactivity that we observed in the temporal cortex of individuals with Alzheimer's disease, and was not seen at all in control mouse brains.

We had previously shown that F5, an antibody to the carboxyterminal end of the amyloid protein precursor that immunolabels pyramidal cell bodies and neuropil in normal human cortex and hippocampus, displays abnormal neuropil staining in Alzheimer's disease hippocampus (Benowitz *et al.*, 1989). Comparable abnormal organization of the neuropil in the CA2/3 region of the hippocampus ipsilateral to the transplant was revealed by F5 immunostaining in the brains of mice transplanted with C100-transfected PC12 cells but not in the brains of control mice. Adjacent Nissl stained sections did not reveal gross morphological abnormalities in the area of decreased F5 staining, suggesting that the disorganization evident in the immunostained section mainly involves neuropil at 4 months post-transplantation.

A recent report (Fukuchi *et al.*, 1994) of work in which APP-C100-transfected P19 mouse embryonic teratocarcinoma cells were transplanted into mouse brains, confirms and extends our transplantation studies. Fukuchi *et al.* neuronally differentiated two independently generated C100 transfected and control P19 cell lines, and then transplanted them into the hippocampal regions of syngeneic mice (C3H/HeNNia) aged 6–8 weeks. Histological analysis of the transplanted mice 2 and 5 months post-transplantation revealed abnormal distortion and shrinkage of the hippocampus on the sides injected with the C100 transfectants, with preferential loss of hippocampal granule cells. Moreover, Aβ antibodies stained granular deposits in the neuropil and hippocampal blood vessel walls. Taken together, the results described in these studies (Neve *et al.*, 1992; Fukuchi *et al.*, 1994) suggest that the carboxyterminal fragment of APP may cause specific neuropathology and neurodegeneration *in vivo*.

12.4 Transgenic mice expressing APP-C100 in the brain display Alzheimer's disease-like neuropathology

To carry out additional *in vivo* tests of our hypothesis that the neurotoxic C100 fragment may play a role both in amyloidogenesis and in the development of the progressive neurodegeneration and loss of cognitive functions in Alzheimer's disease, we introduced into mice a transgene carrying the sequence for APP-C100 under control of the brain dystrophin promoter (Boyce *et al.*, 1991). A 4.65 kb DNA fragment containing a dystrophin brain promoter–APP-C100 fusion gene with the simian virus 40 early region splice and polyadenylylation sequences, was generated and microinjected into the male pronuclei of fertilized eggs from F2 hybrid mice (C57BL/6×SJL; Kammesheidt *et al.*, 1992). Of the 12 of the 36 offspring found to be positive for the transgene, 10 survived for further analysis. Microinjected DNA usually integrates into a genomic locus as multicopy head-to-tail tandem arrays; we used Southern blot analysis to estimate the transgene copy number in each founder line. Three founders carried more than 20 copies of the transgene, and one appeared to carry only one copy, while the remaining lines carried between three and 10 copies each. Seven of these founders produced transgenic offspring when backcrossed with C57BL/6J, and Southern blot analysis of selected F1 and F2 progeny showed that the transgenic DNA was inherited with no observable rearrangements or changes in copy number.

We analyzed brain, skeletal muscle, heart and liver of F1 progeny for expression of the transgene at the age of 4 months. RNA was isolated from these tissues, and reverse transcription was coupled with polymerase chain reaction (RT-PCR) to amplify a segment of the RNA that was predicted to be expressed from the transgene. All of the tested progeny exhibited predominant expression of the transgene in the brain, with brain expression being an order of magnitude greater than expression in any of the other tissues evaluated. The mice were also shown by immunoblot analysis (Neve and Boyce, 1995) to express the APP-C100 protein preferentially in the brain.

We used an affinity-purified polyclonal antibody, E1–42 (Cummings *et al.*, 1992), raised against a peptide representing the 42 amino acid Aβ fragment, to detect Aβ epitopes in the transgenic mouse brains. This antibody does not immunostain normal APP in the human brain, but is specific for pathologic structures in AD brain and, to a lesser extent, in normal aged brain. These include amyloid cores in the neuritic plaques, as well as diffuse amyloid deposits that are not detectable by conventional histological stains for amyloid such as thioflavin S. Immunostaining of control mouse brains with E1-42 showed that this antibody recognizes normal structures to some degree in the mouse brain: very light homogeneous staining of cell bodies was evident in all non-transgenic mice. In contrast, analysis of transgenic brains from nine different lines with this antibody revealed abnormal intraneuronal Aβ immunoreactivity throughout the brain in all of the transgenic mice. In the hippocampus, in particular, the E1-42-immunopositive cells were most clearly visible in Ammon's horn and in the stratum oriens. The E1-42 immunoreactivity occurred as rounded, compact punctate deposits within neurons. Such intracellular accumulation of Aβ immunoreactivity was particularly prominent within the hilar region of the hippocampus. In fact, close examination of the E1-42 staining in this region showed that the immunoreactivity extended beyond the cell body, with punctate deposits visible in some short, somewhat curly, abnormal-appearing fibers. Such occurrence of E1-42 immunoreactivity in the neuropil may represent a later stage of amyloid deposition or pathology than that seen in the cell body, since it was restricted to those mouse lines with the highest transgene expression and was never seen in controls.

Subsequent scrutiny of the mouse brains with a large number of antibodies to Aβ and peptide fragments of Aβ revealed differences in the ability of these antibodies to detect the Aβ immunoreactivity in the transgenic mice. Antibodies generated to conjugated Aβ(1–28) or Aβ(1–42) preferentially detected the Aβ(1–42) monomer on immunoblots and stained only very lightly the punctate deposits in the C100 transgenic mice. However, antibodies generated to unconjugated Aβ(1–42) were immunoreactive with Aβ aggregates but not with monomeric Aβ on immunoblots, and stained darkly the deposits in the transgenic mice. Therefore, the accumulations of Aβ immunoreactivity in the transgenic mouse brains showed darkest immunostaining with antibodies recognizing aggregated Aβ.

We had previously shown (Neve *et al.*, 1992; and see above) that when we transplanted cells expressing APP-C100 into the mouse hippocampus, immunocytochemical examination of the post-transplantation brains with the antibody F5 (characterized in Benowitz *et al.*, 1989), specific for the carboxyterminal nine amino acids of APP, revealed specific neuropil aberrations in the CA2/3 region of the pyramidal cell layer. Hence, we anticipated that we would detect a similar phenomenon in the transgenic mice. Indeed, staining of the brain sections with F5 disclosed a striking change in the subcellular localization of F5 epitope that was particularly evident in the Ca2/3 region of the hippocampus in the transgenic mice. In control mouse brains, F5 immunoreactivity was distributed relatively homogeneously in the neuronal somata of hippocampal pyramidal cells. In contrast, the F5 immunoreactivity in these cells in the transgenic mice took on a punctate appearance in the cell somata, and often extended markedly into the neuronal processes. The hippocampal pyramidal cells from mice in lines 2, 3 and 7 displayed especially dense reaction product in the neuropil, and the F5 immunoreactivity in the soma took the form of larger accumulations, as if the punctate vesicular reactive material was fusing or aggregating. The appearance of the F5 immunoreactivity in these cells was very similar to that in CA1 hippocampal neurons of Sommer's sector in human Alzheimer's disease brains (Benowitz *et al.*, 1989), and may portend developing pathology.

The question of whether or not the abnormal aggregates of F5 immunoreactivity in intracellular organelles and in neuronal processes, as described above, are derived from the product of the APP-C100 transgene or from endogenous APP is important to answer. They are likely to be at least partially derived from the endogenous APP, because the transgene is expressed at less than one-tenth the abundance of endogenous APP; so that the predominent F5 immunoreactivity is most likely due to staining of the endogenous molecule. Nevertheless, to answer this question more definitively, we have constructed a second transgenic mouse, expressing a cDNA that we term FLAGAPP-C100, in which the coding sequence for the hydrophilic 10 amino acid sequence termed FLAG (Pricket *et al.*, 1989) has been fused onto the aminoterminus of APP-C100. This fusion fragment retains its neurotoxicity in culture. FLAG antibodies will be used to distinguish the transgene produce from the endogenous APP gene product in the transgenic mice expressing FLAGAPP-C100. We will use them, in combination with anti-APP antibodies, to determine the location and state of aggregation of FLAGAPP-C100, and of specific epitopes of endogenous APP, in the transgenic mice.

Thioflavin S histochemistry is commonly used in the analysis of Alzheimer's disease brains, because it is an historic amyloid stain which highlights deposition of Aβ in senile plaques and in the cerebral vasculature. Thioflavin S histochemistry showed no abnormal fluorescence of structures in the control mice, or in six of the transgenic mouse lines. However, the mice from lines 2, 3 and 7, with the highest brain expression of the transgene, displayed prominent thioflavin S fluorescence around blood vessels (Kammesheidt *et al.*, 1992), which we also observed in Alzheimer's disease brains stained with thioflavin S. We inferred from this observation that amyloid had accumulated in or around the cerebral blood vessels of these transgenic animals.

We have detected additional abnormalities in the brains of older APP-C100 transgenic mice (Oster-Granite *et al.*, 1992). Analysis of the brains by electron microscopy (EM) disclosed the fact that numerous granule and pyramidal neuronal somata and processes in 11 month transgenic mice contained secondary lysosomal inclusions which were not observed in non-transgenic littermates or in other regions of the brain, such as the cerebellum and spinal cord. Unusual accumulations of intermediate filaments were seen in the hippocampal formation of APP-C100 mice at 11 months of age. These accumulations, particularly noticeable in axonal processes, were not detectable in axonal processes of cerebellum, cortex or spinal cord of the same animals. None of these abnormalities occurred in aged C57BL/6J and SJL mice, or in aged non-transgenic littermates.

12.5 Expression of APP-C100 in the brain leads to massive neurodegeneration in aged transgenic mice

Electron microscopic analysis of the hippocampal formation in APP-C100 mice between 1 and 2 years of age has exposed a progressive neurodegeneration in the brains of these transgenics. At 1 year of age, a significant number of large lucent pyramidal neurons displayed irregular cytoplasmic accumulations of atypical secondary lysosomes (Oster-Granite *et al.*, 1992). In transgenic mice approximately 18 months of age, many more neurons, including granule cell neurons, contained these abnormal inclusions. Such inclusions were rare in the hippocampal neurons of non-transgenic littermates, although they increased somewhat in frequency with age. In 24–28 month old transgenic mice, numerous degenerating pyramidal and granule cells (Figures 1 and 2) and greatly increased numbers of microglia containing phagocytic debris were evident. Examination of the cytoskeleton of hippocampal neurons of 1 year old transgenic animals revealed increased neurofilament concentration in unmyelinated axons, which became progressively disorganized in 2 year old animals (Figure 2). In addition, the synaptic degeneration which is characteristic of Alzheimer's disease was seen throughout the hippocampi of the older transgenic animals (Figure 3). The brains of control mice displayed some ageing pathology, such as secondary lysosomes (Figure 4), but did not show the massive degeneration that distinguished the brains of the older transgenic mice.

12.6 Role of APP and its fragments in the etiology of Alzheimer's disease: what we have learned from genetic intervention studies and animal models

The dual findings that specific point mutations in the APP gene are associated tightly with some forms of familial Alzheimer's disease (reviewed by Mullan and Crawford,

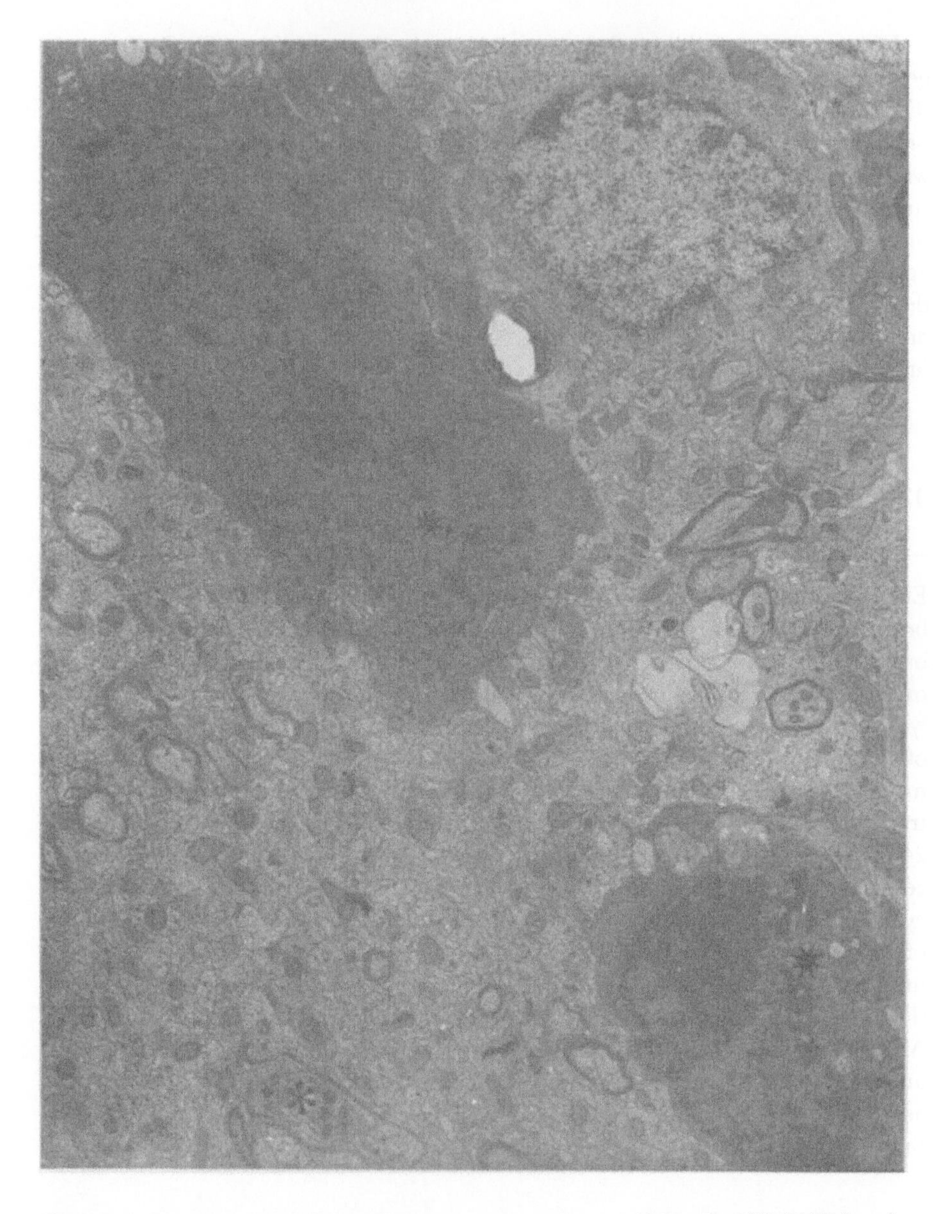

Figure 1 Degenerating dentate granule cell (star) in 2 year old female APP-C100 founder #1005. Observe the condensed cytoplasm and the accumulation of degenerating mitochondria and other cellular inclusions. Axons containing abundant degenerating debris (asterisk) are evident as well are membraneous whorls and inclusions in the neuropil. Note the absence of reactive glial processes in the neuropil. The tissue shown in this and the following figures was fixed with 4% paraformaldehyde, 3% glutaraldehyde in 0.1 M cacodylate buffer. ×8850.

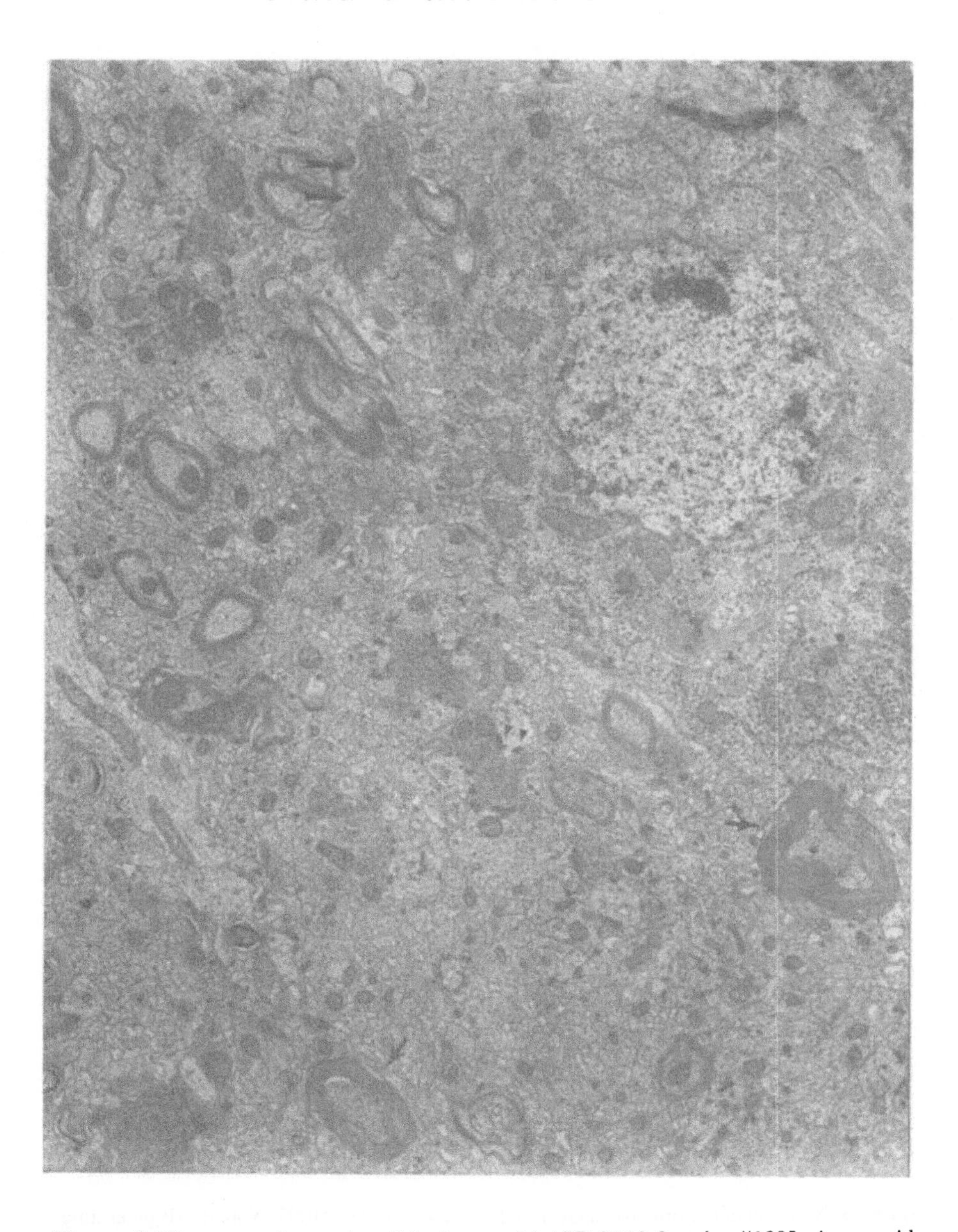

Figure 2 Hippocampal granule cell in 2 year old APP-C100 founder #1005. Axons with increased numbers of neurofilaments are observed in the neuropil (single arrows). Degenerating processes receiving asymmetrical synapses (double arrows) are observed in the neuropil, as well as asymmetric synapses that appear to be intact (double arrowheads). Observe the lack of astrocytic reaction in the neuropil. ×8850.

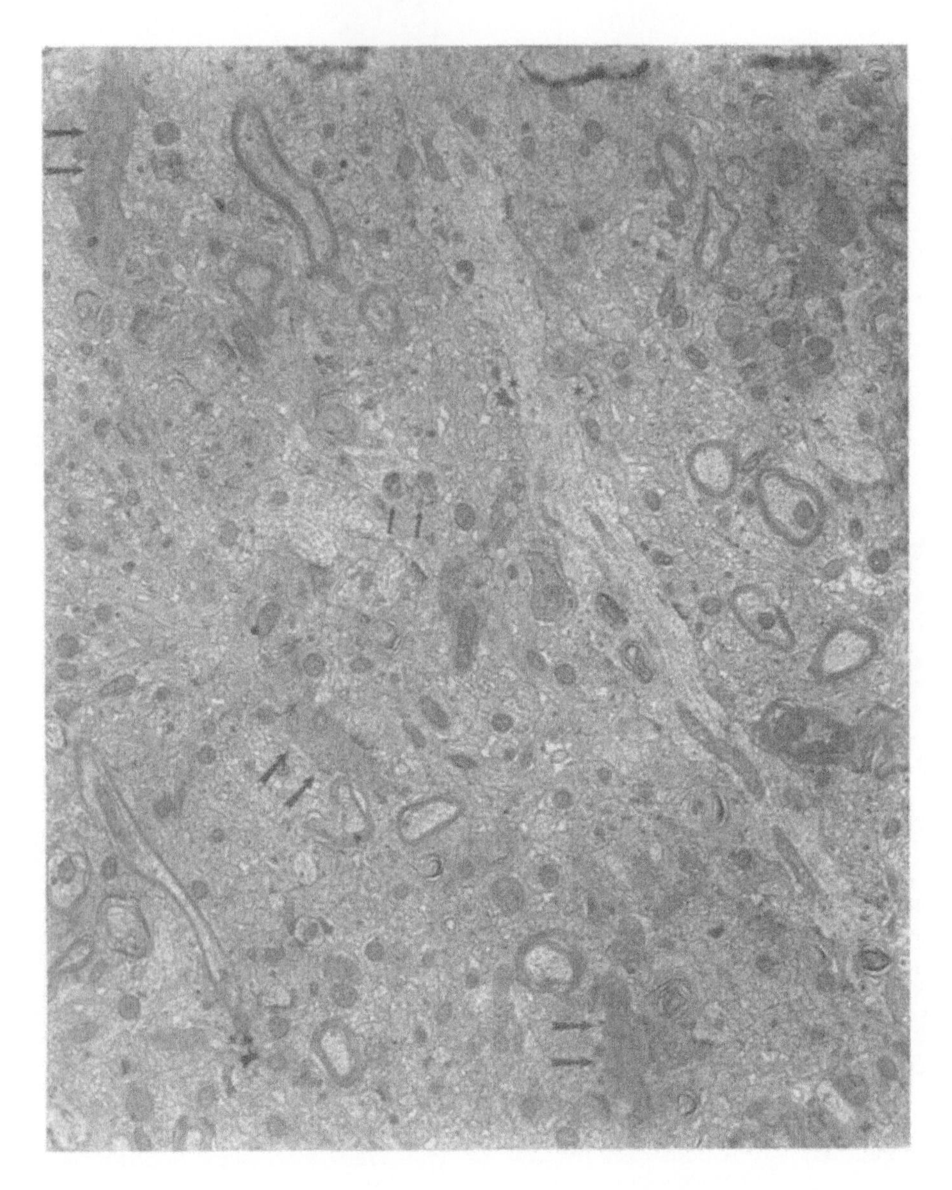

Figure 3 Neuropil of pyramidal cell layer in 2 year old founder #1005. Note the degenerating processes with asymmetric synapses (larger double arrows), the numerous membraneous whorls in the neuropil, and the presence of intact asymmetric synapses on the pyramidal cell dendrite (star). Processes with accumulated debris are also observed (smaller double arrows). ×8700.

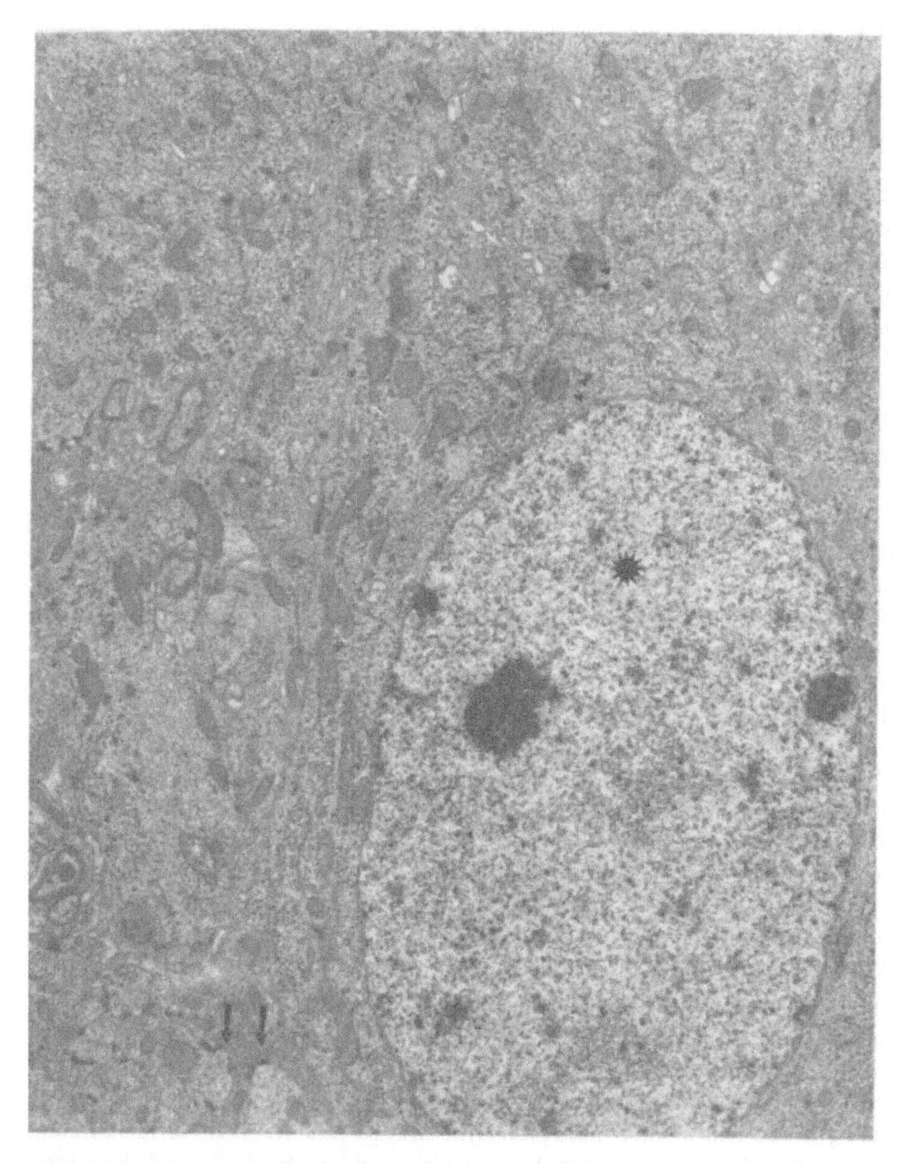

Figure 4 Hippocampal pyramidal cell (star) in control APP-C100 transgenic mouse, aged 1 year. Note the presence of numerous secondary lysosomes within the cytoplasm of this neuron (double arrowheads). Observe that the neuropil contains typical synapses (arrows) and that dendritic processes contain abundant microtubules and intermediate filaments. ×6960.

1993) and that increased deposition of Aβ occurs in the brains of individuals carrying the risk allele of the apolipoprotein E gene (Schmechel *et al.*, 1993) have contributed to the recent increased interest in the role of APP in the disease. An additional and earlier impetus for suspecting an incisive role of APP in the etiology of Alzheimer's disease is that it lies on chromosome 21, which is triplicated in Down's syndrome. Because virtually all individuals with Down's syndrome develop neuropathology almost identical with that of Alzheimer's disease if they live long enough, it has been speculated that simple over-expression of a normal APP gene can cause amyloidosis and neurodegeneration. This speculation has experimental support both *in vivo* and *in vitro*: over-expression of normal APP-751 in transgenic mice leads to extracellular deposition of amyloid in their brains (Quon *et al.*, 1991; see Table 1) and over-expression of APP mRNA in the neuronal cell line P19 causes it to degenerate (Fukuchi *et al.*, 1992; Yoshikawa *et al.*, 1992). In the latter case, amyloidogenic N-terminal-truncated C-terminal fragments accumulated in the cells (Fukuchi *et al.*, 1992), suggesting that over-production of recombinant APP may strain the limit of the 'normal' processing capacity of the cells, with the overflow deflected to an alternative processing pathway in which amyloidogenic fragments are generated. It is not yet known, however, whether APP-C100 or similar fragments do in fact build up in neurons in familial Alzheimer's disease caused by mutations in the APP gene, or in Down's syndrome, in which APP is over-expressed (Neve *et al.*, 1988).

12.6.1 Is APP-C100 neurotoxicity due to Aβ?

Given that APP-C100 is neurotoxic, and that it causes *in vivo* pathology that uncannily resembles in many respects the distinctive neuropathology of Alzheimer's disease, is its neurodegenerative activity a function of the holo-C100 fragment, or of Aβ generated by proteolytic cleavage of C100? Key differences in the mode of neurotoxicity of the two polypeptides suggest that the entire C100 fragment, rather than simply Aβ, comprises the 'smoking gun'. First, published data suggest strongly that the toxicity of APP-C100 is receptor mediated (Kozlowski *et al.*, 1992), whereas the toxicity of Aβ is not (reviewed by Kosik and Coleman, 1992). The binding of APP-C100 to its receptor is specific: Aβ peptides do not complete with C100 for binding to its receptor (M.R. Kozlowski and R.L. Neve, unpublished data). As shown in Kozlowski *et al.* (1992), APP-C100 remains unchanged in size after dissociation from its receptor, providing further confirmation that a proteolytic cleavage product of C100 is not involved in the putatively receptor-mediated neurotoxicity of this molecule. Second, the neurotoxicity of APP-C100 is pH dependent (Kozlowski *et al.*, 1992), whereas that of Aβ fails to show the same pH dependence. Third, APP loses both its neurotoxicity and its ability to bind to its receptor when it aggregates (M.R. Kozlowski *et al.*, unpublished data). In contrast, it is now well established that Aβ is optimally toxic when the peptide is 'aged' and in an aggregated form. Fourth, while we have not rigorously excluded apoptosis as the means of APP-C100 neurotoxity, the manner in which neuronal cells die when exposed to C100 is not overtly apoptotic. We have not observed the cellular 'blebbing' that is characteristic of apoptosis, and we have shown clearly that APP-

Table 1 Animal models for Alzheimer's disease

Experimental manipulation	Phenotype	Reference
Injection of synthetic β(1–40) into rat brain*	Neuronal loss, degenerating neurons and neurites, Alz-50 staining	Kowall *et al.* (1991)
	Impaired spatial learning, decreased choline acetyltransferase activity	Nitta *et al.* (1994)
Injection of Alzheimer β-amyloid cores into rat brain	Neuronal loss, ubiquitin, Alz-50, silver-positive structures	Frautschy *et al.* (1991)
	Phagocytosis and association of β-amyloid with blood vessels	Frautschy *et al.* (1992)
Injection of synthetic β(1–28) into mouse brain	Amnesia for footshock active avoidance training	Flood *et al.* (1991)
Transgenic mice over-expressing APP-751 (NSE promoter)	Extracellular β-amyloid immunoreactive deposits	Quon *et al.* (1991)
Transplants of mouse trisomy 16 hippocampus into normal mouse brain	Immunoreactivity for βA4, ubiquitin, tangle components	Richards *et al.* (1991)
Transgenic mice over-expressing APP-695 (metallothionein promoter)	Impaired spatial memory	Yamaguchi *et al.* (1991)
Transplantation into mouse brain of C100-transfected PC12 cells	Cortical astrophy, Alz-50 staining, neuropil disorganization	Neve *et al.* (1992)
Transgenic mice expressing APP-C100 (dystrophin brain promoter)	Intracellular β/A4 immunoreactivity, thioflavin S-positive blood vessels, accumulation of C-terminal epitopes of APP in lysosomes	Kammesheidt *et al.* (1992)
	Lysosomal inclusions, thickened basement membrane	Oster-Granite *et al.* (1992)
Transgenic mice expressing the entire human APP gene	Normal	Lamb *et al.* (1993) Pearson and Choi (1993) Buxbaum *et al.* (1993)
Transgenic mice expressing APP-C100 (JC viral early gene promoter)	C100 expressed in glia only; normal NMDA and AMPA receptors	Sandhu *et al.* (1993)
Injection of β(1–40)+heparan sulfate proteoglycan into rat brain	Spherical extracellular amyloid deposits	Snow *et al.* (1994)
Transplantation into mouse brain of C100-transfected P19 cells	Neuronal loss, βA4 immunoreactivity in neuropil and vasculature	Fukuchi *et al.* (1994)
Transgenic mice expressing mouse Aβ peptide (NF-L promoter)	Perinuclear zones of unstained cytoplasm in neurons, apoptosis, gliosis	LaFerla *et al.* (1994)

*Giordano *et al.* (1994) later showed that synthetic β(40–1) produced a similar phenotype to β(1–40) when injected into rat brain.

C100 causes lysosomal pathology (Kammesheidt *et al.*, 1992), which is not typical of apoptotic death. In contrast, both in *in vitro* (Loo *et al.*,1993) and the *in vivo* (LaFerla *et al.*, 1995) death of neurons caused by Aβ has been reported to be apoptotic. Interestingly, there have been no reports of apoptotic neuronal death in Alzheimer's disease brain, and a recent paper (Lassmann *et al.*, 1995) provides evidence that degenerating neurons in Alzheimer's disease brain appear to have undergone necrotic rather than apoptotic death.

12.6.2 Genetic animal models for Alzheimer's disease

The most compelling argument for a distinction between C100 and Aβ neurotoxicity is that expression of APP-C100 in the mouse brain *in vivo* causes distinctive Alzheimer's disease-like neuropathology and neurodegeneration that has not been reported in animal models for the *in vivo* action of Aβ (Table 1). Animal models for Aβ neuropathology include non-genetic models utilizing injection of synthetic peptides encoding Aβ or subfragments of it (Kowall *et al.*, 1991; Flood *et al.*, 1991; Nitta *et al.*, 1994; Snow *et al.*, 1994), as well as of Alzheimer's disease β-amyloid cores (Frautschy *et al.*, 1991, 1992). Rodents receiving such injections generally show acute neuropathology in the immediate vicinity of the injection site, and impaired cognitive function has been claimed in some cases. Interestingly, Giordano and colleagues (1994) revealed that synthetic β(40–1), designed to be a negative control peptide, produced a phenotype similar to that caused by β(1–40) when injected into rat brain.

Genetic animal models for Alzheimer's disease include those involving transplantation and also the production of transgenic animals. Embryonic brain tissue from trisomy 16 mice, which do not survive gestation and which are considered a model for human trisomy 21 (Down's syndrome), has been reported to develop some Alzheimer's disease-like pathology when transplanted into normal mouse brain (Richards *et al.*, 1991). APP-C100-transfected cells transplanted into mouse brain caused cortical atrophy, neuronal loss and the appearance of βA4 immunoreactivity and disorganization in the neuropil, much of it occurring at sites distant from the site of the transplantation (Neve *et al.*, 1992; Fukuchi *et al.*, 1994).

Transgenic mice expressing the entire human APP gene were created (Lamb *et al.*, 1993; Pearson and Choi, 1993; Buxbaum *et al.*, 1993) in an effort to duplicate one consequence of trisomy 21, which is the over-expression of APP in the brain (Neve *et al.*, 1988). To date, neuropathological repercussions of such over-expression of the human APP gene in mice have not been reported. However, transgenic mice over-expressing the APP-751 cDNA (Quon *et al.*, 1991) clearly display extracellular β-amyloid immunoreactive deposits that resemble the 'diffuse' amyloid deposits in Alzheimer's disease brain; and transgenic mice over-expressing the APP-695 cDNA are reported to have impairment of spatial memory (Yamaguchi *et al.*, 1991).

Of the many attempts to create a transgenic animal model for Alzheimer's disease by expressing exogenous Aβ cDNA in the mouse brain, that of LaFerla and colleagues (1995) is notable for the altered neuronal morphology that results when exogenous Aβ expression is confined within the cell. These investigators expressed the murine homolog of Aβ under the control of the neuronal promoter for the mouse neurofilament-light (NF-L) gene and discovered morphologically altered cells containing perinuclear zones of unstained cytoplasm throughout the brains of these transgenic mice. Evidence for apoptotic cell death in these brains was reported. Curiously, such abnormalities were not seen when the Aβ transgene product was engineered to be secreted by the neurons in which it was expressed.

12.7 Concluding remarks

If C100 or a similar fragment of APP does indeed account for a significant portion of Alzheimer's disease neuropathology, what are the cellular pathways by which it causes the destruction of cells in the brain? In searching for an answer to this question, it is instructive to recall that synaptic loss and neuronal cell death correlate more strongly with the degree of cognitive impairment in Alzheimer's disease than does amyloid deposition (Terry *et al.*, 1981, 1991; Hamos *et al.*, 1989; DeKosky and Scheff, 1990). Synaptic pathogenesis may even precede the formation of the classical plaques and tangles (Martin *et al.*, 1994). If APP-C100 does, as we hypothesize, play a significant role in Alzheimer's disease neurodegeneration, it must activate (or at least interact with) intracellular pathways that cause the neuronal death, disruption of the neuronal cytoskeleton, loss of cortico-cortical connectivity, synaptic damage and lysosomal abnormalities that are hallmarks of Alzheimer's disease neuropathology. In Figure 5, we present one possible scheme by which neurons may degenerate in this deadly disease. It is clear that the next step is to clone the cDNA for the APP-C100 receptor and to express it in neuronal cells so that we can analyze the signal transduction pathways that are activated when APP-C100 binds to its receptor and begins the cell's deterioration towards neuronal dysfunction.

Acknowledgement

The work described was funded by NIH grant AG12954 (R.L.N.).

References

Benowitz, L.I., Rodriguez, W., Paskevich, P., Mufson, E.J., Schenk, D. & Neve, R.L. (1989) The amyloid precursor protein is concentrated in neuronal lysosomes in normal and Alzheimer disease subjects. *Exp. Neurol.* **106**, 237–250.

Boyce, F.M., Beggs, A.H., Feener, C. & Kunkel, L.M. (1991) Dystrophin is transcribed in brain from a distant upstream promoter. *Proc. Natl. Acad. Sci. USA* **88**, 276–1280.

Buxbaum, J.D., Christensen, J.L., Ruefli, A.A., Greengard, P. & Loring, J.F. (1993) Expression

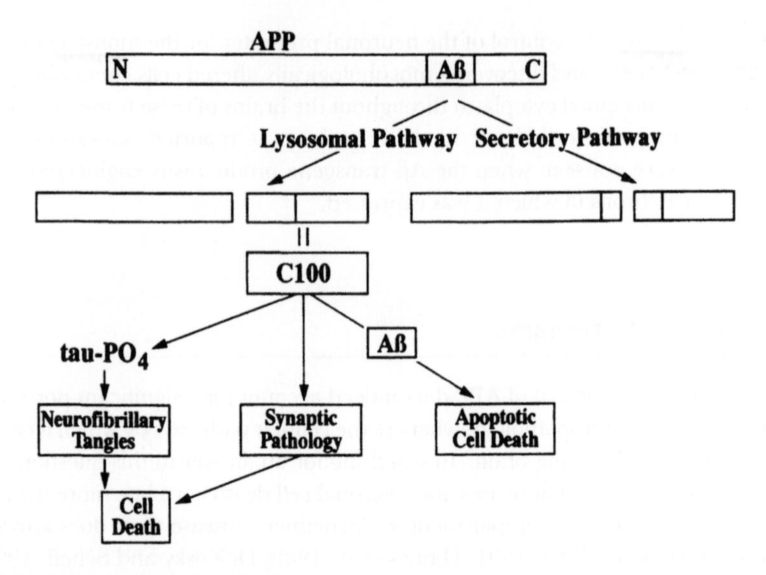

Figure 5 Scheme suggesting the relative roles of APP-C100 and Aβ in Alzheimer's disease. In this scheme, lysosomal processing pathway for APP becomes overloaded with C100 due to any of a number of insults. C100 is directly toxic to neurons; further processing of C100 produces Aβ, which causes secondary toxicity.

of APP in brains of transgenic mice containing the entire human APP gene. *Biochem. Biophys. Res. Commun.* **197**, 639–645.

Cataldo, A.M. & Nixon, R.A. (1990) Enzymatically active lysosomal proteases are associated with amyloid deposits in Alzheimer brain. *Proc. Natl. Acad. Sci. USA* **87**, 3861–3865.

Cummings, B.J., Su, J.H., Geddes, J.W., Van Nostrand, W.E., Wagner, S.L., Cunningham, D.D. & Cotman, C.W. (1992) Aggregation of the amyloid precursor protein within degenerating neurons and dystrophic neurites in Alzheimer's disease. *Neuroscience* **48**, 763–777.

DeKosky, S.T. & Scheff, S.W. (1990) Synapse loss in frontal cortex biopsies in Alzheimer's disease: correlation with cognitive severity. *Ann. Neurol.* **27**, 457–464.

Flood, J.F., Morley, J.E. & Roberts, E. (1991) Amnestic effects in mice of four synthetic peptides homologous to amyloid β protein from patients with Alzheimer disease. *Proc. Natl. Acad. Sci. USA* **88**, 3363–3366.

Frautschy, S.A., Baird, A. & Cole, G.M. (1991) Effects of injected Alzheimer β-amyloid cores in rat brain. *Proc. Natl. Acad. Sci. USA* **88**, 8362–8366.

Frautschy, S.A., Cole, G.M. & Baird, A. (1992) Phagocytosis and deposition of vascular β-amyloid in rat brains injected with Alzheimer β-amyloid. *Am. J. Pathol.* **140**, 1389–1399.

Fukuchi, K., Karnino, K., Deeb, S.S., Smith, A.C., Dang, T. & Martin, G.M. (1992) Overexpression of amyloid precursor protein alters its normal processing and is associated with neurotoxicity. *Biochem. Biophys. Res. Commun.* **182**, 165–173.

Fukuchi, K., Sopher, B., Furlong, C.E., Smith, A.C., Dang, N.T. & Martin, G.M. (1993) Selective neurotoxicity of COOH-terminal fragments of the β-amyloid precursor protein. *Neurosci. Lett.* **154**, 145–148.

Fukuchi, K., Kunkel, D.D., Schwartzkroin, P.A., Kamino, K., Ogburn, C.E., Furlong, C.E. & Martin, G.M. (1994) Overexpression of a C-terminal portion of the β-amyloid precursor protein in mouse brains by transplantation of transformed neuronal cells. *Exp. Neurol.* **127**, 253–264.

Giordano, T., Pan, J.B., Monteggia, L.M., Holzman, T.F., Snyder, S.W., Krafft, G., Ghanbari, H. & Kowall, N.W. (1994) Similarities between β amyloid peptides 1–40 and 40–1: effects on aggregation, toxicity *in vitro*, and injection in young and aged rats. *Exp. Neurol.* **125**, 175–182.

Hamos, J.E., DeGennaro, L.J. & Drachman, D.A. (1989) Synaptic loss in Alzheimer's disease and other dementias. *Neurology* **39**, 355–361.

Kammesheidt, A., Boyce, F.M., Spanoyannis, A.F., Cummings, B.J., Ortegon, M., Cotman, C.W., Vaught, J.L. & Neve, R.L. (1992) Amyloid deposition and neuronal pathology in transgene mice expressing the carboxyterminal fragment of the Alzheimer amyloid precursor in the brain. *Proc. Natl. Acad. Sci. USA* **89**, 10857–10861.

Kosik, K.S. & Coleman, P. (1992) Is β-amyloid neurotoxic? *Neurobiol. Aging* **13**, 535–630.

Kowall, N.W., Beal, M.F., Busciglio, J., Duffy, L.K. & Yankner, B.A. (1991). An *in vivo* model for the neurodegenerative effects of β amyloid and protection by substance P. *Proc. Natl. Acad. Sci. USA* **88**, 7247–7251.

Kozlowski, M.R., Spanoyannis, A.F., Manly, S.P., Fidel, S.A. & Neve, R.L. (1992) The neurotoxic carboxyterminal fragment of the Alzheimer amyloid precursor binds specifically to a neuronal cell surface molecule: pH dependence of the neurotoxicity and the binding. *J. Neurosci.* **12**, 1679–1687.

LaFerla, F.M., Tinkle, B.T., Bieberich, C.J., Haudenschild, C.C. & Jay, G. (1995) The Alzheimer's Aβ peptide induces neurodegeneration and apoptotic cell death in transgenic mice. *Nature Genet.* **9**, 21–29.

Lamb, B.T., Sisodia, S.S., Lawler, A.M., Slunt, H.H., Kitt, C.A., Kearns, W.G., Pearson, P.L., Price, D.L. & Gearhart, J.D. (1993) Introduction and expression of the 400 kilobase *precursor amyloid protein* gene in transgenic mice. *Nature Genet.* **5**, 22–29.

Lassmann, H., Bancher, C., Breitschopf, H., Wegiel, J., Bobinski, M., Jellinger, K. & Wisniewski, H.M. (1995) Cell death in Alzheimer's disease evaluated by DNA fragmentation in situ. *Acta Neuropathol.* **89**, 35–41.

Lee, V.M.-Y., Valin, B.J., Otvos, L. & Trojanowski, J.Q. (1991) A68: a major subunit of paired helical filaments and derivatized forms of normal tau. *Science* **251**, 675–678.

Loo, D.T., Copani, A., Pike, C.J., Whittemore, E.R., Walencewicz, A.J. & Cotman, C.W. (1993) Is β-amyloid neurotoxic? *Proc. Natl. Acad. Sci. USA* **90**, 7951–7955.

Mann, D.M.A. & Yates, P.O. (1974) The quantitative assessment of lipofuscin pigment, cytoplasmic RNA and nucleolar volume in senile dementia. *Neuropathol. Appl. Neurobiol.* **4**, 129–135.

Martin, L.J., Pardo, C.A., Cork, L.C. & Price, D.L. (1994) Synaptic pathology and glial responses to neuronal injury precede the formation of senile plaques and amyloid deposits in the aging cerebral cortex. *Am. J. Pathol.* **145**, 1358–1381.

Martin, T.L., Felsenstein, K.M. & Baetge, E.E. (1991) Cells transfected with amyloid precursor protein DNA fragment AB-1 produce medium toxic to differentiated PC12 cells. *Soc. Neurosci. Abstr.* **17**, 1072.

Maruyama, K., Terakado, K., Usami, M. & Yoshikawa, K. (1990) Formation of amyloid-like fibrils in COS cells overexpressing part of the Alzheimer amyloid protein precursor. *Nature* **347**, 566–569.

Mullan, M. & Crawford, F. (1993) Genetic and molecular advances in Alzheimer's disease. *Trends Neurosci.* **16**, 398–403.

Neve, R.L. & Boyce, F.M. (1995) Construction and analysis of transgenic mice expressing amyloidogenic fragments of the Alzheimer amyloid protein precursor. *Methods Neurosci.* (in press).

Neve, R.L., Finch, E.A. & Dawes, L.R. (1988) Expression of the Alzheimer amyloid precursor gene transcripts in the human brain. *Neuron* **1**, 669–677.

Neve, R.L., Kammesheidt, A. & Hohmann, C.F. (1992) Brain transplants of cells expressing the carboxyterminal fragment of the Alzheimer amyloid protein precursor cause specific neuropathology *in vivo*. *Proc. Natl. Acad. Sci. USA* **89**, 3448–3452.

Nitta, A., Itoh, A., Hasegawa, T., Nabeshima, T. (1994) β-amyloid protein-induced Alzheimer's disease animal model. *Neurosci. Lett.* **170**, 63–66.
Oster-Granite, M.L., Greenan, J. & Neve, R.L. (1992) Structural alterations in the brains of mice transgenic for the carboxy terminal portion of the amyloid precursor protein gene. *Soc. Neurosci.* Abstr. **18**, 730.
Pearson, B.E. & Choi, T.K. (1993) Expression of the human β-amyloid precursor protein gene from a yeast artificial chromosome in transgenic mice. *Proc. Natl. Acad. Sci. USA* **90**, 10578–10582.
Prickett, K.S., Amberg, D.C., Hopp, T.P. (1989) A calcium-dependent antibody for indentification and purification of recombinant proteins. *Biotechniques* **7**, 580–589.
Quon, D., Wang, Y., Catalano, R., Scardina, J.M., Murakami, K. & Cordell, B. (1991) Formation of β-amyloid protein deposits in brains of transgenic mice. *Nature* **352**, 239–241.
Richards, S.-J., Waters, J.J., Beyreuther, K., Masters, C.L., Wischik, C.M., Sparkman, D.R., White, C.L. III, Abraham, C.R. & Dunnett, S.B. (1991) Transplants of mouse trisomy 16 hippocampus provide a model of Alzheimer's disease neuropathology. *EMBO J.* **10**, 297–303.
Sandhu, F.A., Porter, R.H.P., Eller, R.V., Zain, S.B., Salim, M. & Greenamyre, J.T. (1993) NMDA and AMPA receptors in transgenic mice expressing human β-amyloid protein. *J. Neurochem.* **61**, 2286–2289.
Schmechel, D.E., Saunders, A.M., Strittmatter, W.J., Crain, B.J., Hulette, C.M., Joo, S.H., Pericak-Vance, M.A., Goldgaber, D. & Roses, A.D. (1993) Increased amyloid β-peptide deposition in cerebral cortex as a consequence of apolipoprotein E genotype in late-onset Alzheimer disease. *Proc. Natl. Acad. Sci. USA* **90**, 9649–9653.
Snow, A.D., Sekiguchi, R., Nochlin, D., Fraser, P., Kimata, K., Mizutani, A., Arai, M., Schreier, W.A. & Morgan, D.G. (1994) An important role of heparan sulfate proteoglycan (perlecan) in a model system for the deposition and persistence of fibrillar Aβ-amyloid in rat brain. *Neuron* **12**, 219–234.
Terry, R.D., Peck, A., DeTeresa, R., Schechter, R. & Horoupian, D.S. (1981) Some morphometric aspects of the brain in senile dementia of the Alzheimer type. *Ann. Neurol.* **10**, 184–192.
Terry, R.D., Masliah, E., Salmon, D.P., Butters, N., DeTeresa, R., Hill, R., Hansen, L.A. & Katzman, R. (1991) Physical basis of cognitive alterations in Alzheimer's disease: Synapse loss is the major correlate of cognitive impairment. *Ann. Neurol.* **30**, 572–580.
Ueda, K., Masliah, E., Saitoh, T., Bakalis, S.L., Scoble, H. & Kosik, K.S. (1990) Alz-50 recognizes a phosphorylated epitope of tau protein. *J. Neurosci.* **10**, 3295–3304.
Wolf, D., Quon, D., Wang, Y. & Cordell, B. (1990) Identification and characterization of C-terminal fragments of the β-amyloid precursor produced in cell culture. *EMBO J.* **9**, 2079–2084.
Wolozin, B.L., Pruchnicki, A., Dickson, D.W. & Davies, P. (1986) A neuronal antigen in the brains of Alzheimer patients. *Science* **232**, 8–650.
Yamaguchi, F., Richards, S.-J., Beyreuther, K., Salbaum, M., Carlson, G.A. & Dunnett, S.B. (1991) Transgenic mice for the amyloid precursor protein 695 isoform have impaired spatial memory. *NeuroReport* **2**, 781–784.
Yankner, B.A., Dawes, L.R., Fisher, S., Villa-Komaroff, L., Oster-Granite, M.L. & Neve, R.L. (1989) Neurotoxicity of a fragment of the amyloid precursor associated with Alzheimer's disease. *Science* **245**, 417–420.
Yoshikawa, K., Aizawa, T. & Hayashi, Y. (1992) Degeneration *in vitro* of post-mitotic neurons overexpressing the Alzheimer amyloid protein precursor. *Nature* **359**, 64–67.

CHAPTER 13

NEURONAL PRECURSORS IN THE BRAIN OF ADULT MAMMALS: BIOLOGY AND APPLICATIONS

Carlos Lois and Arturo Alvarez-Buylla

The Rockefeller University, 1230 York Avenue, New York, NY 10021, USA

Table of Contents

13.1 Introduction

It is generally assumed that the generation of neurons ceases before or soon after birth, and consequently, that neurons are not replaced in the brain of adult animals (Jessell, 1991; Jacobson, 1991). Indeed, these are considered to be two of the most distinctive characteristics of the nervous system, since its functioning is thought to depend on the constancy of its circuits (Cowan, 1973; Rakic, 1985; Eckenhoff and Rakic, 1988). According to this hypothesis, neurons must be long-lived for their synaptic connections to be able to encode information over long periods of time.

Genetic Manipulation of the Nervous System
ISBN 0-12-437165-5

In the last three decades, however, several studies have shown that new neurons continue to be generated in the brain of adult animals (Altman, 1962; Straznicky and Gaze, 1971; Kaplan and Hinds, 1977; Bayer *et al.*, 1982; Raymond and Easter, 1983; Goldman and Nottebohm, 1983; Paton and Nottebohm, 1984; Lopez-Garcia *et al.*, 1988; Alvarez-Buylla *et al.*, 1990a). Here we will review studies about neural precursors present in the brain of adult mammals and the mechanisms by which they migrate and differentiate *in vitro* and *in vivo.* In addition, we will discuss new experimental opportunities opened up by the presence of putative neural stem cells in the adult mammalian brain: neural stem cells could be used to introduce genes into specific populations of neurons; cell lines could be generated from the adult brain; neural precursors derived from adult brain could be used for transplantation.

13.2 Neurogenesis in adult animals

During development, neural precursors proliferate in the walls of the ventricles and give rise to cells that migrate to their destinations where they differentiate into neurons (reviewed in McConnell, 1988; Rakic, 1990; Jacobson, 1991). This mechanism of neurogenesis, whereby cells that are born in the walls of the ventricles migrate to their targets where they differentiate into neurons, continues in the brain of adult fish (Kirsche, 1967), reptiles (Garcia-Verdugo *et al.*, 1986) and birds (Alvarez-Buylla and Nottebohm, 1988). In adult rodents neurogenesis has been described in the hippocampal dentate gyrus and olfactory bulb. In contrast to the other genera in which adult neurogenesis has been reported in the ventricular zone (VZ) and is followed by long migration, newly generated neurons in mammals were thought to be derived from local precursors that reside within these structures (Altman, 1970; Mair *et al.*, 1982; Bayer, 1983; Kaplan *et al.*, 1985; Corotto *et al.*, 1993; Cameron *et al.*, 1993).

In the hippocampus, Altman and Das (1965) identified [^{3}H]thymidine ([^{3}H]T)-labeled cells in the basal zone of the dentate gyrus, and suggested that these cells were the precursors of the newly formed granule neurons. A later study by Kaplan suggested that in the hilus of the dentate gyrus, proliferating neuroblasts could be identified by combining electron microscope analysis and thymidine autoradiography. Surprisingly, this work suggested that some of the dividing neuronal precursors received synapses (Kaplan and Bell, 1984). The cells that proliferate within the hippocampus of adult mammals, however, have not been characterized molecularly. Moreover, it is not known whether they can give rise to other cell types in addition to the granule neurons of the dentate gyrus.

The location of the precursors of the newly generated neurons in the olfactory bulb of adult animals is still more controversial. Whereas Altman suggested in 1969 that newly generated neurons in the olfactory bulb originated from cells that proliferated in the subventricular zone (SVZ) of the lateral ventricle, other studies indicated that they were derived from neuronal precursors located within the olfactory bulb (Mair *et al.*, 1982; Corotto *et al.*, 1993).

13.3 Cell proliferation in the brain of adult rodents: the subventricular zone cells

In mammals, proliferation in the VZ ceases in most parts of the neural tube around birth. Cells located in the ventricular layer are thought to differentiate into ependymal cells that line the ventricles of the brain and the central canal of the spinal cord. There is, however, a region of the ventricular walls in the forebrain, the SVZ (also called the subependymal zone or layer), where cell proliferation continues after birth. Shortly after birth the murine SVZ is a band of tissue about 20 cells thick around the anterior part of the lateral ventricles. In newborn rodents, cells from the SVZ gave rise to glial cells (LeVine and Goldman, 1988; Levison and Goldman, 1993) and small interneurons (Altman, 1970; Fentress *et al.*, 1981; Luskin, 1993) in different structures of the anterior forebrain. As animals mature, the SVZ thins out and in adult animals it persists as a layer 2–5 cells thick in the lateral wall of the lateral ventricle, overlying the ependymal cells. SVZ cells continue to proliferate in the brain of adult mice (Smart, 1961), rats (Privat and Leblond, 1972), cats (Altman, 1963), dogs (Blakemore and Jolly, 1972) and monkeys (McDermott and Lantos, 1990). Several competing hypotheses have been proposed regarding the fate of SVZ cells in the brain of adult mammals.

13.4 The controversial fate of the subventricular zone cells in adult mammals

Proliferating cells in the SVZ of adult mammals were described as early 1912 by Allen, but the fate of their progeny has remained a matter of controversy ever since. Smart studied the proliferation of cells in the SVZ using [^{3}H]T and concluded that the fate of SVZ cells is to die after mitosis (Smart, 1961). Altman initially suggested that SVZ cells migrate laterally into the white matter and corpus callosum adjacent to the lateral ventricle, where they would differentiate into glial cells (Altman, 1966). In contrast, in a later study he suggested that SVZ cells differentiate into neurons in the olfactory bulb, neocortex and caudate nucleus (Altman, 1969). Many other studies, however, support his initial proposition and suggest that the fate of dividing SVZ cells in adult mice is to differentiate into glial cells in the neighboring parenchyma (Lewis, 1968a; Privat and Leblond, 1972; Imamopto *et al.*, 1978; Morshead and Van der Kooy, 1992).

More recently, the hypothesis that death is the fate of proliferating cells in the SVZ was revived (Morshead and Van der Kooy, 1992) by a report in which multiple methods [bromodeoxyuridine (BrdU), [^{3}H]T and retroviral labeling] were used to trace these cells. They confirmed previous reports and showed that several days after [^{3}H]T or BrdU administration, no labeled cells are found in the SVZ. The authors reasoned that the inability to detect labeled cells several days after administration of these markers could be due to two possibilities: (1) the progeny of the dividing cells dies; or (2) the label is diluted by continuous division as it is split between the progeny. A marker of cell division that is genetically inherited and thus not diluted by successive divisions can resolve this problem. Genetically engineered retroviruses can be used for this purpose (reviewed in Sanes, 1989; Cepko *et al.*, 1993). Morshaed and van der Kooy reported that the progeny of dividing SVZ cells do not differentiate but instead die shortly after mitosis. This conclusion was based on two observations: (1) The number of retrovirally labeled SVZ cells does not increase in the SVZ with increasing survival times; and (2) no retrovirally labeled cells can be detected in the vicinity of the SVZ. The first observation suggested that the progeny of dividing SVZ cells does not remain in the SVZ, since otherwise labeled cells should accumulate. The second observation suggested that the progeny of SVZ cells does not migrate to the neighboring parenchyma (striatum, cortex and corpus callosum) in contrast to earlier hypotheses (Altman, 1966; Paterson *et al.*, 1973). In this study, however, only frontal sections between the rostral genu of the corpus callosum and the start of the third ventricle were examined. Thus, retrovirally labeled cells could not have been detected should they have migrated to other brain regions not included in the histological sections examined.

13.5 Subventricular zone cells from the brain of adult mice can differentiate into neurons

The differentiation potential of proliferating SVZ cells from adult mice has recently been investigated *in vitro*. When cells that divide constitutively in the SVZ of adult mice are labeled *in vivo* with [^{3}H]T and then explanted and cultured, labeled neurons and glial cells form *in vitro* (Lois and Alvarez-Buylla, 1993; Kirschenbaum and Goldman, 1995). The vast majority of the new neurons formed *in vitro* are derived from cells that undergo their last division in the brain. In contrast, with these same culture conditions, very few neurons are labeled when [^{3}H]T is added to the culture medium, indicating that the neuronal precursors do not divide *in vitro*. In contrast, glial cells derived from the adult SVZ continue to divide *in vitro* (Figure 1). This indicates that proliferating SVZ cells in the brain of adult mice have the potential to generate neurons and glial cells, raising the possibility that under some conditions these cells may generate new neurons *in vivo*.

The fate of SVZ cells in the brain of adult mice has been recently reinvestigated (Lois and Alvarez-Buylla, 1994). First, by using local [^{3}H]T microinjection, it has been shown that SVZ cells that divide in the lateral ventricle migrate to the olfactory bulb. Second, microinjection of the lipophilic dye, DiI reveals that SVZ cells that migrate to the olfactory bulb differentiate into neurons in the granule cell and periglomerular layers. Third, transplantation of brain tissues from neuron-specific enolase (NSE) gene transgenic mice demonstrates that SVZ cells are able to migrate to the olfactory bulb only when they are present in the SVZ, but not when they are placed in neighboring regions. The NSEp transgenic mice carry the reporter gene β-galactosidase under the

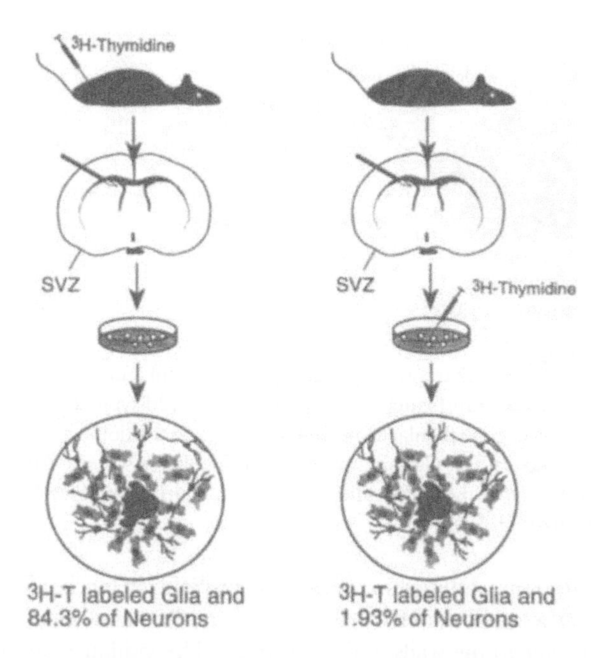

Figure 1 SVZ cells that divide in the walls of the lateral ventricle of adult mice generate neurons and glia *in vitro*. The experiment on the left illustrates that after SVZ cells are labeled by an injection of [^{3}H]T *in vivo*, the majority of the neurons generated *in vitro* are [^{3}H]T-labeled. In contrast, very few neurons are labeled with [^{3}H]T when this marker is added to the culture medium as illustrated on the right.

transcriptional control of the promoter of the NSE gene, and this transgene is only expressed in mature neurons (Forss-Petter *et al.*, 1990). These experiments indicate that in adult mice SVZ cells are neuronal precursors that divide in the walls of the lateral ventricle, migrate long distances (up to 5 mm) and differentiate into neurons in the olfactory bulb (Lois and Alvarez-Buylla, 1994; Figure 2 and Plate 2). Thus, contrary to the established dogma, long distance neuronal migration continues in the brain of adult mammals. The migration of SVZ cells from the lateral ventricle into the olfactory bulb in adult mice is similar, if not identical, to the migration of neuronal precursors recently demonstrated by Luskin (1993) in newborn rats and originally suggested by Altman (1969). Thus, neuronal birth, long range migration and neuronal differentiation in this part of the mammalian brain starts during development and persists into adult life. This long migration suggests that the primary stem cells for adult neurogenesis are located on the walls of the lateral ventricle, similarly to what happens during development.

In accordance with the initial nomenclature used by Altman (1969) we refer to this migratory pathway as the rostral migratory stream (RMS). It is interesting that cells in the RMS move parallel to the orientation of the walls of the ventricles. This form

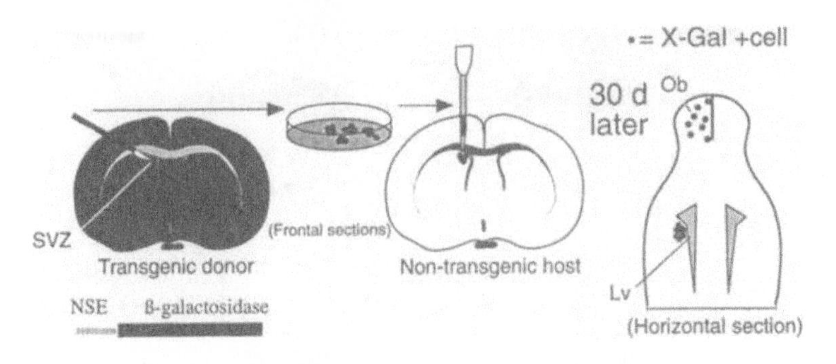

Figure 2 SVZ cells from adult transgenic mice migrate 4–6 mm and differentiate into neurons in the olfactory bulb after transplantation into the SVZ of wild-type adult hosts. Transgenic donors carry the reporter gene β-galactosidase under the transcriptional control of the neuron-specific enolase gene promoter. Cells that migrate to the olfactory bulb express the β-galactosidase gene indicating that they have differentiated into neurons (see Plate 2). Modified from Lois and Alvarez-Buylla (1994).

of tangential migration is different in bearing and probably in mechanism to the better studied radial migration (Rakic, 1972; Rakic *et al.*, 1994), that takes young neurons in a path perpendicular to the walls of the ventricles. The guidance and translocation mechanism used by neuronal precursors in the RMS in neonates and adults remains largely unknown. There are two features of the tangential migration of SVZ cells which are in contrast to radial migration. First, cells moving in the RMS apparently migrate as groups of cells aligned in linear structures ('chains'). Migrating cells in these chains maintain a close association to each other until they reach the olfactory bulb where they disperse and migrate radially to reach peripheral layers, Lois *et al.*, 1995. Second, cells migrating in the rostral migratory stream continue to divide during their journey. Interestingly, both these phenomena have also been observed in the migration of neural crest cells (Bronner-Fraser, 1993). It is not known whether proliferative neural precursors migrating along the RMS towards the olfactory bulb retain the self-renewal properties of stem cells.

Migrating cells in the SVZ and along the RMS express high levels of the polysialylated form of neuron cell adhesion protein (PSA-NCAM; Bonfanti and Theodosis, 1994; Rousselot *et al.*, 1995). Interestingly, mice with a targeted mutation of the NCAM gene show clear alterations of the olfactory bulb (Tomasiewicz *et al.*, 1993). In these mice there is an accumulation of proliferating SVZ cells close to the lateral ventricle and a reduction of cells in the RMS within the olfactory bulb. In addition, this alteration is replicated by local injection of the enzyme endosialidase which removes the sialic acid residues from PSA-NCAM (Ono *et al.*, 1994). Together these results indicate that PSA-NCAM probably plays an important role in the migration of neuronal precursors along these pathways in the juvenile and adult mammalian brain. Further understanding of the molecular mechanism of chain migration of neural precursors in the brain of adult mice will be important to direct these cells to other brain regions.

13.6 What is the identity of the neural precursors in the brain of adult mice?

The experiments described above showed that cells that proliferate in the SVZ of adult mice are neural precursors. The SVZ, however, is a complex structure that contains several different cell types (Privat and Leblond, 1972; Blakemore and Jolly, 1972), and it is still not known what the identity of the precursors that reside in the SVZ of adult mice is.

Several studies have reported that certain cells from the CNS from embryos or adult animals can proliferate and generate neurons when cultured *in vitro* under different conditions (Temple, 1989; Cattaneo and McKay, 1990; Reynolds and Weiss, 1992; Reynolds *et al.*, 1992; Richards *et al.*, 1992). In addition, several markers have been described that are expressed by neural precursors in the brain, such as nestin (Bignami *et al.*, 1982; Lendahl *et al.*, 1990) and MASH-1 (Johnson *et al.*, 1990). Interestingly, studies in mammalian (Frederiksen and McKay, 1988; McKay, 1989) and avian (Gray and Sanes, 1992) embryos, and in adult birds (Alvarez-Buylla *et al.*, 1990b), suggest that radial glia may be the neuronal precursors in the brain. The identity and location of the neural stem cells in the brain, however, are still unknown.

Reynolds and Weiss (1992) showed recently that cells present in the brain of adult mice divide *in vitro* when exposed to epidermal growth factor (EGF), and that these dividing cells can give rise to neurons and glia. In this work, the striata of adult mice (including the SVZ) were dissected and enzymatically dissociated into single cell suspensions. Cells were plated onto uncoated culture dishes in the presence of serum-free medium plus EGF. Under these culture conditions, the authors observed that a small percentage of plated cells begin to proliferate. These proliferating cells do not express glial fibrillary acidic protein (GFAP), microtubule associated protein-2 (MAP-2), NSE or neurofilaments (NF), markers characteristic of mature astrocytes and neurons. Most of them, however, express nestin, an intermediate filament known to be present in undifferentiated neuronal precursors (Lendahl *et al.*, 1990). Interestingly, when these proliferating cells are cultured in the presence of horse serum and are allowed to attach to the surface of the culture dish, they differentiate into neurons and glial cells (Figure 3). The neurons derived from EGF-responsive cells are immunopositive for NSE, NF 168 kD, MAP-2, GABA and substance P, whereas the astrocytes are positive for GFAP. By using this culture technique, Reynolds and Weiss have been able to subculture proliferating EGF-responsive cells for over 30 passages (Rubenstein *et al.*, 1994). Since these cells are able to proliferate, self-renew and generate differentiated progeny, they have been classified as neuronal stem cells.

Recent experiments indicate that the SVZ is the region that contains the cells that proliferate *in vitro* and give rise to neurospheres under the influence of EGF (Morshead *et al.*, 1994). First, cells in the SVZ are immunopositive for nestin and the EGF receptor. Second, microdissection and subsequent culture of different regions of the brain indicates that only tissue containing the SVZ gives rise to neurospheres. In addition, this study shows that the ablation of dividing cells in the brain by high doses of [^{3}H]T does not affect the number of neurospheres generated *in vitro*. Hence, it has been

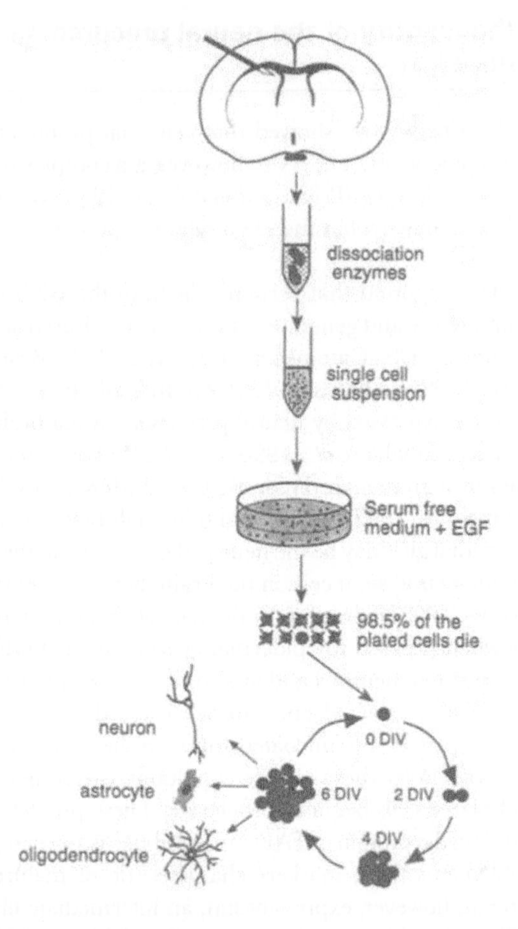

Figure 3 Generation of neurospheres from SVZ of adult mice. Some of the cells that survive proliferate *in vitro* under the influence of EGF and generate clusters of dividing cells called neurospheres. These neurospheres can give rise to neurons, astrocytes and oligodendrocytes when allowed to attach to the substrate. In addition, individual cells can be subcultured to give rise to other neurospheres. Modified from Reynolds and Weiss (1993).

suggested that the cells that constitutively proliferate in the SVZ of adult mice are not the source of EGF-responsive cells *in vitro*. Instead, it has been proposed that a relatively quiescent population of SVZ cells gives rise to the EGF-responsive cells.

Another study shows that cells dissociated from the brain of adult mice proliferate when exposed to basic fibroblast growth factor (bFGF) or EFG, and then differentiate into neurons (Richards *et al.*, 1992). In this study, the authors indicated that proliferating neural precursors can be obtained from the diencephalon, hippocampus, cerebral cortex and tissue containing striatum and septum.

13.7 What is the developmental potential of subventricular zone cells from adult animals?

As of now, SVZ cells from adult mice have only been shown to generate interneurons in the olfactory bulb in the brain (Lois and Alvarez-Buylla, 1994). In addition, the neuronal precursors described by Reynolds and Weiss generate neurons that contain the same neurotransmitters and peptides as do the interneurons derived from SVZ cells *in vivo*. So far, there is no evidence that heterotopic transplantation of SVZ cells results in the alteration of their developmental potential (unpublished observations). Should these cells give rise to other type of neurons it would indicate that SVZ cells are true pluripotential neural stem cells. In addition, it would be of great clinical interest to be able to generate different types of neurons, possessing both specific neurotransmitters and axonal and dendritic connections. These cells could be used to replace lost neurons and to restore function after neurological deficits resulting from disease or injury. Perhaps, given the appropriate environmental conditions, these neural precursors may generate other types of cells in addition to the ones observed spontaneously under natural conditions.

Neuronal loss due to disease or injury may result in severe neurological deficits which in most cases are permanent. Any recovery that follows is believed not to be due to neuronal replacement, but to a compensatory response of the undamaged neurons. In the last two decades, there has been great interest in neuronal cell transplantation as a means to replace cells lost by injury or degeneration (Björklund, 1991). The utility of this technique as a therapy for humans, however, is still a matter of controversy. In addition, there are several problems that limit the applicability of the technique. Most importantly, only brain tissue from embryos has been shown to survive when grafted into the brain of adults (both in experimental animals and in humans). The use of embryos as a source of cells for transplantation has started a controversial polemic, with complex legal and ethical problems. In addition, the small size of the embryonic brain limits the number of cells that can be obtained for each transplant. As described before, cells from the brain of adult animals can migrate and differentiate into neurons when grafted into the brain of adult hosts (Lois and Alvarez-Buylla, 1994). This observation suggests that therapies based on transplantation and adult neurogenesis could be useful for the treatment of neurological diseases. Basically, two approaches can be imagined. First, neural precursors could be expanded *in vitro* and used as a source for transplantation into the brain. Second, neural precursors could be induced to proliferate, migrate and differentiate in the adult brain.

In addition, neural precursors in the adult CNS could be used as vehicles to introduce genes into the brain. There are several features that make the SVZ cells an advantageous system for this purpose. First, because these cells proliferate constitutively, genes can be transferred into them with retroviral vectors (Morshead and Van der Kooy, 1992). Second, SVZ cells are located in a well-defined region of the brain so that they can be transfected by single intracerebral injections. Third, these cells have the ability to migrate to and differentiate into neurons in brain regions distant from their proliferation site. Understanding the mechanisms that regulate the migration and differentiation of these cells may permit the directed delivery of transfected cells to specific regions of the brain and to tailor the phenotype of their progeny to suit specific needs.

Obviously, there are many biological and technical problems that need to be solved before neurogenesis could be used to repair brain damage by cell replacement in adults. Most importantly, neurogenesis has never been reported to occur in the brain of adult humans. In addition, there is no evidence that in any homeotherm vertebrate CNS neuronal replacement can occur after injury. The goal of using adult neurogenesis for brain repair, however, may not be so far out of reach. First, SVZ cells continue to proliferate in the brain of adult primates (Lewis, 1968b: McDermott and Lantos, 1990; Rakic and Kornack, 1993). Second, a recent study indicates that the periventricular temporal lobe from adult humans can generate new neurons *in vitro* (Kirschenbaum *et al.*, 1994). Third, adult lizards can regenerate their cortex after massive surgical or toxic injury by the production of new neurons (Minelli *et al.*, 1977; Font *et al.*, 1991). In adult birds and mammals, newly generated neurons do replace older ones in the avian telencephalon (Nottebohm, 1985; Alvarez-Buylla *et al.*, 1990a; Kirn and Nottebohm, 1993) and in the mammalian olfactory bulb (Kaplan and Hinds, 1977; Kaplan *et al.*, 1985). Several studies, however, have shown that in these same brain regions, experimentally induced neuronal loss is not followed by neuronal replacement (Monti-Graziadei and Graziadei, 1992; and unpublished observations). This observation suggests that the inability to replace brain neurons lost by injury in higher vertebrates may be due to the effect of putative inhibitory substances that prevent the neuronal replacement that occurs under normal conditions (Silver and Steindler, 1990; Laywell and Steindler, 1991). Alternatively, it is possible that the experimental lesions eliminate other elements of the neuropil such as blood vessels, glial cells or synaptic sites that may be necessary for the differentiation and/or survival of the newly generated neurons.

13.8 Conclusions

The brain of adult mammals retains a population of constitutively dividing cells called the SVZ cells. These cells have the capacity to generate neurons both *in vivo* and *in vitro*. Methods have been developed to grow these neuronal precursors *in vitro*. SVZ cells in the brain of adult mice migrate long distances and differentiate into neurons *in vivo*. In addition, these cells can successfully migrate and generate neurons when grafted between adult animals. These characteristics make SVZ cells a useful vehicle to introduce genes into the brain. Thus, SVZ cells in the brain of adult mice are an ideal system in which to study the mechanisms involved in neuronal birth, migration and differentiation. In addition, further developments of this system may permit the use of adult neurogenesis to repair brain damage resulting from disease or injury.

Acknowledgements

Part of the work presented here was supported by NIH grant NS 24478 and a Sinsheimer award.

References

Allen, E. (1912) Cessation of mitosis in central nervous system of the albino rat. *J. Comp. Neurol.* **22**, 547–568.

Altman, J. (1962) Are new neurons formed in the brains of adult mammals? *Science* **135**, 1127–1128.

Altman, J. (1963) Autoradiograhic investigation of cell proliferation in the brains of rats and cats. *Anat. Rec.* **145**, 573–591.

Altman, J. (1966) Proliferation and migration of undifferentiated precursor cells in the rat during postnatal gliogenesis. *Exp. Neurol.* **16**, 263–278.

Altman, J. (1969) Autoradiographic and histological studies of postnatal neurogenesis. IV. Cell proliferation and migration in the anterior forebrain, with special reference to persisting neurogenesis in the olfactory bulb. *J. Comp. Neurol.* **137**, 433–458.

Altman, J. (1970) Postnatal neurogenesis and the problem of neural plasticity. In *Developmental Neurobiology* (ed. Himwich, W.A.) pp. 197–237. Springfield, IL, C.C. Thomas.

Altman, J. & Das, G.D. (1965) Autoradiographic and histological evidence of postnatal hipocampal neurogenesis in rats. *J. Comp. Neurol.* **124**, 319–336.

Alvarez-Buylla, A. & Nottebohm, F. (1988) Migration of young neurons in adult avian brain. *Nature* **335**, 353–354.

Alvarez-Buylla, A., Kirn, J.R. & Nottebohm, F. (1990a) Birth of projection neurons in adult avian brain may be related to perceptual or motor learning. *Science* **249**, 1444–1446.

Alvarez-Buylla, A., Theelen, M. & Nottebohm, F. (1990b) Proliferation 'hot spots' in adult avian ventricular zone reveal radial cell division. *Neuron* **5**, 101–109.

Bayer, S.A. (1983) 3H-Thymidine-radiographic studies of neurogenesis in the rat olfactory bulb. *Exp. Brain Res.* **50**, 329–340.

Bayer, S.A., Yackel, J.W. & Puri, P.S. (1982) Neurons in the rat dentate gyrus granular layer substantially increase during juvenile and adult life. *Science* **216**, 890–892.

Bignami, A., Raju, T. & Dahl, D. (1982) Localization of Vimentin, the nonspecific intermediate filament protein, in embryonal glia and in early differentiating neurons. *Devel. Biol.* **91**, 286–295.

Björklund, A. (1991) Neural transplantation – an experimental tool with clinical possibilities. *Trends Neurosci.* **14**, 319–322.

Blakemore, W.F. & Jolly, D.R. (1972) The subependymal plate and associated ependyma in the dog. An ultrastructural study. *J. Neurocytol.* **1**, 69–84.

Bonfanti, L. & Theodosis, D.T. (1994) Expression of polysialylated neural cell adhesion molecule by proliferating cells in the subependymal layer of the adult rat, in its rostral extension and in the olfactory bulb. *Neuroscience* **62**, 291–305.

Bronner-Fraser, M. (1993) Mechanisms of neural crest cell migration. *BioEssays* **15**, 221–230.

Cameron, H.A., Wooley, C.S., McEwen, B.S. & Gould, E. (1993) Differentiation of newly born neurons and glia in the dentate gyrus of the adult rat. *Neuroscience* **56**, 337–344.

Cattaneo, E. & McKay, R. (1990) Proliferation and differentiation of neuronal stem cells regulated by nerve growth factor. *Nature* **347**, 762–765.

Cepko, C.L., Ryder, E.F., Austin, C.P., Walsh, C. & Fekete, D.M. (1993) Lineage analysis using retrovirus vectors. *Methods Enzymol.* **225**, 933–960.

Corotto, F.S., Henegar, J.A. & Maruniak, J.A. (1993) Neurogenesis persists in the subependymal layer of the adult mouse brain. *Neurosci. Lett.* **149**, 111–114.

Cowan, W.M. (1973) Neuronal death as a regulative mechanism in the control of cell number in the nervous system. In *Development and Aging in the Nervous System* (eds Rockstein, M. & Sussman, F.D.), pp. 19–43. New York, Academic Press.

Eckenhoff, M.F. & Rakic, P. (1988) Nature and fate of proliferative cells in the hippocampal dentate gyrus during the life span of the Rhesus monkey. *J. Neurosci.* **8**, 2729–2747.

Fentress, J.C., Stanfield, B.B. & Cowan, W.M. (1981) Observations on the development of the striatum in mice and rats. *Anat Embryol.* **163**, 275–298.

Font, E., García-Verdugo, J.M., Alcántara, S. & López-García, C. (1991) Neuron regeneration reverses 3-acetylpyridine-induced cell loss in the cerebral cortex of adult lizards. *Brain Res.* **551**, 230–235.

Forss-Petter, S., Danielson, P.E., Catsicas, S., Battenberg, E., Price, J., Nerenberg, M. & Sutcliffe, J.G. (1990) Transgenic mice expressing b-galactosidase in mature neurons under neuron-specific enolase promoter control. *Neuron* **5**, 187–197.

Frederiksen, K. & McKay, R.D.G. (1988) Proliferation and differentiation of rat neuroepithelial precursor cells *in vivo*. *J. Neurosci* **8**, 1144–1151.

Garcia-Verdugo, J.M., Farinas, I., Molowny, A. & Lopez-Garcia. C. (1986) Ultrastructure of putative migrating cells in the cerebral cortex of *Lacerta galloti*. *J. Morphol.* **189**, 189–197.

Goldman, S.A. & Nottebohm, F. (1983) Neuronal production, migration, and differentiation in a vocal control nucleus of the adult female canary brain. *Proc. Natl. Acad. Sci. USA* **80**, 2390–2394.

Gray, G.E. & Sanes, J.R. (1992) Lineage of radial glia in the chicken optic tectum. *Development* **114**, 271.

Imamoto, K., Paterson, J.A. & Leblond, C.P. (1978) Radioautographic investigation of gliogenesis in the corpus callosum of young rats. I. Sequential changes in oligodendrocytes. *J. Comp. Neurol.* **180**, 115–138.

Jacobson, M. (1991) *Developmental Neurobiology*. New York, Plenum Press.

Jessell, T.M. (1991) Reactions of neurons to injury. In *Principles of neural science* (eds. Kandel, E.R., Schwartz, J.H. & Jessell, T.M.) pp. 258–269. New York, Elsevier.

Johnson, J.E., Birren, S.J. & Anderson, D.J. (1990) Two rat homologues of Drosophila achaete-scute specifically expressed in neuronal precursors. *Nature* **346**, 858–861.

Kaplan, M. & Bell, D. (1984) Mitotic neuroblasts in the 9 day old and 11 month old rodent hippocampus. *J. Neurosci.* **4**, 1429–1441.

Kaplan, M.S. & Hinds, J.W. (1977) Neurogenesis in the adult rat: electron microscopic analysis of light radioautographs. *Science* **197**, 1092–1094.

Kaplan, M.S., McNelly, N.A. & Hinds, J.W. (1985) Population dynamics of adult-formed granule neurons of the rat olfactory bulb. *J. Comp. Neurol.* **239**, 117–125.

Kirn, J.R. & Nottebohm, F. (1993) Direct evidence for loss and replacement of projection neurons in adult canary brain. *J. Neurosci.* **13**, 1654–1663.

Kirsche, W. (1967) Uber postembryonale matrixzonen in Gehirn verschiedener Vertebrate und deren Beziehung zur Hirnbauplanlehre. *Z. Mikrosk. Anat. Forsch.* **77**, 313–406.

Kirschenbaum, B. & Goldman, S.A. (1995) Brain-derived neutrophic factor promotes the survival of neurons arising from the adult rat forebrain subependymal zone. *Proc. Natl. Acad. Sci. USA* **92**, 210–214.

Kirschenbaum, B., Nedergaard, M., Preuss, A., Barami, K., Fraser, R.A.R. & Goldman, S.A. (1994) *In vitro* neuronal production and differentiation by precursor cells derived from the adult human forebrain. *Cerebral Cortex* **6**, 576–589.

Laywell, E.D. & Steindler, D.A. (1991) Boundaries and wounds, glia and glycoconjugates: cellular and molecular analyses of developmental partitions and adult brain lesions. *Ann. NY Acad. Sci.* **633**, 122–141.

Lendahl, U., Zimmerman, L.B. & McKay, R.D.G. (1990) CNS stem cells express a new class of intermediate filament protein. *Cell* **60**, 585–595.
LeVine, S.M. & Goldman, J.E. (1988) Embryonic divergence of oligodendrocyte and astrocyte lineages in developing rat cerebrum. *J. Neurosci.* **8**, 3992–4006.
Levison, S.W. & Goldman, J.E. (1993) Both oligodendrocytes and astrocytes develop from progenitors in the subventricular zone of postnatal rat forebrain. *Neuron* **10**, 201–212.
Lewis, P.D. (1968a) The fate of the subependymal cell in the adult rat brain, with a note on the origin of microglia. *Brain* **91**, 721–736.
Lewis, P.D. (1968b) Mitotic activity in the primate subependymal layer and the genesis of gliomas. *Nature* **217**, 974–975.
Lois, C. & Alvarez-Buylla, A. (1993) Proliferating subventricular zone cells in the adult mammalian forebrain can differentiate into neurons and glia. *Proc. Natl. Acad. Sci. USA* **90**, 2074–2077.
Lois, C. & Alvarez-Buylla, A. (1994) Long-distance neuronal migration in the adult mammalian brain. *Science* **264**, 1145–1148.
Lopez-Garcia, C., Molowny, A., Garcia-Verdugo, J.M. & Ferrer, I. (1988) Delayed postnatal neurogenesis in the cerebral cortex of lizards. *Dev. Brain Res.* **43**, 167–174.
Luskin, M.B. (1993) Restricted proliferation and migration of postnatally generated neurons derived from the forebrain subventricular zone. *Neuron* **11**, 173–189.
Mair, R.G., Gellman, R.L. & Gesteland, R.C. (1982) Postnatal proliferation and maturation of olfactory bulb neurons in the rat. *Neuroscience* **7**, 3105–3116.
McConnell, S.K. (1988) Development and decision-making in the mammalian cerebral cortex. *Brain Res. Rev.* **13**, 1–23.
McDermott, K.W.G. & Lantos, P.L. (1990) Cell proliferation in the subependymal layer of the postnatal marmoset, *Callithrix jacchus. Devel. Brain Res.* **57**, 269–277.
McKay, R.D.G. (1989) The origins of cellular diversity in the mammalian central nervous system. *Cell* **58**, 815–821.
Minelli, G., Del Grande, P. & Mambelli, M.C. (1977) Preliminary study of the regenerative process of the dorsal cortex of the telencephalon of *Lacerta viridis. Z. Mikrosk. Anat. Forsch.* **91**, 241–246.
Monti-Graziadei, A.G. & Graziadei, P.P. (1992) Sensory reinnervation after partial removal of the olfactory bulb. *J. Comp. Neurol.* **316**, 32–44.
Morshead, C.M. & Van der Kooy, D. (1992) Postmitotic death is the fate of constitutively proliferating cells in the subependymal layer of the adult mouse brain. *J. Neurosci.* **12**, 249–256.
Morshead, C.M., Reynolds, B.R., Craig, C.G., McBurney, M.W., Staines, W.A., Morsassutti, D., Weiss, S. & Van der Kooy, D. (1994) Neural stem cells in the adult mammalian forebrain: a relatively quiescent subpopulation of subependymal cells. *Neuron* **13**, 1071–1082.
Nottebohm, F. (1985) Neuronal replacement in adulthood. In *Hope for a new neurology.* (ed. Nottebohm, F.) pp. 143–161, New York, New York Academy of Sciences.
Ono, K. & Kawamura, K. (1990) Mode of neuronal migration in the pontine stream of fetal mice. *Anat. Embryol.* **182**, 11–19.
Ono, K., Tomasiewickz, H., Magnuson, T. & Rutischauser, U. (1994) N-CAM mutation inhibits tangential neuronal migration and is phenocopied by enzymatic removal of polysialic acid. *Neuron* **13**, 595–609.
Paterson, J.A., Privat, A., Ling, E.A. & Leblond, C.P. (1973) Investigation of glial cells in demithin sections. III. Transformation of subependymal cells into glial cells, as shown by radioautography after ^{3}H-Thymidine injection into the lateral ventricle of the brain of young rats. *J. Comp. Neur.* **149**, 83–102.
Paton, J.A. & Nottebohm, F. (1984) Neurons generated in the adult brain are recruited into functional circuits. *Science* **225**, 1046–1048.
Privat, A. & Leblond, C.P. (1972) The subependymal layer and neighboring region in the brain of the young rat. *J. Comp. Neur.* **146**, 277–302.

Rakic, P. (1972) Mode of cell migration to the superficial layers of fetal monkey neocortex. *J. Comp. Neurol.* **145**, 61–84.
Rakic, P. (1985) Limits of neurogenesis in primates. *Science* **227**, 1054–1056.
Rakic, P. (1990) Principles of neural cell migration. *Experientia* **46**, 882–891.
Rakic, P. & Kornack, D.R. (1993) Constraints on neurogenesis in adult primate brain: An evolutionary advantage? In *Neuronal Death and Repair* (ed. Cuello, A.C.) pp. 257–266. Amsterdam, Elsevier.
Rakic, P., Cameron, R.S. & Komuro, H. (1994) Recognition, adhesion, transmembrane signaling and cell motility in guided neuronal migration. *Curr. Opin. Neurobiol.* **4**, 63–69.
Raymond, P.A. & Easter, S.S. (1983) Post-embryonic growth of the optic tectum in goldfish. I. Location of germinal cells and numbers of neurons produced. *J. Neurosci.* **3**, 1077–1091.
Reynolds, B. & Weiss, S. (1992) Generation of neurons and astrocytes from isolated cells of the adult mammalian central nervous system. *Science* **255**, 1707–1710.
Reynolds, B.A. & Weiss, S. (1993) EGF-Responsive stem cells in the mammalian central nervous system. In *Neuronal Cell Death and Repair.* (ed. Cuello, C.A.) pp. 247–255. Amsterdam, Elsevier.
Reynolds, B.A., Tetzlaff, W. & Weiss, S. (1992) A multipotent EGF-responsive striatal embryonic progenitor cell produces neurons and astrocytes. *J. Neurosci.* **12**, 4565–4574.
Richards, L.J., Kilpatrick, T.J. & Bartlett, P.F. (1992) De novo generation of neuronal cells from the adult mouse brain. *Proc. Natl. Acad. Sci. USA* **89**, 8591–8595.
Rousselot, P., Lois, C. & Alvarez-Buylla, A. (1995) Embryonic (PSA) N-CAM reveals chains of migrating neuroblasts between the lateral ventricle and the olfactory bulb of adult mice. *J. Comp. Neurol.* **351**, 51–61.
Rubenstein, J.L.R., Martinez, S., Shimamura, K. & Puelles, L. (1994) The embryonic vertebrate forebrain: the prosomeric model. *Science* **266**, 578–580.
Sanes, J.R. (1989) Analysing cell lineage with a recombinant retrovirus. *Trends Neurosci.* **12**, 21–28.
Silver, J. & Steindler, D.A. (1990) Axonal boundaries and inhibitory mechanisms during neural development and regeneration. *Exp. Neurol.* **109**, 1–139.
Smart, I. (1961) The subependymal layer of the mouse brain and its cell production as shown by radioautography after thymidine-H3 injection. *J. Comp. Neurol.* **116**, 325–348.
Straznicky, K. & Gaze, R.M. (1971) The growth of the retina in *Xenopus laevis*: an autoradiographic analysis. *J. Embryol. Exp. Morphol.* **26**, 67–79.
Temple, S. (1989) Division and differentiation of isolated CNS blast cells in microculture. *Nature* **340**, 471–473.
Tomasiewicz, H., Ono, K., Yee, D., Thompson, C., Goridis, C., Rutishauser, U. & Magnuson, T. (1993) Genetic deletion of a neural cell adhesion molecule variant (N-CAM 180) produces distinct defects in the central nervous system. *Neuron* **11**, 1163–1174.

CHAPTER 14

DIRECT INJECTION OF PLASMID DNA INTO THE BRAIN

Masaaki Tsuda* and Takashi Imaoka†

*Department of Microbiology, Faculty of Pharmaceutical Sciences, Okayama University, Tsushima-naka 1-1-1., Okayama 700, and †Department of Neurological Surgery, Okayama University Medical School, Shikata-cho 2-5-1, Okayama 700, Japan

Table of Contents

14.1 Introduction

It is desirable to develop an efficient and safe method for gene transfer into the mammalian brain, not only for clinical application of gene therapy for neurological diseases of the central nervous system (CNS) but also for a basic study of gene function in the brain. Numerous advances have been made to improve the methods for gene transfer into animal tissues, and two primary approaches have emerged. The *ex vivo* approach depends upon genetic transduction of cultured cells which are subsequently implanted into a host organism. Successful methods using this approach to express transgenes in targeted regions of the brain have been reported by several groups (Gage *et al.*, 1987; Tsuda *et al.*, 1990). On the other hand, the *in vivo* approach concentrates on direct transfer of genetic materials to the tissue cells using viral infection or DNA transfection with chemical DNA-delivering reagents. Infection with recombinant retroviruses (Turner and Cepko, 1987) and adenoviruses (Le Gal La Salle *et al.*, 1993) is available for gene transfer into brain cells. As an alternative to this virus-mediated method, direct injection of expression vector plasmid DNAs into the brain with cationic lipids is now also being evaluated (Ono *et al.*, 1990; Roessler and Davidson, 1994).

Genetic Manipulation of the Nervous System
ISBN 0-12-437165-5

Conceptually, this cationic lipid-mediated DNA delivery into the brain, without the use of viral infection, has several advantages. First, construction of plasmid DNAs containing functional genes and adequate enhancers/promoters is easily performed and less time consuming than generating viral vectors. Second, the use of plasmid DNAs to deliver genetic materials is not associated with risks of deleterious side effects such as cytotoxicity, autoimmune encephalitis or productive virus infection. Third, persistence of introduced plasmid DNAs in the chromosomal state would reduce the unpredictable effects of integration of transgenes into random sites within chromosomal DNA after injection. Fourth, multiple transgenes can be incorporated simultaneously and expressed by the same cells, and therefore it is easy to exert multiple transgene effects on brain functions. Finally, from the clinical viewpoint, transient expression of transgenes by *in vivo* DNA transfection would be advantageous in that it would allow cessation of the treatment regimen if undesirable side effects occurred in patients.

Our previous study (Ono *et al.*, 1990) showed successful cationic liposome-mediated gene transfer into the brain cells of neonatal mice with the expression of reporter genes for up to 9 days after injection. Although several cases of successful gene transfer into vertebrate brains have been reported (Holt *et al.*, 1990; Li-Xin *et al.*, 1992; Roessler and Davidson, 1994), the usefulness of this method has been limited because of the low efficiency of DNA transfection in brain cells. To circumvent this low transfection efficiency, we investigated gene transfer into the adult rat brain by continuous injection of cationic liposome–DNA complexes. A marked increase in DNA delivery efficiency could be obtained reproducibly with glial cells of the brain. In this review, we discuss several technical factors which might be required for maximal gene transfer into brain cells using cationic liposome–DNA complexes.

14.2 Materials and methods

14.2.1 Plasmid DNAs

Plasmid DNAs constructed for DNA transfection in culture are commonly available for direct gene transfer into the brain. To examine the uptake of plasmid DNAs by the tissue cells after injection, adequate reporter genes such as firefly luciferase, chloramphenicol acetyltransferase (CAT) and *Eschenichia coli* β-galactosidase (*lacZ*) genes can be used, which are generally fused with virus enhancers/promoters such as those of simian virus (SV) 40, Raus sarcoma virus (RSV) and cytomegalovirus (CMV) for effective survey of expression of transgenes in the exposed tissues.

When the β-galactosidase gene is fused with the nuclear location signal derived from the SV40 T-antigen gene as with the plasmid L7RHβ-gal (Kalderon *et al.*, 1984), the β-galactosidase protein produced can be transported into the nuclei of cells and, therefore, the nuclei show dense staining with X-gal (5-bromo-4-chloro-3-indolyl-Dgalactopyranoside; (Ono *et al.*, 1990) as shown in Figure 2. When the brain is injured, some types of glial cells, probably derived from the monocyte/macrophage lineage, appear and produce false β-galactosidase positive signals because of their lysosomal β-galactosidase activities (Sheng *et al.*, 1993; Roessler and Davidson, 1994). By identifying these nuclear or cytoplasmic signals stained with X-gal, we can easily distinguish cells expressing the exogenous *E. coli* β-galactosidase from those expressing endogenous β-galactosidase activity. Nevertheless, it seems important to confirm the production of *E. coli* β-galactosidase proteins in exposed tissues by immunohistochemical analyses using anti-*E. coli* β-galactosidase antiserum. Using anti-CAT antiserum, it is also possible to detect cells expressing the CAT protein in exposed brains (Ono *et al.*, 1990). In this case, however, certain groups of cells cross-reactive to anti-CAT antiserum are sometimes recognized in the brain in a developmentally and regionally specific fashion (our unpublished data).

14.2.2 DNA delivery mediators

A wide variety of methods, including the use of DEAE dextran, calcium phosphate, liposome fusion, recombinant virus infection, microinjection and electroporation, have been developed to facilitate the uptake of exogenous DNA by cultured cells. However, most of these methods have several disadvantages related to either cytotoxicity, variable efficiency or inconvenience. To circumvent such problems, new DNA transfection protocols have been developed that make use of synthetic cationic lipids.

N-[1-(2,3-Dioleyloxy)propyl]-*N*,*N*,*N*-trimethylammonium chloride (DOTMA; Figure 1A) forms positively charged unilamellar liposomes that interact with DNA to form liposome–DNA complexes due to ionic interactions between the positively charged group on the DOTMA molecule and the negatively charged phosphate groups on the DNA (Felgner *et al.*, 1987). These DOTMA–DNA complexes with a net positive charge can fuse with the negatively charged surface membrane of the cells. Another synthetic cationic lipid, lipopolyamine dioctadecylamidoglycylspermine (DOGS; Figure 1B), is composed of a cationic head group (spermine residue) linked to two hydrocarbon chains through a spacer allowing a strong interaction in the DNA minor groove (Behr *et al.*, 1989; Loeffler *et al.*, 1990). Such lipospermines readily compact and mask negatively charged DNA, thus promoting efficient gene transfer through the cell membrane and subsequent expression.

Transfection efficiencies with such cationic liposomes tend to be one or more orders of magnitude higher than with other transfection methods such as those using calcium phosphate and DEAE dextran for established cell lines (Felgner *et al.*, 1987; Behr *et al.*, 1989). Although the cytotoxicities of these reagents vary with different cell types, duration of exposure to cationic lipids and density of cell culture, both reagents appear to be less toxic than calcium phosphate and DEAE dextran.

The cationic liposome–DNA complexes, with 100% entrapment of the DNA, contrast with the conventional liposome encapsulation (Fraley *et al.*, 1980) which usually entraps less than 10% of the DNA and requires an additional purification step to

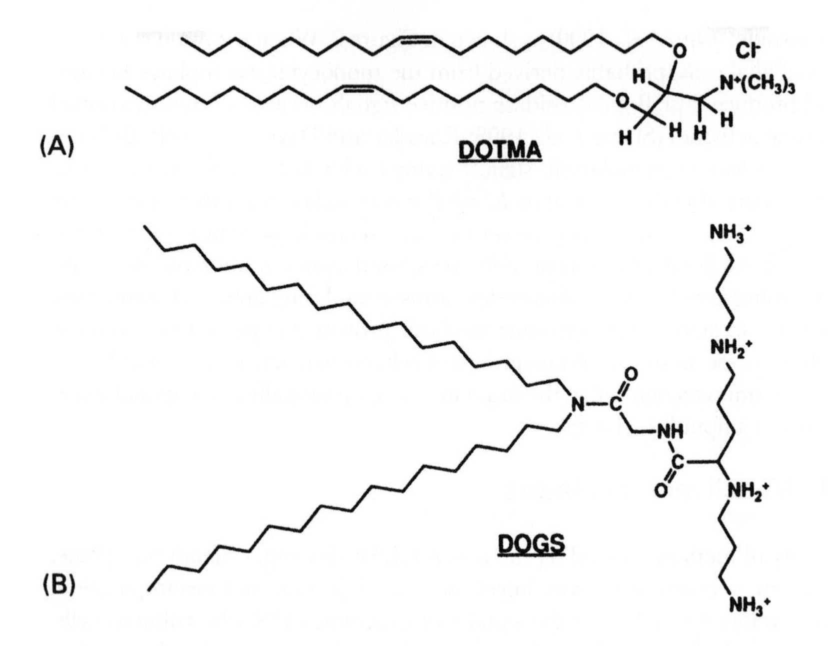

Figure 1 Chemical structures of cationic lipids used for gene transfer. (A) *N*-[1-(2,3-dioleyloxy)propyl]-*N*,*N*,*N*,-trimethylammonium chloride (DOTMA; Felgner *et al.*, 1987). (B) Dioctadecylamidoglycylspermine (DOGS; Behr *et al.*, 1989). Both cationic lipids form liposome–DNA complexes with plasmid DNA and serve to deliver this DNA into mammalian cells.

remove unencapsulated DNA. Particles formed with lipospermines cannot be regarded as conventional liposomes since their interior is not aqueous but rather resembles that of a virus, in which the nucleocapsid is filled with protein, polyamines and genetic material (Loeffler *et al.*, 1990). These reagents are commercially supplied as Lipofectin (GIBCO/BRL New York, USA) and Transfectam (IBF Inc. Maryland USA). Taken together, it seems convenient to use these cationic lipids for direct gene transfer into the brain. Non-lipid-mediated gene transfer is available for mouse skeletal muscles using sucrose (Wolff *et al.*, 1990) and other mediators including fructose, galactose and sodium chloride (Ono *et al.*, 1994). However, such non-lipid mediators do not appear to work for delivering DNAs into brain cells *in vivo* (our unpublished data).

14.2.3 Injection

For maximal gene transfer into animal tissues using *in vivo* transfection, an adequate ratio of cationic liposomes to DNA appears to be required (Zhu *et al.*, 1993; Philip *et al.*, 1993). When plasmid DNAs were injected intravenously into mice or directly into muscle or brain tissue, incremental expression of transgenes was observed with increasing amounts of DNA. The amount of cationic lipid should be increased in parallel with this increase in amount of injected DNA to produce a net positive charge for the cationic liposome–DNA complexes which is required for attachment of the complexes to cellular membranes and for protection of the encapsulated DNAs from possible cleavage by several types of nucleases after injection. However, increasing the amount of lipid too much may increase the cytotoxic effects, which was suggested in the case of cultured cells (Felgner *et al.*, 1987).

After mixing plasmid DNAs with cationic lipids, the solution containing liposome–DNA complexes can be injected stereotactically under anesthesia to various brain regions of adult rats using cannulae. In most cases tested so far, a single injection has been used to deliver DNA not only to the brain but also to other tissues. Quite recently, however, continuous injection into the adult rat brain was found to be much more effective in delivering DNA to brain cells than a single injection which occurs over only a short period (Imaoka *et al.*, 1995), as described below. Following an appropriate recovery period, immunohistochemical or biochemical analysis can be performed to detect the expression of reporter genes in the exposed tissues.

14.3 Results

14.3.1 Expression of transgenes in brain cells

Possible utilization of expression plasmid vectors as DNA delivery vehicles for gene transfer into rodent brains was initially demonstrated with neonatal mice (1 week old) in our laboratory (Ono *et al.*, 1990). Since the skull of neonatal mice is still thin and soft, it is easy to penetrate with a microsyringe. In addition, a high efficiency of gene transfer into brain cells could be expected because populations of dividing cells, which are highly competent for DNA transfection, are still abundant in the brain during the neonatal stages of development. Injection of L7RH-βgal plasmid DNA harboring the *E. coli* β-galactosidase gene fused with the nuclear localization signal of the SV40 T-antigen gene with DOTMA resulted in a pattern of scattered nuclear-like rounded β-galactosidase-positive signals on brain slices, but detection of such positive signals was very rare (Figure 2a,b). Low efficiency of gene transfer by direct injection of plasmid DNA into the brain was also reported in the embryonic brain of *Xenopus laevis* (Holt *et al.*, 1990). Luciferase activity in the brain following direct injection of DOTMA–DNA complexes into the ventricles was about two orders of magnitude less than that with isolated brain preparations incubated for 8 h in transfection medium including DOTMA–DNA complexes. These observations indicated that a cationic lipid could be used to deliver DNA into brain cells but its delivery efficiency was very low.

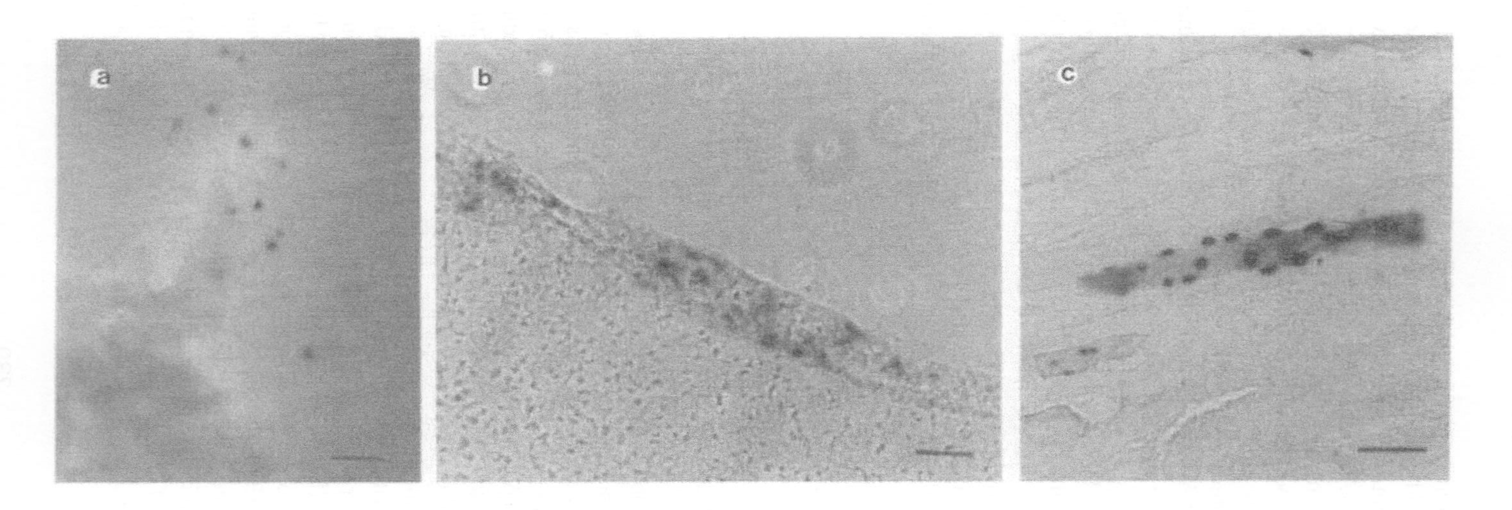

Figure 2 Detection of exogenous *E. coli* β-galactosidase-positive cells in mouse brain or muscle after direct injection of plasmid DNA with DOTMA. (a,b) L7RHβ-gal plasmid DNA was injected into the brain of 1 week old mice with DOTMA, and X-gal-stained cells on 1 mm (a) or 20 μm (b) thick brain sections were observed by light microscopy (Ono *et al.*, 1990). Cell types of these β-galactosidase-positive cells were not identified. (c) L7RHβ-gal and pRSVβ-gal plasmid DNAs were injected simultaneously into the quadriceps with 20% fructose solution (Ono *et al.*, 1994). Scale bars: (a) 100 μm; (b,c) 25 μm.

14.3.2 Improvement of delivery efficiency

One of the reasons for this lowered transfection efficiency might be an inhibitory effect of extracellular materials on tissues or charged materials in the cerebrospinal fluid on the interaction of cationic liposome–DNA complexes with the surface of brain cells (Holt *et al.*, 1990). To exclude such barriers in the brain, it might be necessary to flush the ventricles or targeted brain regions with enzymatic solutions to degrade such molecules just before injection of liposome–DNA complexes.

As a more important factor for improving the DNA delivery efficiency, however, the length of time which the liposome–DNA complexes remain in contact with brain cells may be considered. Direct gene transfer into mouse skeletal muscles is successful with a highly reproducible transfection efficiency (Wolff *et al.*, 1990; Ono *et al.*, 1994). In this case, the solution including DNA with cationic lipid or non-lipid mediators does not readily diffuse out of muscles but remains in the tissue for a relatively long period after injection; such prolonged contact might be responsible for the observed high frequencies of DNA incorporation into muscle cells. Therefore a single injection, as used to date for DNA delivery by direct injection into brain, does not seem to allow sufficient time for efficient DNA transfection of brain cells. Therefore, we examined the efficiency of continuous injection of plasmid DNA into the adult rat brain.

pRSVβ-gal plasmid DNA (30 μg) mixed with DOGS at a ratio of 3:1 (DOGS:DNA) was injected into the right striatum of the adult rat brain using an osmotic pump (Alzet; Palo Alto, USA), which allowed continuous injection for up to 14 days (Imaoka *et al.*, 1995). In addition to this advantage in prolonging the contact of cationic liposome–DNA complexes with brain cells, application of continuous injection would allow the total volume injected to be larger than that of a single injection, limited to 5–10 μl by the intracranial size of the recipient animals and because injection of a large volume into restricted brain regions would carry associated risks of regional trauma and non-specific brain damage. The larger the volume injected, the more DNA can be delivered into the brain (our unpublished data). As a result, a dramatic increase in the number of cells immunoreactive to anti-*E. coli* β-galactosidase antiserum was observed around the traces of the inserted cannulae after injection for 7 days, compared to that observed after injection for only 1 day (Figure 3 and Table 1). No immunoreactive cells were detected in the contralateral hemispheres of these animals. Thus, continuous direct injection of liposome–DNA complexes seems to facilitate highly efficient gene transfer into the brain.

14.3.3 Cell types expressing transgenes

Most of the cells immunoreactive to anti-*E. coli* β-galactosidase antiserum which were obtained by continuous injection were thought to be astrocytes because they were also stained with anti-glial fibrillary acidic protein (GFAP) antiserum (our unpublished data). We observed no neuronal cells among the *E. coli* β-galactosidase-immunoreactive cells. By observing the ultrastructure of β-galactosidase-positive cells using transmission electron microscopy, Roessler and Davidson (1994) also identified multiple

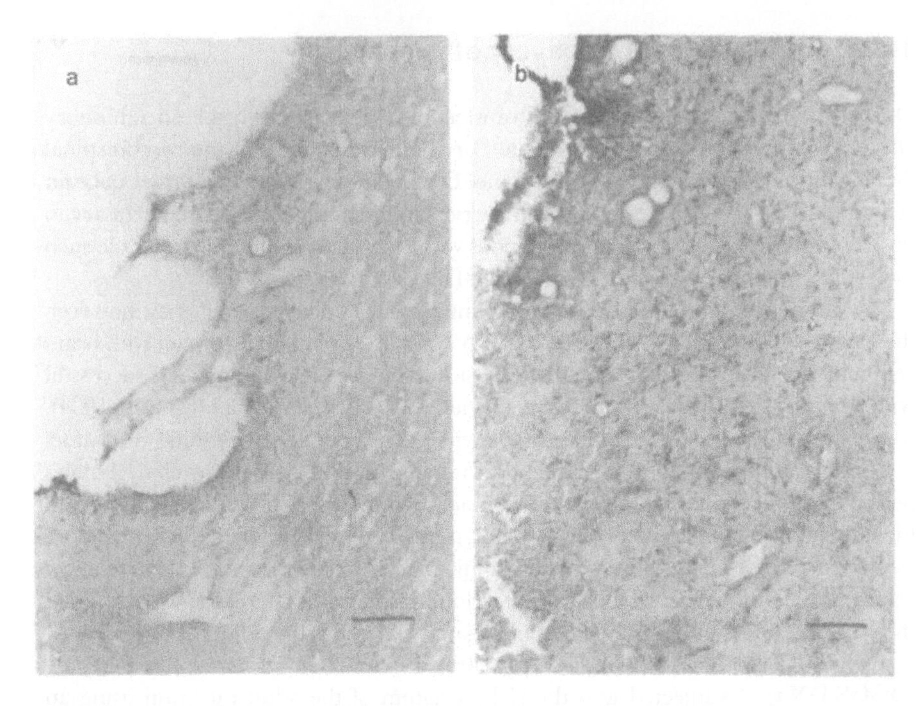

Figure 3 Immunohistochemical staining of *E. coli* β-galactosidase-expressing cells after continuous injection of plasmid DNA. The continuous injection of pRSVβ-gal plasmid DNA with DOGS for 1 day (a) or 7 days (b) was carried out using an osmotic pump, and then sections were subjected to immunohistochemical analysis using anti-*E. coli* β-galactosidase antiserum. A large number of cells immunoreactive to β-galactosidase antiserum were observed in sections from animals treated for 7 days, while only a few such cells were observed with 1 day injection. Scale bars: (a,b) 150 μm.

neuroglia with signals for *E. coli* β-galactosidase after a single injection of DOTMA–DNA complexes.

The possibility of gene transfer into post-mitotic neurons of the adult rat brain was recently suggested by Kato *et al.* (1994) using the hemagglutinating virus of Japan (HVJ)–liposome method. In this case, however, the transfection efficiencies not only of neuronal cells but also of glial cells appear to be very low, probably because of the use of the conventional liposome method. Therefore, it seems reasonable to conclude that at least until now cationic liposome–DNA complexes predominantly served to deliver DNA to glial cells in brain, which is probably due to their active endocytotic activities.

14.3.4 Delivery of functional genes to brain cells

It is possible to deliver functional genes into brain cells with an appropriate expression vector. Roessler and Davidson (1994) delivered the human β-glucuronidase gene

Table 1 Number of *E. coli* β-galactosidase-positive cells

	Duration of injection	
	1 day (n=7)	7 days (n=7)
No. of cells	4.7±2.4	307.0±59.2

Following continuous injection for the indicated days, the animals were sacrificed and the expression of *E. coli* β-galactosidase gene was analyzed by immunohistochemical analyses using anti-*E. coli* β-galactosidase antibody. Values are means ±SD.

with SV40 enhancer/promoter into the caudate putamen of 7 week old heterozygous deficient $gus^{mps/+}$ mice, a prototypical animal model of the mucopolysaccharidoses (MPS), a group of lysosomal storage diseases characterized by the accumulation of glycosaminoglycans within the neuroglial cells. They found successful expression of the human β-glucuronidase gene in the neuroglial cells largely confined to the area along the needle trace. On the other hand, Li-Xin *et al.* (1992) reported delivery of the cholecystokinin (CCK) gene by a single intracerebroventricular injection of DOTMA–DNA complexes into the brains of P77PMC rats, which are deficient in cholecystokinin octapeptide (CCK-8) in the CNS and highly sensitive to audiogenic seizure. Surprisingly, injection of CCK gene reduced audiogenic seizures in the P77PMC rat, most obviously at 3 and 4 days after injection, the time course of which was almost identical with expression of the *E. coli* β-galactosidase gene. However, it is a primary requirement to raise the transfection efficiency of direct gene transfer for obtaining reproducible effects on the recovery of functionally deficient phenotypes.

14.3.5 Peristence and distribution of transgene expression in the brain

Expression of transgene increased markedly within about 5 days after direct injection into the brain, and remained for at least 3 weeks, although at a much decreased level (Roessler and Davidson, 1994). This is consistent with results with intravenous or intraperitoneal DNA delivery into adult mice, in which the transgene was highly expressed for 7 days but still expressed in multiple tissues for at least 9 weeks after a single injection (Zhu *et al.*, 1993; Philip *et al.*, 1993). Such prolonged transgene expression for up to 2 months was also detected with mouse skeletal muscles by direct gene transfer using non-lipid mediators (Wolff *et al.*, 1990; Ono *et al.*, 1994). During the continuous expression of transgene, most of the injected DNA exists extrachromosomally in a non-integrated and circular form that does not replicate (Wolff *et al.*, 1990; Zhu *et al.*, 1993; Philip *et al.*, 1993).

The β-galactosidase-positive cells were confined to a certain area along the edge of the needle trace in the brain even with continuous injection of sample solution (about 210 l) for 7 days (Figure 3a,b). No signs of brain dysfunction were observed as a result of continuous injection throughout the period tested (at least 3 weeks). For local gene transfer in animal brains, therefore, it is convenient to use cationic lipids as DNA delivery mediators.

14.3.6 Co-transfection

Most transfected cells in culture co-express the transgene when two plasmid DNAs, each of which carry different genes, are simultaneously added to the culture medium. Such co-transfection of plasmid DNAs can be performed with *in vivo* transfection into brain cells. Holt *et al.* (1994) identified the cells expressing both luciferase and CAT genes after corresponding plasmid DNAs were transfected into embryonal brain slices with DOTMA. Ono *et al.* (1994) also tried to co-transfect two types of β-galactosidase genes, whose products were distributed to the nucleus and cytoplasm, respectively, in adult mouse skeletal muscle cells by direct injection of DNA, and found the co-expression of these gene products in confined regions of myofibers (Figure 2c). Thus, it seems possible to deliver multiple genes together to the same brain cells using direct injection of plasmid DNA with cationic liposomes.

14.4 Discussion

The main disadvantage with cationic liposome-mediated DNA delivery has been the low efficiency of DNA transfection into brain cells. However, continuous injection of cationic liposome–DNA complexes into adult rat brain using an osmotic pump has increased the transfection efficiency by up to about two orders of magnitude compared to that obtained by a single injection. This technical advancement promises wide application of this cationic liposome-mediated DNA delivery system, not only for clinical treatment of the CNS disorders but also for basic analysis of gene functions in the brain.

At present, however, glial cells but not neuronal cells are the major cell populations in the brain to which DNA is delivered by the cationic liposome-mediated method. This supports the idea that the highly efficient DNA transfection attained by continuous injection could be due to reactive gliosis which may be a result of injection trauma (Sheng *et al.*, 1993). From such a limited cell population for DNA delivery, the transgenes might be limited at present to those encoding secreted proteins such as neurotrophic factors and cytokines or to those encoding some enzymes such as catalase and superoxide dismutase, which can interact with reactive oxygen species, because proteins synthesized from such transgenes in glial cells could be expected to show their effects on surrounding regions in the brain.

To develop this liposome-mediated method further in future, it is necessary to improve the transfection efficiency to non-dividing cells such as neuronal cells. For this purpose, further modification of cationic lipids as delivery reagents is essential.

Cationic lipopolyamine is known to be effective for DNA transfection into non-dividing cells in culture (Behr *et al.*, 1989; Loeffler *et al.*, 1990). However, it serves well to deliver DNA not only to neuronal cells but also to glial cells in primary cultures of brain cells (our unpublished data). The majority of *E. coli* β-galactosidase-positive cells after continuous injection into the brain with DOGS were glial and not neuronal cells (Figure 3b and our unpublished data). However, Philip *et al.* (1993) reported that the choice of cationic lipids in the liposomes conferred some degree of tissue-specific transgene expression *in vivo*, probably due to specific interactions of individual cationic lipids with specific cell types. Therefore, it may be possible to develop modified cationic lipids or to find adequate compositions of cationic lipids for the liposomes, which might be more adequate in delivering DNA to non-dividing cells.

In addition, it is important to investigate the molecular mechanisms which regulate the existence of plasmid DNAs in an extrachromosomal state in cells of several tissues. It is still unknown what mechanisms maintain the plasmid DNA within the cells for such a relatively long period of time (2 or more months) without any evidence of their replication. This persistence of plasmid DNA in mammalian cells might, at least in part, be attributable to the transcriptionally active state of transgenes under the control of viral enhancers/promoters. A gradual inactivation of viral enhancers/promoters could occur following continuous expression of active transgene, and result in a loss of plasmid DNA by enzymatic cleavage or functional inactivation of the transgene by formation of chromatin structures. Thus, it would be of interest to use a cell type-specific enhancer/promoter for glial or neuronal cells instead of those from viral genes. Selection of such an adequate enhancer/promoter would prolong the transcriptionally active state of the transgene and result in longer maintainance of the plasmid DNA in tissue cells. This use of cell type-specific enhancer/promoter would also confer cell type specificity on the expression of transgene in the exposed tissue. In any case, understanding of the mechanisms of the relatively long maintenance of plasmid DNA may allow us to control the length of expression of transgenes carried on the plasmid DNA.

A series of recent studies concerning gene transfer into brain cells using replication-deficient adenovirus vectors seem to guarantee efficient and safe gene transfer to a variety of brain cells (Neve, 1993; Davidson *et al.*, 1993). Le Gal La Salle *et al.* (1993) reported that stereotactic injection of recombinant adenoviruses into the rat hippocampus and the substantia nigra resulted in the detection of β-galactosidase-positive cells for 2 months, which were identified not only as glial cells but also as neuronal cells by morphological and immunohistochemical analyses. In addition, more extensive infection of brain cells can be achieved with the injection of increasing amounts of recombinant adenoviruses. At present, as a DNA delivery system, the adenovirus vector seems to be superior to the cationic liposome-mediated method with respect to the DNA delivery efficiency and the application to neuronal cells.

Nevertheless, several disadvantages seem to still be associated with adenovirus vectors with regard to clinical treatment (Neve, 1993). First, the long-term safety of adenoviruses residing in the brain has not been determined with respect to possible formation of replicative adenoviruses in infected cells. Second, cytopathogenicity may occur around the injection site when high titer virus suspensions (over 10^9 pfu ml^{-1}) are injected into the brain (Akli *et al.*, 1993). Third, a humoral immune response to the adenovirus vectors has been evaluated in experimental animals in different organs such as lungs and peritoneum (Bajocchi *et al.*, 1993), which may interfere with repeated administration of the vector depending upon the organ.

In contrast, such disadvantages may be avoided with the cationic liposome-mediated method, probably as a result of the biodegradable characteristics of liposome–DNA complexes. One of the advantages with this method is the capability of repeated administration of liposome–DNA complexes to the tissues without cytotoxicity or an immune response. In addition, it is easy to administer several different genes simultaneously and express them in the same brain cells, and construction of plasmid DNA containing transgenes is not a time-consuming process compared to that of recombinant adenoviruses. Furthermore, if reactive gliosis can be stimulated by some neurotrophic factors, it would be easy to administer such protein factors exogenously with plasmid DNA to raise the *in vivo* transfection efficiency during the period of continuous injection. Such advantages suggest that the liposome-mediated method may be more suitable for gene therapy of CNS diseases, in which functional genes can be administered in a drug-like manner.

Overall, the success in increasing the transfection efficiency of brain cells by continuous injection using an osmotic pump seems to provide an efficient and safe alternative for gene transfer into brain cells. Owing to this technical advancement, it may be possible to apply this cationic liposome-mediated method for gene therapy of acquired CNS diseases such as Parkinson's and Alzheimer's diseases. Such studies are now in progress using experimental mice. On the other hand, it also seems useful to apply this method for investigating gene functions in restricted regions of the brain. Genetically defective knock-out mice are available for the study of several brain functions, including learning and memory (Li *et al.*, 1994; Jones *et al.*, 1994). As a supplementary method to the use of knock-out mice, direct gene transfer into brain cells can be used to recover the expression of disrupted genes in a restricted region of knock-out mouse brain by local delivery of corresponding intact genes. Since the distribution of transfected cells was rigidly confined to the regions surrounding the site of injection, the cationic liposome-mediated method is mostly suitable for the local recovery of disrupted gene function. Such utilization of direct gene transfer with knock-out mice would facilitate understanding of gene functions in the brain. While several conditions for maximal gene transfer into brain cells remain to be established, the cationic liposome-mediated gene transfer reviewed here is the most controllable and manageable alternative reported to date, having several unique advantages which allow cloned genes to be handled like drugs.

References

Akli, S., Caillaud, C., Vigne, E., Stratford-Perricaudet, I.D., Poenaru, L., Perricaudet, M., Kahn, A. & Peschanski, M.R. (1993) Transfer of a foreign gene into the brain using adenovirus vectors. *Nature Genet.* **3**, 224–228.

Bajocchi, G., Feldman, S.H., Crystal, R.G. & Mastrangeli, A. (1993) Direct *in vivo* gene transfer to ependymal cells in the central nervous system using recombinant adenovirus vectors. *Nature Genet.* **3**, 229–234.

Behr, J.P., Demeneix, B., Loeffler, J.P. & Perez-Mutul, J. (1989) Efficient gene transfer into mammalian primary endocrine cells with lipopolyamine coated DNA. *Proc. Natl. Acad. Sci. USA* **86**, 6982–6986.

Davidson, B.L., Allen, E.D., Kozarsky, K.F., Wilson, J.M. & Roessler, B.J. (1993) A model system for *in vivo* gene transfer into the central nervous system using an adenoviral vector. *Nature Genet.* **3**, 219–223.

Felgner, P.L, Gadek, T.R., Holm, M., Roman, R., Chan, H.W., Wenz, M., Northrop, J.P., Ringold, G.M. & Danielsen, M. (1987) Lipofection: a highly efficient, lipid-mediated DNA-transfection procedure. *Proc. Natl. Acad. Sci. USA* **84**, 7413–7417.

Fraley, R., Subramani, S., Berg, P. & Papahadjopoulos, D. (1980) Introduction of liposome-encapsulated SV40 DNA into cells. *J. Biol. Chem.* **255**, 10431–10435.

Gage, F.H., Wolff, J.A., Rosenberg, M.B., Xu, L., Yee, J.-K., Shults, C. & Friedman, T. (1987) Grafting genetically modified cells to the brain: possibilities for the future. *Neuroscience* **23**, 795–807.

Holt, C.E., Garlick, N., Cornel, E. (1990) Lipofection of cDNA in the embryonic vertebrate central nervous system. *Neuron* **4**, 203–214.

Imaoka, T., Date, I., Miyoshi, Y., Ono, T., Furuta, T., Asari, S., Ohmoto, T., Yasuda, T. & Tsuda, M. (1995) Preliminary results of gene transfer to central nervous system by continuous injection of DNA-liposome complex. *Cell Transplant.* **4**, (Suppl. 1), s23–s26.

Jones, K.R., Farinas, I., Backus, C. & Reichardt, F. (1994) Targeted disruption of the BDNF gene perturbs brain and sensory neuron development but not motor neuron development. *Cell* **76**, 989–999.

Kalderon, D., Roberts, B.L., Richardson, W.D. & Smith, A. (1984) A short amino acid sequence able to specify nuclear location. *Cell* **39**, 499–509.

Kato, K., Yoneda, Y., Okada, Y., Kiyama, H. & Shiosaka, S. (1994) Gene transfer and the expression of a foreign gene *in vivo* in post-mitotic neurons of the adult rat brain using the hemagglutinating virus of the Japan-liposome method. *Mol. Brain Res.* **25**, 359–363.

Le Gal La Salle, G., Robert, J.J., Berrard, S., Ridoux, V., Stratford-Perricaudet, L.D., Perricaudet, M. & Mallet, J. (1993) An adenovirus vector for gene transfer into neurons and glia in the brain. *Science* **259**, 988–990.

Li, Y., Erzurumlu, R.S., Chen, C., Jhaveri, S. & Tonegawa, S. (1994) Whisker-related neuronal patterns fail to develop in the trigerminal brainstem nuclei of NMDAR1 knockout mice. *Cell* **76**, 427–437.

Li-Xin, Z., Min, W. & Ji-Sheng, H. (1992) Suppression of audiogeneic epileptic seizures by intracerebral injection of a CCK gene vector. *NeuroReport* **3**, 700–702.

Loeffler, J.P., Barthel, F., Feltz, P., Behr, J.P., Sassone-Corsi, P. & Feltz, A. (1990) Lipopolyamine-mediated transfection allows gene expression studies in primary neuronal cells. *J. Neurochem.* **54**, 1812–1815.

Neve, R.L. (1993) Adenovirus vectors enter the brain. *TINS* **16**, 251–253.

Ono, T., Fujino, Y., Tsuchiya, T. & Tsuda, M. (1990) Plasmid DNAs directly injected into mouse brain with lipofectin can be incorporated and expressed by brain cells. *Neurosci. Lett.* **117**, 259–263.

Ono, T., Ono, K., Mizukawa, K., Ohta, T., Tsuchiya, T. & Tsuda, M. (1994) Limited diffusibility of gene products directed by a single nucleus in the cytoplasm of multinucleated myofibres. *FEBS Lett.* **337**, 18–22.

Philip, R., Liggitt, D., Philip, M., Dazin, P. & Debs, R. (1993) *In vivo* gene delivery: efficient transfection of T lymphocytes in adult mice. *J. Biol. Chem.* **268**, 16087–16090.

Roessler, B.J. & Davidson, B.L. (1994) Direct plasmid mediated transfection of adult murine brain cells *in vivo* using cationic liposomes. *Neuroscience Lett.* **167**, 5–10.

Sheng, J.G., Shirabe, S., Nishiyama, S., Schwartz, J.P. (1993) Alterations in striatal glial fibrillary acidic protein expression in response to 6-hydroxydopamine-induced denervation. *Exp. Brain Res.* **95**, 450–456.

Tsuda, M., Yuasa, S., Fujino, Y., Sekikawa, M., Ono, K., Tsuchiya, T. & Kawamura, K. (1990) Retrovirus-mediated gene transfer into mouse cerebellar primary culture and its application to the neural transplantation. *Brain Res. Bull.* **24**, 787–792.

Turner, D.L. & Cepko, C.L. (1987) A common progenitor for neurons and glia persists in rat retina late in development. *Nature* **328**, 131–136.

Wolff, J.A., Malone, R.W., Williams, P., Chong, W., Acsadi, G., Jani, A. & Felgner, P.L. (1990) Direct gene transfer into mouse muscle *in vivo*. *Science* **247**, 1465–1468.

Zhu, N., Liggitt, D., Liu, Y. & Debs, R. (1993) Systemic gene expression after intravenous DNA delivery into adult mice. *Science* **261**, 209–211.

CHAPTER 15

ANTISENSE OLIGODEOXYNUCLEOTIDES AS NOVEL NEUROPHARMACOLOGICAL TOOLS FOR SELECTIVE EXPRESSION BLOCKADE IN THE BRAIN

Markus Heilig* and Karl-Hermann Schlingensiepen†

*Department of Clinical Neuroscience, Karolinska Institute, Stockholm, Sweden, and †Max-Planck Institut für Biophysikalische Chemie, Göttingen, Germany

Table of Contents

15.1 Introduction

In the short time which has passed since the first reports of its successful *in vivo* application in the brain, the use of antisense oligodeoxynucleotides (AS-ODNs) has developed from a disputed novelty to an everyday tool in neuropharmacology. Few fundamentally novel approaches have been adopted with equal speed. Here, we outline the combination of utility and accessibility which in our opinion accounts for this development. Yet in its short history, antisense methodology has also acquired something of a reputation for being inconsistent, if not outright capricious. On the basis of our numerous failures and some successes, we discuss methodological considerations relevant to issues of efficacy and specificity.

The concept of antisense-mediated gene inhibition was introduced by Stephenson and Zamecnik (Stephenson and Zamecnik, 1978; Zamecnik and Stephenson, 1978).

Genetic Manipulation of the Nervous System
ISBN 0-12-437165-5

The development of automated chemical DNA synthesis, and increasingly successful cloning and sequencing efforts enabled a widespread application of this principle about a decade later. AS-ODNs are oligomers – typically 14–30 nucleotides long – of single-stranded DNA, or DNA analogues chemically modified for the purpose of increasing stability. They are designed complementary ('antisense') to strategically chosen sequences within a target transcript, with the goal of decreasing (or, ideally, eliminating) the levels of the corresponding protein. Upon cellular entry – a still debated issue, accounted for in a separate section – AS-ODNs are thought to Watson–Crick base-pair with the target mRNAs to transiently inhibit expression with a high degree of specificity (Helene and Toulme, 1990; Akhtar and Juliano, 1992). The mechanisms of action are still being investigated, and will be reviewed.

Transiently blocking the expression of specific genes in the living brain provides a valuable investigational tool in systems where conventional pharmacological tools are lacking, such as numerous neuropeptide systems and intracellular signalling pathways. Furthermore, the specificity of conventional drugs may not be sufficient in systems with a high degree of heterogeneity, such as dopaminergic or serotonergic receptors. The specificity inherent in the genetic code may here allow for a better functional analysis. Antisense studies can thereby bypass costly and often capricious drug development efforts, and point to systems where such efforts may later be justified. Another intriguing prospect is that DNA analogues employing the antisense principle may become therapeutically useful drugs themselves. While present costs of synthesis are prohibitive, an extrapolation from the immediate past suggests that AS-ODNs could come well within reach as pharmaceuticals. The critical issues are more likely to be those of stability, specificity of action, lack of toxicity and possibility for proper targeting; these issues are reviewed.

15.2 Selected applications

If translation, in one manner or another, is inhibited by AS-ODNs, proteins expressed at low levels and subjected to a continuous turnover should constitute targets well suited for antisense inhibition. Another suitable type of target would be proteins which are not at all expressed constitutively, but only transiently induced. On the other hand, structural proteins with a slow turnover rate might not be easily targeted with AS-ODNs at all. In agreement with these theoretical considerations, successful targeting of (1) neurotransmitter receptors, (2) neuropeptides and (3) inducible transcription factors has been most consistently reported.

15.2.1 Receptors

Receptor antagonists are only available for a minority of receptor active brain messengers, mainly those belonging to the small-molecular transmitter families, or the 'classical transmitters': the monamines, acetylcholine and the amino acids. For a large number of neuropeptides, cytokines and neurotrophic factors, antagonists have long been lacking. Also, the selectivity of available antagonists is often not sufficient to distinguish with precision between receptor subtypes, illustrated by the difficulties to functionally characterize the role of recently cloned dopamine (DA) and serotonin (5-HT) receptor subtypes.

The concept that blocking the expression of a membrane-associated receptor could constitute a functional equivalent of blocking the receptor seems to have arisen in immunology. In a seminal study (Zheng *et al.*, 1989), AS-ODNs corresponding to transcripts encoding the V-α and V-β chains of the T cell receptor (TCR) were used to suppress TCR expression in T cell hybridomas. When AS-ODNs targeting appropriate V genes were added after protease treatment (which temporarily removes functional TCR), little or no recovery of TCR expression or function was observed over 48 h. This effect was specific for the TCR V genes utilized by the T cell. Further evidence in the same direction was obtained using the interleukin-1 (IL-1) receptor as target (Burch and Mahan, 1991). In the concentration range 3–30 μM, both phosphodiester and phosphorothioate (see Section 15.4.2) 18-mer AS-ODNs targeting the murine IL-1 receptor message inhibited IL-1-stimulated prostaglandin E_2 (PGE_2) synthesis by cultured murine fibroblasts in a time- and concentration-dependent manner. A control ODN and an AS-ODN targeting the human IL-1 receptor were without effect. Target specificity was also demonstrated by showing that the AS-ODN did not affect PGE_2 synthesis induced by other cytokines. The attentuation of the cellular response to IL-1 caused by the AS-ODN correlated with a loss in cell surface receptors for IL-1, without any change in the number of bradykinin receptors on these cells. Finally, in mice, subcutaneous injection with an AS-ODN to the murine IL-1 receptor markedly inhibited the infiltration of neutrophils in response to subsequent injection of IL-1, demonstrating the potential *in vivo* utility of the antisense approach.

Meanwhile, in a line of studies on the functional role of central neuropeptide Y (NPY), we had obtained evidence that this messenger acts as an endogenous anti-anxiety signal by activating a Y1 subtype of NPY receptors in the amygdala (Heilig *et al.*, 1989, 1992a, 1993a, 1994). These results were, however, based on agonist administration studies. Useful receptor antagonists were not available, preventing attempts to establish whether the anti-anxiety actions of exogenous NPY and its analogues reflect a physiological role of the endogenous peptide. Around this time, the Y1 receptor had been cloned from several species (Eva *et al.*, 1990; Herzog *et al.*, 1992; Larhammar *et al.*, 1992), providing the sequence information necessary for AS-ODN design. In a combined *in vitro/in vivo* approach, we could then demonstrate (Heilig *et al.*, 1992b; Wahlestedt *et al.*, 1993a) that:

(1) an 18-mer phosphodiester AS-ODN, when added for 5 days at 0.2 μM to cortical neurons from rat embryos cultured under serum-free conditions, selectively decreased the density of Y1 receptors by approximately 60%;
(2) the decrease in Y1 receptor density was accompanied by a comparable decrease in the second messenger response to NPY (i.e. inhibition of forskolin-induced cAMP accumulation);

(3) intracerebroventricular (i.c.v.) administration of 50 μg of the same AS-ODN every 12 h for 2 days produced a selective loss of Y1 binding density by approximately 60%;
(4) the loss of Y1 receptors *in vivo* was followed by a marked increase in experimental anxiety as tested in the elevated plus-maze, a model where central NPY administration has the opposite effect.

In a follow-up study (Heilig, 1995), this approach was used to help resolve whether food intake stimulation, another prominent effect of central NPY (Stanley, 1993), is mediated by Y1 receptors. Since Y1 AS-ODN treatment not only did not suppress but even increased food intake, it seems likely that another NPY receptor subtype is responsible for the latter action of the peptide. Thus the utilization of AS-ODNs has here helped clarify a heterogeneity which might be beneficial for future drug design efforts.

Expression blockade with AS-ODNs is an attractive strategy in systems where it circumvents a lack of conventional antagonists. There has, however, also been considerable value to validating antisense methodology by applying it in systems where, on the contrary, an extensive conventional pharmacological characterization is available. One step in this direction was taken by demonstrating an effect similar to that seen with the non-competitive *N*-methyl-D-aspartate (NMDA) receptor antagonist MK-801, i.e. a cytoprotective action following experimental cerebral ischaemia, by inhibiting NMDA-R1 protein expression with a combination of two phosphodiester AS-ODNs (Wahlestedt *et al.*, 1993b). Perhaps more convincing evidence of this type has been obtained by two different laboratories for dopaminergic D2 receptors. Creese and colleagues (Zhang and Creese, 1993) found that i.c.v. infusion of a phosphorothioate AS-ODN corresponding to the rat D2 receptor mRNA reduced rat striatal D2 receptors by approximately 50%, while D1, muscarinic and serotonin 5-HT_2 receptors were unaffected. D2 receptor autoradiography indicated a D2 receptor down-regulation of about 50% throughout the striatum and over 70% in the nucleus accumbens; reversibility after 5 days was also demonstrated. A random control ODN was inactive. The antisense treatment inhibited locomotion induced by the D2 receptor agonist quinpirole, without altering behaviour induced by a D1 receptor agonist. Antisense treatment also elicited catalepsy and reduced spontaneous locomotor activity. Thus the profile of D2 antisense action was virtually identical to that observed with classical competitive D2 antagonists. Similar data have been obtained in mice by Weiss and collaborators, although the reduction in receptor density was lower in these studies (Weiss *et al.*, 1993; Zhou *et al.*, 1994). A 20-mer phosphorothioate D2 AS-ODN, administered by this group i.c.v. to mice with unilateral 6-hydroxydopamine lesions of the corpus striatum, inhibited rotations induced by D2 dopamine receptor agonists but did not block rotations induced by D1 or muscarinic cholinergic receptor stimulation. This behavioural effect was seen with 1 day of repeated injections of D2 antisense, and almost complete inhibition was seen after 6 days of treatment. The effect of D2 antisense was reversible also in this study: behavioural recovery occurred by 2 days after cessation of antisense treatment. Repeated D2 AS-ODN administration specifically reduced the levels of D2 dopamine receptors and D2 dopamine receptor mRNA; D1 receptors were unaffected.

Five distinct subtypes of DA receptors have been cloned over the recent years. The DA D3 receptor subtype (Sokoloff *et al.*, 1990) has attracted particular interest due to its largely limbic distribution, and its possible involvement in mediating the abuse potential of central stimulants (Caine and Koob, 1993). The lack of selective, high affinity D3 antagonists has, however, prevented the latter hypothesis from being examined. Recently, we developed a phosphorothioate 15-mer targeted at the D3 transcript, which seems to inhibit D3 receptor expression both *in vitro* (in stably transfected CHO cells) and *in vivo* in the rat striatum (Nissbrandt *et al.*, 1995). Although a lack of selective D3 ligands prevents a precise quantitation of the receptor loss, conservative indirect estimation suggests a D3 B_{max} reduction of approximately 50% in ventral striatum, accompanied by profound effects on DA synthesis.

Among numerous other receptor systems which are currently being targeted with antisense, we choose to review two simply because at the time this is written, consistent and elegant results have been obtained in both by more than one laboratory. First, the expression of angiotensin-1 (AT1) receptors has been effectively inhibited both *in vitro* and *in vivo* using intraventricular administration of phosphorothioate AS-ODN (Sakai *et al.*, 1994; Gyurko *et al.*, 1993). As expected, this treatment attenuated drinking induced by i.c.v. administration of angiotensin II (Sakai *et al.*, 1994). Perhaps more importantly, a long-lasting decrease in blood pressure was produced by AT1 antisense in spontaneously hypertensive rats (Gyurka *et al.*, 1993). Secondly, the expression of progesterone receptors has been consistently blocked with AS-ODN by several groups, with profound effects on sexual behaviour (Pollio *et al.*, 1993; Mani *et al.*, 1994; Ogawa *et al.*, 1994). Numerous other reports of antisense expression blockade of neurotransmitter receptors have been presented, including opioid (Standifer *et al.*, 1994) and serotonergic receptor subtypes; for the latter, this strategy is expected to help bring order into the physiology of 14 cloned receptor species.

In summary, extensive evidence is available to support that neurotransmitter receptor expression can be consistently inhibited by relatively similar treatment schedules of AS-ODNs, and the dosages and administration schedules required are emerging.

15.2.2 Neuropeptides

Successful inhibition of interleukin-6 (IL-6) expression obtained in culture (Levy *et al.*, 1991) demonstrated the potential for targeting also the 'presynaptic' component of peptidergic systems. Accordingly, inhibition of NPY expression has been obtained in the rat hypothalamus, with an expected decrease in feeding as a result (Akabayashi *et al.*, 1994). Expression blockade of corticotropin-releasing hormone (CRH) has also been successfully achieved, and produced behavioural effects similar to those seen with a CRH antagonist (Skutella *et al.*, 1994a). In an elegant study, the synthesis of the pro-opiomelanocortin (POMC)-derived peptides adrenocorticotropin (ACTH) and β-endorphin (β-END) was markedly reduced by an ODN complementary to a region of β-END mRNA both in AtT-20 cells and *in vivo* (Spampinato *et al.*, 1994). We have observed a dose- and sequence-dependent inhibition of pro-dynorphin synthesis after intrastriatal administration of phosphorothioate AS-ODNs (Georgieva *et al.*, 1995). Also, using intraventricular rather than localized administration, inhibition of vasopressin expression has been reported, and was accompanied by a transient diabetes insipidus (Skutella *et al.*, 1994b). Finally, highly sequence-specific effects of oxytocin antisense infusion into the hypothalamus on suckling have recently reported. This was, however, observed within 4–5 h of infusion, and in the absence of a detectable decrease in the tissue content of the peptide. These findings may reflect a depletion of a rapidly releasable peptide pool, which comprises too small a fraction of the total tissue content of peptide to be detected (Neumann *et al.*, 1994). Neuropeptides can thus be targeted with AS-ODNs, but tend in view of the latter data and our own experience to be technically more complicated. Also, the efficiency may be lower than that seen in numerous receptor systems. The more complex results with peptide antisense may be related to turnover kinetics of the peptide in question; in a case where *de novo* synthesis of NPY occurs, i.e. preceding the progesterone-induced LH surge in female rats, brief i.c.v. administration of unprotected AS-ODNs was highly effective, allowing the clarification of a causal relation between the two events (Kalra *et al.*, 1994).

15.2.3 Inducible transcription factors

Inducible transcription factors are, from a kinetic point of view, ideal as AS-ODN targets. Targeting proteins of this kind opens up a window on cellular processes normally not accessible for a pharmacological analysis. In fact, successful local antisense inhibition of transcription factor expression was reported in a peripheral system prior to its application in the brain (Simons *et al.*, 1992).

c-*fos* and related transcription factors are expressed in the brain in response to a wide range of stimuli, and c-*fos* induction has been extensively used as a marker of neuronal activation. However, the possible functional role of this phenomenon in the living brain has remained poorly understood, largely due to a lack of tools which would make it possible to block it (Morgan and Curran, 1991). For example, it has been established that c-*fos* expression is induced in the striatum by cocaine and amphetamine through an activation of dopaminergic D1 receptors (e.g. Moratalla *et al.*, 1993), and in the central nucleus of amygdala (CNA) by various stressors (Campeau *et al.*, 1991; Arnold *et al.*, 1992); yet the functional relevance of these phenomena for reward and anxiety has remained unknown. The situation has been similar in most other systems where c-*fos* induction occurs, such as cerebral ischaemia.

The feasibility of time- and dose-dependently blocking c-*fos* expression in the brain by a single localized injection of a phosphorothioate AS-ODN directly into the target area was demonstrated by Robertson and colleagues (Chiasson *et al.*, 1992; Dragunow *et al.*, 1993). This group reported a sequence-specific 70–80% reduction in the number of Fos-positive nuclei when c-*fos* induction with amphetamine was preceded by AS-ODN administration. Using the same methodology, we could show that inhibiting c-*fos* expression acutely (10 h post-injection) blocked locomotion induced by cocaine (Heilig *et al.*, 1993b). This somewhat surprising effect of blocking the expression of a transcription factor has been consistently confirmed in subsequent studies with amphetamine and rotational activation (Dragunow *et al.*, 1993; Sommer *et al.*, 1993). Hence, we were less surprised to then find a reduction in experimental anxiety when stress-induced c-*fos* expression in the rat amygdala was blocked with AS-ODNs (Möller *et al.*, 1994). Elegant results of blocking ischaemia-induced c-*fos* expression in the hippocampus have also recently been published, although the functional consequences were not addressed in that study (Liu *et al.*, 1994).

Using the antisense technique it has been demonstrated that the three members of the Jun transcription factor family have at least partly antagonistic effects. The transactivation properties and binding affinities of the three Jun proteins differ from each other (Schütte *et al.*, 1989; Chiu *et al.*, 1989; Ryseck and Bravo, 1991). However, until recently, little was known about the function of JunB and JunD. Moreover, hardly anything is known about their role in complex processes like plastic adaptations in the brain, although their expression during such processes is known to occur frequently.

c-*jun* is a proto-oncogene which promotes cell growth. In contrast, experiments with AS-ODN have shown that *junB* and *junD* are inhibitors of cell proliferation (Schlingensiepen and Brysch, 1992). The inhibition of cell growth by *junD* was recently confirmed by over-expression experiments (Pfarr *et al.*, 1994). Using antisense technology we next found that JunB is not only a negative regulator of proliferation, but also plays an important role in neuronal differentiation. *junB* expression was required for neuronal differentiation of PC-12 phaeochromocytoma cells (Schlingensiepen *et al.*, 1993), and of hippocampal neurons (Schlingensiepen *et al.*, 1994). Inhibition of *junB* expression with an AS-ODN prevented neuronal differentiation in both cell types. In contrast, suppression of c-Jun protein promoted neurite outgrowth. Western blotting showed that each antisense sequence inhibited expression of the targeted inducible transcription factor, but not of the other family members.

In vivo, JunB and Fos are the predominantly expressed AP-1 (activator protein 1) transcription factor components in the suprachiasmatic nucleus (SCN) of the rat brain after light stimulation that resets the circadian clock (Kornhauser *et al.*, 1992). Such a phase shift of the circadian rhythm can be induced in rats housed in darkness by a light pulse during the subjective night phase (Wollnik, 1992). Co-suppression of Fos and JunB in rat SCN with AS-ODNs completely prevented the light-induced phase shift, with no effect of a randomized control. The *junB* AS-ODN selectively reduced JunB, but not c-Jun expression (Schlingensiepen *et al.*, 1994; Wollnik *et al.*, 1995). Finally, in a further *in vivo* model, the c-*jun* AS-ODN specifically inhibited the acquisition and retention of a learning task in rats, with neither a mismatch c-*jun* nor a randomized ODN having any effects (Tischmeyer *et al.*, 1994).

Studies successfully employing AS-ODNs to block the expression of transcriptional factors are already becoming too numerous to be covered exhaustively. Interestingly, studies of mechanisms behind AP-1 factor induction have recently also been aided by antisense technology, by the demonstration that an antisense blockade of CREB (cAMP-responsive element binding protein) expression blocks c-*fos* induction by amphetamine (Konradi *et al.*, 1994).

15.3 Delivery, cellular uptake and mode of action

While the basic principle of antisense action has been recognized for some time in studies employing intracellular injection in cultured cells, the potential for *in vivo* efficacy is intimately related to the realization that AS-ODNs can penetrate brain tissue, and enter cells.

15.3.1 Modes of delivery, and tissue distribution in the brain

Unmodified and phosphodiester antisense ODNs, ranging from 14- to 30-mers, are polyvalent negatively charged molecules with molecular weights sometimes exceeding 10 000 Da. It is thus not surprising that naked ODNs do not pass the blood–brain barrier (BBB). Strategies such as biotinylation, or coupling to carriers have been suggested to enhance ODN uptake across the BBB (e.g. Pardridge and Boado, 1991), and this clearly represents an area of critical importance for the future success of AS-ODNs as therapeutics in brain disease. Presently, this problem awaits its solution, and ODNs have to be delivered inside the BBB.

Both intermittent injection and continuous infusion of ODNs into the CSF compartment have been successfully utilized in numerous studies reviewed in Section 15.2. Best data regarding elimination and distribution kinetics are available with the latter mode of delivery (Whitesell *et al.*, 1993). For natural, phosphodiester ODNs, concentrations of up to 15 mM delivered at 1 μl h^{-1} seem to be well tolerated. Although most of the material was rapidly degraded, small amounts of intact ODNs could be recovered from CSF samples. For phosphorothioate ODNs, concentrations of up to 1.5 mM were well tolerated at the same delivery rate for a week, but the 'therapeutic window' was narrow: delivery of a 3 mM solution was lethal. CSF ODN levels could readily be maintained at 0.1% of the delivery solution. For comparison, the half-life of phosphorothioate ODNs after bolus injection was reported to be around 20 min.

Brain tissue penetration of phosphorothioate ODNs is slow, but unexpectedly efficient for molecules of this size and charge. After a single bolus injection of 40 nmol labelled ODNs, a distribution gradient was apparent on tissue sections, with the highest signal noted at the ependymal ventricular lining. After continuous infusion for 1 week, however, the gradient was less apparent, and a relatively even distribution was seen even at long distances from the CSF compartment (Whitesell *et al.*, 1993). Finally, several studies reviewed above have successfully utilized localized administration of ODNs into target brain areas, but little information is available about resulting tissue concentrations and distribution. With phosphorothioate ODNs, repeated localized phosphorothioate injection, or continuous local infusion may be toxic (Chiasson and Robertson, 1994).

15.3.2 Cellular uptake and intracellular fate

Expression blockade has been reported in cultured cells upon intracellular injection of antisense sequences (e.g. Smith *et al.*, 1988; Dagle *et al.*, 1990; Woolf *et al.*, 1992), but a requirement for this type of delivery mode would obviously make antisense methodology useless for all but a few select applications; also, the specificity of expression blockade upon intracellular injection in a commonly used system, the *Xenopus* oocyte, has been questioned (Smith *et al.*, 1990). It has been a matter of some debate whether, and if so how, AS-ODNs can enter cells when delivered extracellularly. Our pragmatic attitude is that, if functional effects are observed and found to be sequence specific, cellular entry must be assumed to have occurred. What then remains to be clarified is the underlying mechanism.

Most hitherto published studies of cellular ODN uptake have been performed in non-neuronal cells. Two early ambitious efforts yielded similar results. Loke *et al.* (1989) found that ODNs were taken up by cultured cells in a saturable, size-dependent manner compatible with receptor-mediated endocytosis. Polynucleotides of any length behaved as competitive inhibitors of ODN transport across cell membranes providing they had a 5′-phosphate moiety. Using oligo(dT)-cellulose for affinity purification, an 80 kDa surface protein which seems to mediate transport was identified. Similarly, in experiments with cultured mouse fibroblast and ascites carcinoma cells, Yakubov *et al.* (1989) found that cellular uptake of ODN derivatives is achieved by an endocytosis mechanism. Uptake was considerably more efficient at low ODN concentrations (less than 1 μM), seemingly because at this concentration a significant percentage of the total ODN pool was absorbed on the cell surface and internalized by a more efficient absorptive endocytosis process. The binding of oligomers to two candidate proteins was inhibited by other ODNs, single- and double-stranded DNA, and RNA. Other polyanions, such as heparin and chondroitin sulphates A and B, did not inhibit binding. These and other observations suggest that receptor-mediated endocytosis is involved in mediating uptake of phosphodiester and phosphorothioate ODNs to mammalian cells at low concentrations, while the contribution of fluid phase endocytosis increases as ODN concentrations rise. Methylphosphonates, in contrast, do not seem to utilize receptor-mediated endocytosis at all (Stein *et al.*, 1988; Akhtar and Juliano, 1992; Budker *et al.*, 1992; Stein and Cheng, 1993).

Uptake data are beginning to appear from endocrine and/or neuronally derived cells as well. In a recent study employing the POMC-expressing cell line AtT-20, which retains many of the differentiated phenotypic features of corticotrophs, a phosphorothioate AS-ODN was stable in the cell culture medium for 24 h, and cellular uptake, while low (approximately 2.5% of the added ODN), produced intracellular levels of the ODN sufficient to form a ribonuclease-resistant duplex with complementary cellular mRNA. As would be predicted, this was paralleled by an inhibited expression of a POMC-derived peptide, β-endorphin (Spampinato *et al.*, 1994). A similar time course of uptake has been observed in our laboratories with fluorescein- and BrdU-labelled phosphorothioate ODN; an association with membranes of cultured cells was seem within 1 h, and significant intracellular uptake within 24 h (Brysch *et al.*, 1994). *In vitro*, uptake has also been described into human glioma cells (Nitta and Sato, 1994). *In vivo*, prolonged (7 day) infusion of phosphorothioate ODN resulted in the presence of labelled ODNs in both neuronal and glial cells throughout the brain (Whitesell *et al.*, 1993). Although time course data are beginning to appear both in the studies reviewed here and others, it is our experience that the compounded time course of uptake and expression blockade differs widely with cell type, oligo length, chemistry etc. Even *in vivo*, using the same probe design and administration conditions, the time course of c-*fos* expression blockade can vary between 3 and 10 h depending on the brain region targeted. Clearly, there is no substitute for systematically testing out the conditions in each specific system, and it is our experience that failure to do so accounts for many instances of 'negative results' with AS-ODNs in neurobiology.

Strategies for enhancing cellular uptake of ODNs are being developed, and may in the future become important for the successful application of this methodology. Coupling to biotin (Pardridge and Boado, 1991) or delivery in complexes with cationic lipids (Felgner *et al.*, 1987; Bennett *et al.*, 1992) are two uptake-enhancing strategies which have been suggested. The latter method has demonstrated a remarkable capability to mediate intracellular delivery of plasmid DNA *in vivo* both in the periphery (Zhu *et al.*, 1993) and in the brain (Roessler and Davidson, 1994; Heilig *et al.*, unpublished), and appears to possess a yet unexplored potential for delivery of antisense ODNs in the brain. Finally, the intracellular distribution and fate of ODNs is not clear yet. The punctate accumulation of phosphorothioate ODNs is suggestive of a lysosomal localization, but no ODN could be recovered from lysosomal fractions of cells incubated with phosphorothioate ODNs (Stein and Cheng, 1993).

15.3.3 Mode of action

The exact mechanism by which AS-ODNs inhibit the translation of the targeted mRNA has not been definitively determined, but steric hindrance of the ribosome during translation appears to be a major factor, and is supported by elegant data where progressively truncated receptor protein could be detected with several ODNs targeting different regions of the LH receptor transcript (West and Cooke, 1991). A further mode of action which is often postulated is the degradation of the DNA-RNA duplex by RNase H. This enzyme specifically cleaves the RNA strand of DNA–RNA hybrids. *In vitro* experiments show that phosphorothioate ODNs hybridized to RNA also form substrates for RNase H. In contrast, duplexes of methylphosphonate DNA–RNA are not accepted as RNase H substrates.

In mammalian cells, however, RNase H plays a minor if any role for the activity of AS-ODNs. Numerous reports demonstrate that AS-ODNs can down-regulate a specific gene product without causing a degradation of the corresponding mRNA in a variety of mammalian cell types, including neurons *in vivo* (Chiang *et al.*, 1991; Jachimczak *et al.*, 1993; Wahlestedt *et al.*, 1993b; Brysch *et al.*, 1994). Others have, however, found a decrease of mRNA to coincide with the decrease in protein (Zhou *et al.*, 1994; Robertson, personal communication). Thus a possible role of RNase H awaits further exploration, but it is clear that efficient antisense-mediated expression blockade may be achieved without it.

Additional mechanisms may mediate antisense effects. AS-ODNs have successfully been targeted to the 5′ and 3′ untranslated regions of mRNA, suggesting that binding of the antisense molecule may cause conformational changes which prevent translation. Alternatively, such conformational changes may lead to destabilization of the target mRNA. AS-ODNs may also disrupt secondary and tertiary structures in the mRNA which form recognition and binding sites for non-ribosomal proteins involved in RNA processing, RNA transport from the nucleus to the cytoplasm, regulation of translation, or stabilization of the mRNA (Vickers *et al.*, 1991).

Our incomplete understanding of mechanisms by which antisense ODNs affect neuronal cells *in vivo* is underscored by recent observations of virtually immediate, yet highly sequence specific electrophysiological effects of oxytocin antisense infused directly into hypothalamic tissue on oxytocinergic neurons (Neumann and Pittman, 1994).

15.4 Technical issues

The numerous reports of successful antisense application in the brain have been greeted with a considerable scepticism by workers who have had a predominantly negative experience of attempts to employ this strategy, or have encountered non-specific effects. This reflects the highly real fact that factors influencing efficacy and specificity of AS-ODN action have not been well known, and, while some antisense designs work well, others simply do not. Although by no means completely predictive, some key factors in this context are beginning to emerge.

15.4.1 Target selection and efficacy

15.4.1.1 Location of target sequence within the message

Most *in vivo* and *in vitro* antisense experiments are performed with the mRNA as the primary target in mind. For this approach, which is the antisense strategy in the strict sense, the ODN sequence can be complementary to: (1) part of the coding region; (2) untranslated parts of an RNA molecule, i.e. the 5′ and 3′ untranslated, or intron sequences; or (3) to the border region between these and the coding sequence. A target region that is often proposed is the start codon and surrounding bases in order to interfere with initiation of protein translation (Strauss *et al.*, 1992; Kitajima *et al.*, 1993). While this approach has worked for some targets, many others can be inhibited equally or more efficiently by ODNs targeted to sequences located further downstream in the coding region (West and Cooke, 1991; Jachimczak *et al.*, 1993; Schlingensiepen *et al.*, 1993; Brysch *et al.*, 1994; Bennett *et al.*, 1994; Spampinato *et al.*, 1994). The intron–exon border of pre-mRNA has been targeted with the aim of interfering with RNA splicing. ODNs targeted to such splice acceptor sites have been effective in down-regulating herpes simplex virus replication, but whether this was due to antisense interference with splicing remained unclear (Kulka *et al.*, 1989).

15.4.1.2 Affinity and secondary structure

Among factors which, independently of target location, influence efficacy is the affinity of ODNs, expressed as free binding energy, or melting temperature. Experiments in our laboratory suggest that ODNs in which the free energy of binding based on nucleotide nearest neighbour estimates (Freier *et al.*, 1986; Breslauer *et al.*, 1986) is evenly distributed along the length of the molecule, offer the best compromise between high affinity to the designated target and low unspecific effects on other sequences. Next, secondary structure of both the ODN and the target message appear to be of importance. It is our experience that ODNs with a high potential for homodimer or hair-pin formation are less likely to succeed. Prediction of potential secondary structures of target RNA has also been useful in our hands for defining regions that might be accessible, e.g. potential loop structures. Unfortunately, available algorithms for RNA secondary structure formation (Freier *et al.*, 1986) are limited in their prediction to transcripts of less than 1200 bp.

15.4.1.3 Choice of target transcript

Since AS-ODNs are generally employed with the goal of reducing cellular levels of a key functional protein, the kinetics of the protein chosen are critical. Even if a near complete inhibition of translation were to be achieved, proteins which are constitutively expressed and only slow turned over may not be at all accessible for this approach. At the other end of the spectrum, inducible proteins are obviously better suited.

15.4.2 Stability and chemical modifications

Short nucleic acid chains such as DNA or RNA ODNs are rapidly degraded by ubiquitous nucleases in serum as well as intracellularly. The half-life of unmodified, phosphodiester DNA–ODN is only around 30 min in serum (Wickstrom, 1986). Unmodified ODNs have been tested in a number of systems, and the concentrations used to achieve adequate down-regulation of target proteins were 10 to 100-fold higher than with nuclease-resistant chemical analogues (Gewirtz and Calabretta, 1988; Caceres and Kosik, 1990; Morrison, 1991). Accordingly, data supporting degradation of unmodified ODN within minutes after i.c.v. administration have been obtained (Whitesell *et al.*, 1993). Due to their instability, unmodified ODN are primarily useful in serum-free culture systems, or in compartments with a very low nuclease content. The CSF, while being such a compartment, ceases to be so the moment the BBB is broken down by trauma or infection. Even in these systems, intracellular stability is low, and therefore the use of unmodified ODNs is rarely justified. A notable exception might be chronic localized treatment directly into target brain areas, a mode of delivery where repeated or prolonged administration of currently available stable analogues seems to be toxic (Chiasson and Robertson, 1994).

The short half-life of phosphodiester ODNs can be overcome by introducing chemical modifications to the DNA structure. Such modifications are primarily located in the phosphate backbone (Miller *et al.*, 1981; Eckstein, 1983; Dagle *et al.*, 1990; Huang *et al.*, 1991, 1993), but modifications of the sugar moieties and bases have also been described (e.g. Morvan *et al.*, 1987; Perbost *et al.*, 1989). For studies in cultured cells and *in vivo*, methylphosphonate and phosphorothioates have been the most widely used chemical analogues to date. In methylphosphonate ODNs, the charge-bearing oxygen of the phosphodiester bond has been replaced by a neutral methyl group. These ODNs are non-ionic, and highly resistant to nuclease degradation. Their lipophilic nature was expected to facilitate their passive diffusion across the cell membrane, but proved to lead to great solubility problems and a low efficiency compared to other analogues. Most cell culture and *in vivo* experiments, and particularly clinical studies are now being carried out with phosphorothioates. In this analogue, one of the non-bridging oxygens is replaced by a sulphur atom. Phosphorothioates are highly stable against nuclease degradation (Campbell *et al.*, 1990; Manoharan *et al.*, 1992). Non-sequence-specific effects and toxicity which are sometimes attributed to this modification are in our experience to a large extent due to impurities in the ODN preparation, and a lack of specificity of the particular sequence used.

15.4.3 Establishing efficacy and specificity, and choice of controls

The specificity of antisense effects must be carefully controlled. A homology search of known sequences is mandatory, and it is obvious that complete matches with sequences other than the intended target exclude a sequence as a target for antisense molecules. It must be kept in mind, however, that only a fraction of the genome is sequenced for both humans, mice and rats. Statistically, specificity is related to ODN length, although not in straightforward manner. The shortest sequence which is likely to be unique within a vertebrate mRNA pool is 13 bases, but already this ODN length contains several internal 10-mers, each of which might be fully complementary to transcripts other than those intended, thus increasing the risk for non-specific effects. Increasing ODN length further will further increase the number of possible internal oligomers, and therefore the risk. Experimental results in the *Xenopus* oocyte seem to support these considerations, and may mean that AS-ODNs, like most drugs, can only achieve relative specificity (Woolf *et al.*, 1992). It must, however, be kept in mind that mechanisms of antisense action might differ in mammalian cells.

Functional studies using antisense-mediated gene inhibition are only valid if adequate assays and controls show that an observed biological effect is due to the specific inhibition of a particular gene. Thus, reduction of the protein product must be demonstrated. This is usually done by immunochemical methods, i.e. Western blotting, ELISA or immunohistochemistry. Binding density, however, is the standard method for quantitating functional receptor protein, while enzyme proteins can be estimated by activity assays. If other proteins are also decreased, the reason can be insufficient specificity, but downstream regulatory effects are also a possible explanation. Inhibiting a transcription factor such as Fos or Jun will certainly influence downstream genes, but in a different time frame from the inhibition of the target protein itself. Carefully chosen time points for assaying protein levels can distinguish between these two possibilities. When looking for specificity at the appropriate time point it is also important to assay for proteins with similar induction properties and turnover rates, e.g. in the case of transcription factors, inducible factors such as c-Jun and JunB can yield a fair comparison, while a constitutively expressed factor such as JunD is less useful (Schlingensiepen *et al.*, 1993).

It is obviously impossible to assay for changes in the expression levels of all cellular proteins. Thus, experiments with control ODNs provide a parallel means of establishing specificity. There is much controversy regarding the choice of the ideal control.

(1) The sense counterpart of the antisense sequence is commonly chosen. Sense ODNs have the same predicted melting temperature as the antisense sequence. However, sense ODNs may produce specific effects in some cells, by interfering with transcription rather than translation, or with naturally occurring antisense molecules (see e.g. Helene and Toulme, 1990). We as well as others have observed sequence-specific sense effects in the brain (Georgieva *et al.*, 1995; Robertson, personal communication); interestingly, it appears that these effects are most likely obtained with prolonged treatment utilizing chemically stabilized ODNs, consistent with the possibility that ODN transport to the nucleus is required.

(2) Randomized sequences, also referred to as 'scrambled ODNs' contain the same bases as the AS-ODNs, but arranged in a randomized order. They usually have melting temperatures that are different from those of the antisense sequence.

(3) Another type of control ODNs are mismatched sequences. Here, 2–4 bases within the antisense sequence are exchanged. In order to maintain melting temperatures, C and G should be exchanged for each other, as should A and T. Depending on the number and spacing of the mismatches, these controls exhibit decreased efficacy, or no effect at all, indicating the specificity of the original antisense ODN (e.g. Tischmeyer *et al.*, 1994).

Each of the above controls has its limitations. It has therefore been suggested that at least two different types of controls should be used to determine the sequence specificity of antisense effects (Stein and Krieg, 1994). Another elegant demonstration of specificity would be obtained if a lack of AS-ODN activity could be demonstrated in cells or animals which have a deletion or a null mutation of the target gene, i.e. a 'mismatched target control'. However, for *in vivo* studies, such 'target mismatch systems' would have to be knockout animals or mutant species, which are mostly not available.

15.5 Summary and prospects for the future

Specific antisense blockade of expression can presently be regarded as an established neuropharmacological tool for targeting neurotransmitter receptors, neuropeptides and inducible transcription factors. When used with appropriate caution and adequate controls, this strategy offers a powerful tool for targeting systems otherwise not accessible for pharmacological analysis. Antisense methodology is an attractive complement to knockout animals, another strategy utilizing genetic sequence information for functional studies. In comparison, knockouts have the advantage of completely ablating the expression of the target gene, while only partial inhibition can commonly be expected with antisense. The practical advantages of antisense methodology in this comparison include its relative simplicity, speed and lower cost. From a theoretical point of view, transient antisense-mediated inhibition of expression is also more likely to reveal the functional role of target genes, since compensatory mechanisms activated during development, or developmental aberrations cannot cloud the picture. Perhaps the greatest promise of antisense strategies, however, is in an area where genetically engineered animals do not compete. If problems of delivery can be solved, and stable DNA analogues with little toxicity developed, AS-ODNs may open up a whole new pharmacology for therapeutic applications.

References

Akabayashi, A., Wahlestedt, C., Alexander, J.T. & Leibowitz, S.F. (1994) Specific inhibition of endogenous neuropeptide Y synthesis in arcuate nucleus by antisense oligonucleotides suppresses feeding behavior and insulin secretion. *Brain Res. Mol. Brain Res* **21**, 55–61.

Akhtar, S. & Juliano, R.L. (1992) Cellular uptake and intracellular fate of antisense oligonucleotides. *Trends Cell Biol.* **2**, 139–144.

Arnold, F.J., De Lucas Bueno, M., Shiers, H., Hancock, D.C., Evan, G.I. & Herbert, J. (1992) Expression of c-*fos* in regions of the basal limbic forebrain following intracerebroventricular corticotropin-releasing factor in unstressed or stressed male rats. *Neuroscience* **51**, 377–390.

Bennett, C.F., Chiang, M.Y., Chan, H., Shoemaker, J.E. & Mirabelli, C.K. (1992) Cationic lipids enhance cellular uptake and activity of phosphorothioate antisense oligonucleotides. *Mol. Pharmacol.* **41**, 1023–1033.

Bennett, C.F., Condon, T.P., Grimm, S., Chan, H. & Chiang, M.Y. (1994) Inhibition of endothelial cell adhesion molecule expression with antisense oligonucleotides. *J. Immunol.* **152**, 3530–3540.

Breslauer, K.J., Frank, R., Blocker, H. & Marky, L.A. (1986) Predicting DNA duplex stability from the base sequence. *Proc. Natl. Acad. Sci. USA* **83**, 3746–3750.

Brysch, W., Magal, E., Louis, J.C., Kunst, M., Klinger, I., Schlingensiepen, R. & Schlingensiepen, K.H. (1994) Inhibition of p185/c-erb-2 proto-oncogene expression by antisense oligodeoxynucleotides down-regulates p185-associated tyrosine-kinase activity and strongly inhibits mammary tumor-cell proliferation. *Cancer Gene Ther.* **1**, 99–105.

Brysch, W., Rifai, A., Tischmeyer, W. & Schlingensiepen, K.H. (1995) Rational drug design, pharmacokinetics and organ uptake of antisense phosphorothioate oligodeoxynucleotides *in vivo*. In *Antisense Oligonucleotide Therapy: Current Status* (ed. Agrawa, S.). Totowa, Humana Press (in press).

Budker, W., Knorre, D. & Vlassov, V. (1992) Cell membranes as barriers for antisense constructions. *Antisense Res. Devel.* **2**, 177–184.

Burch, R.M. & Mahan, L.C. (1991) Oligonucleotides antisense to the interleukin 1 receptor mRNA block the effects of interleukin 1 in cultured murine and human fibroblasts and in mice. *J. Clin. Invest.* **88**, 1190–1196.

Caceres, A. & Kosik, K.S. (1990) Inhibition of neurite polarity by tau antisense oligonucleotides in primary cerebellar neurons. *Nature* **343**, 461–463.

Caine, S.B. & Koob, G.F. (1993) Modulation of cocaine self-administration in the rat through D-3 dopamine receptors. *Science* **260**, 1814–1816.

Campbell, J.M., Bacon, T.A. & Wiskstrom, E. (1990) Oligodeoxynucleoside phosphorothioate stability in subcellular extracts, culture media, sera and cerebrospinal fluid. *J. Biochem. Biophys. Methods* **20**, 259–267.

Campeau, S., Hayward, M.D., Hope, B.T., Rosen, J.B., Nestler, E.J. & Davis, M. (1991) Induction of the c-*fos* proto-oncogene in rat amygdala during unconditioned and conditional fear. *Brain Res.* **565**, 349–352.

Chiang, M.Y., Chan, H., Zounes, M.A., Freier, S.M., Lima, W.F. & Bennett, C.F. (1991) Antisense oligonucleotides inhibit intercellular adhesion molecule 1 expression by two distant mechanisms. *J. Biol. Chem.* **266**, 18162–18171.

Chiasson, B.J. & Robertson, H.A. (1994) Toxicity following *in vivo* infusion of antisense oligonucleotide: role of the interinfusion interval. *Soc. Neurosci. Abstrs.* 426.3.

Chiasson, B.J., Hooper, M.L., Murphy, P.R. & Robertson, H.A. (1992) Antisense oligonucleotide eliminates *in vivo* expression of c-*fos* in mammalian brain. *Eur. J. Pharmacol.* **227**, 451–453.

Chiu, R., Angel, P. & Karin, M. (1989) Jun-B differs in its biological properties from, and is a negative regulator of, c-Jun. *Cell* **59**, 979–986.

Dagle, J.M., Walder, J.A. & Weeks, D.L. (1990) Targeted degradation of mRNA in *Xenopus* oocytes and embryos directed by modified oligonucleotides: studies of An2 and cyclin in embryogenesis. *Nucleic Acid Res.* **18**, 4751–4757.

Dragunow, M., Lawlor, P., Chiasson, B. & Robertson, H. (1993) C-*fos* antisense generates apomorphine and amphetamine-induced rotation. *NeuroReport* **5**, 305–306.

Eckstein, F. (1983) Phosphorothioate analogues of nucleotides – tools for the investigation of biochemical processes. *Angew. Chem.* **22**, 423–506.

Eva, C., Keinanen, K., Monyer, H., Seeburg, P. & Sprengel, R. (1990) Molecular cloning of a novel G protein-coupled receptor that may belong to the neuropeptide receptor family. *FEBS Lett.* **271**, 81–84.

Felgner, P.L., Gadek, T.R., Holm, M., Roman, R., Chan, H.W., Wenz, M., Northrop, J.P., Ringold, G.M. & Danielsen, M. (1987) Lipofection: a highly efficient, lipid-mediated DNA-transfection procedure. *Proc. Natl. Acad. Sci. USA* **84**, 7413–7417.

Freier, S.M., Kierzek, R., Jaeger, J.A., Sugimoto, N., Caruthers, M.H., Neilson, T. & Turner, D.H. (1986) Improved free-energy parameters for predictions of RNA duplex stability. *Proc. Natl. Acad. Sci. USA* **83**, 9373–9377.

Georgieva, J., Heilig, M., Nylander, I., Herrera-Marschi, M. & Terenius, L. (1995) *In vivo* antisense inhibition of prodynorphin expression in rat striatum: dose-dependence and sequence specificity. *Neurosci. Lett.* (in press.)

Gewirtz, A.M. & Calabretta, B. (1988) A c-*myb* antisense oligodeoxynucleotide inhibits normal human hematopoiesis *in vitro*. *Science* **242**, 1303–1306.

Gyurko, R., Wielbo, D. & Phillips, M.I. (1993) Antisense inhibition of AT1 receptor mRNA and angiotensinogen mRNA in the brain of spontaneously hypertensive rats reduces hypertension of neurogenic origin. *Regul. Pept.* **49**, 167–174.

Heilig, M. (1995) Antisense inhibition of neuropeptide Y (NPY) Y1 receptor expression blocks the anxiolytic-like action of NPY in amygdala and paradoxically increases feeding. *Regul. Pep.* (in press).

Heilig, M., Söderpalm, B., Engel, J.A. & Widerlöv, E. (1989) Centrally administered neuropeptide Y (NPY) produces anxiolytic-like effects in anxiety models. *Psychopharmacology (Berl)* **98**, 524–529.

Heilig, M., McLeod, S., Koob, G.K. & Britton, K.T. (1992a) Anxiolytic-like effect of neuropeptide Y (NPY), but not other peptides in an operant conflict test. *Regul. Pept.* **41**, 61–69.

Heilig, M., Pich, E.M., Koob, G.F., Yee, F. & Wahlestedt, C. (1992b) *In vivo* downregulation of neuropeptide Y (NPY) Y1 receptors by i.c.v. antisense oligodeoxynucleotide administration is associated with signs of anxiety in rats. *Soc. Neurosci. Abstr.* 642.18.

Heilig, M., McLeod, S., Brot, M., Heinrichs, S.C., Menzaghi, F., Koob, G.F. & Britton, K.T. (1993a) Anxiolytic-like action of neuropeptide Y: mediation by Y1 receptors in amygdala, and dissociation from food intake effects. *Neuropsychopharmacology* **8**, 357–363.

Heilig, M., Engel, J.A. & Söderpalm, B. (1993b). C-*fos* antisense in the nucleus accumbens blocks the locomoter stimulant action of cocaine. *Eur. J. Pharmacol.* **236**, 339–340.

Heilig, M., Koob, G.F., Ekman, R. & Britton, K.T. (1994) Corticotropin-releasing factor and neuropeptide Y: role in emotional integration. *Trends Neurosci.* **17**, 80–85.

Helene, C. & Toulme, J.J. (1990) Specific regulation of gene expression by antisense, sense and antigene nucleic acids. *Biochim. Biophys. Acta* **1049**, 99–125.

Herzog, H., Hort, Y.J., Ball, H.J., Hayes, G., Shine, J. & Selbie, L.A. (1992) Cloned human neuropeptide Y receptor couples to two different second messenger systems. *Proc. Natl. Acad. Sci. USA* **89**, 5794–5798.

Hunag, Z., Schneider, K.C. & Benner, S.A. (1991) Building blocks for analogues of ribo- and deoxyribonucleotides with dimethylene-sulfide, -sulfoxide, and -sulfone groups replacing phosphodiester linkages. *J. Org. Chem.* **56**, 3869–3882.

Hunag, Z., Schneider, K.C. & Benner, S.A. (1993) Oligonucleotide analogs with dimethylene-sulfide, -sulfoxide, and -sulfone groups replacing phosphodiester linkages. *J. Org. Chem.* **56**, 3869–3882

Huang, Z., Schneider, K.C. & Benne, S.A. (1993). Oligonucleotide analogs with dimethylene-sulfide, -sulfoxide, and -sulfone groups replacing phosphodiester linkages. *Methods Mol. Biol.* **20**, 315–353.

Jachimczak, P., Bogdahn, U., Schneider, J., Behl, C., Meixensberger, J., Apfel, R., Dorries, R., Schlingensiepen, K.H. & Brysch, W. (1993) The effect of transforming growth factor-beta 2-specific phosphorothioate-anti-sense oligodeoxynucleotides in reversing cellular immunosuppression in malignant glioma. *J. Neurosurg.* **78**, 944–951.

Kalra, P.S., Bonavera, J.J., Dube, M.G., Crowley, W.R. & Kalra, S.P. (1994) Inhibition of endogenous neuropeptide Y synthesis by antisense oligodeoxynucleotide administration suppresses the progesteron-induced LH-surge. *Soc. Neurosci. Abstr.* 436.8.

Kitajima, I., Shinohara, T., Bilakovics, J., Brown, D.A., Xu, X. & Nerenberg, M. (1992) Ablation of transplanted HTLV-I Tax-transformed tumors in mice by antisense inhibition of NF-kappa B. *Science* **258**, 1792–1795.

Konradi, C., Cole, R.L., Heckers, S. & Hyman, S.E. (1994) Amphetamine regulates gene expression in rat striatum via transcription factor CREB. *J. Neurosci.* **14**, 5623–534X.

Kornhauser, J.M., Nelson, D.E., Mayo, K.E. & Takahashi, J.S. (1992) Regulation of jun-B messenger RNA and AP-1 activity by light and a circadian clock. *Science* **255**, 1581–1584.

Kulka, M., Smith, C.C., Aurelian, L., Fishelevich, R., Meade, K., Miller, P. & Tso, P.O. (1989) Site specificity of the inhibitory effects of oligo(nucleoside methylphosphonate)s complementary to the acceptor splice junction of herpes simplex virus type 1 immediate early mRNA 4. *Proc. Natl. Acad. Sci. USA* **86**, 6868–6872.

Larhammar, D., Blomqvist, A.G., Yee, F., Jazin, E., Yoo, H. & Wahlestedt, C. (1992) Cloning and functional expression of a human neuropeptide Y/peptide YY receptor of the Y1 type. *J. Biol. Chem.* **267**, 10935–10938.

Levy, Y., Tsapis, A. & Brouet, J.C. (1991) Interleukin-6 antisense oligonucleotides inhibit the growth of human myeloma cell lines. *J. Clin. Invest.* **88**, 696–699.

Loke, S.L., Stein, C.A., Zhang, X.H., Mori, K., Nakanishi, M., Subasinghe, C., Cohen, J.S. & Neckers, L.M. (1989) Characterization of oligonucleotide transport into living cells. *Proc. Natl. Acad. Sci. USA* **86**, 3474–3478.

Mani, S.K., Allen, J.M., Clark, J.H., Blaustein, J.D. & O'Malley, B.W. (1994) Convergent pathways for steroid hormone- and neurotransmitter-induced rat sexual behavior. *Science* **265**, 1246–1249.

Manoharan, M., Johnson, L.K., McGee, D.P., Guinosso, C.J., Ramasamy, K., Springer, R.H., Bennett, C.F., Ecker, D.J., Vickers, T., Cowsert, L. et al. (1992) Chemical modifications to improve uptake and bioavailability of antisense oligonucleotides. *Ann NY Acad Sci*, **660**, 306–309.

Miller, P.S., McFarland, K.B., Jayaraman, K. & Tso, P.O.P. (1981) Biochemical and biological effects of non-ionic nucleic acid methylphosphonates. *Biochemistry* **20**, 1874–1880.

Moratalla, R., Vickers, E.A., Robertson, H.A., Cochran, B.H. & Graybiel, A.M. (1993) Coordinate expression of c-*fos* and jun B is induced in the rat striatum by cocaine. *J. Neurosci.* **13**, 423–433.

Morgan, J.I. & Curran, T. (1991) Stimulus-transcription coupling in the nervous system: involvement of the inducible proto-oncogenes *fos* and *jun*. *Annu. Rev. Neurosci.* **14**, 421–451.

Morrison, R.S. (1991) Suppression of basic fibroblast growth factor expression by antisense oligodeoxynucleotides inhibits the growth of transformed human astrocytes. *J. Biol. Chem.* **266**, 728–734.

Morvan, F., Rayner, B., Imbach, J.L., Thenet, S., Bertrand, J.R., Paoletti, J., Malvy, C. & Paoletti, C. (1987) alpha-DNA II. Synthesis of unnatural alpha-anomeric oligodeoxyribonucleotides containing the four usual bases and study of their substrate activities for nucleases. *Nucleic Acid. Res.* **15**, 3421–3437.

Möller, C., Bing, O. & Heilig, M. (1994) C-*fos* expression in the amygdala: *in vivo* antisense modulation and role in emotionality. *Cell Mol. Neurobiol.* **14**, 415–423.

Neumann, I. & Pittman, Q.J. (1994) An oxytocin antisense oligonucleotide infused into the suprachiasmatic nucleus rapidly alters the excitability of OXT neurons in lactating rats. *Soc. Neurosci. Abstr.* 631.6.

Neumann, I., Porter, W.F., Landgraf, R. & Pittman, Q.J. (1994) Rapid effect on suckling of an oxytocin antisense oligonucleotide administered into rat supraoptic nucleus. *Am. J. Physiol.* **267**, R852–R858.

Nissbrandt, H., Ekman, A., Eriksson, E. & Heilig, M. (1995) Dopamine D3 receptor antisense influences dopamine synthesis in rat brain. *NeuroReport* **6**, 573–576.

Nitta, T. & Sato, K. (1994) Specific inhibition of c-*sis* protein synthesis and cell proliferation with antisense oligodeoxynucleotides in human glioma cells. *Neurosurgery* **34**, 309–314.

Ogawa, S., Olazabal, U.E., Parhar, I.S. & Pfaff, D.W. (1994) Effects of intrahypothalamic administration of antisense DNA for progesterone receptor mRNA on reproductive behavior and progesterone receptor immunoreactivity in female rat. *J. Neurosci.* **14**, 1766–1774.

Pardridge, W.M. & Boado, R.J. (1991) Enhanced cellular uptake of biotinylated antisense oligonucleotide or peptide mediated by avidin, a cationic protein. *FEBS Lett.* **288**, 30–32.

Perbost, M., Lucas, M., Chavis, C., Pompon, A., Baumgartner, H., Rayner, B., Griengl, H. & Imbach, J.L. (1989) Sugar modified oligonucleotides. I. Carbo-oligodeoxynucleotides as potential antisense agents. *Biochem. Biophys. Res. Commun.* **165**, 742–747.

Pfarr, C.M., Mechta, F., Spyrou, G., Lallemand, D., Carillo, S. & Yaniv, M. (1994) Mouse JunD negatively regulates fibroblast growth and antagonizes transformation by ras. *Cell* **76**, 747–760.

Pollio, G., Xue, P., Zanisi, M., Nicolin, A. & Maggi, A. (1993) Antisense oligonucleotide blocks progesterone-induced lordosis behavior in ovariectomized rats. *Brain Res. Mol. Brain Res.* **19**, 135–139.

Roessler, B.J. & Davidson, B.L. (1994) Direct plasmid mediated transfection of adult murine brain cells *in vivo* using cationic liposomes. *Neurosci. Lett.* **167**, 5–10.

Ryseck, R.P. & Bravo, R. (1991) c-JUN, JUNB, and JUN D differ in their binding affinities to AP-1 and CRE consensus sequences: effect of FOS proteins. *Oncogene* **6**, 533–542.

Sakai, R.R., He, P.F., Yang, X.D., Ma, L.Y., Guo, Y.F., Reilly, J.J., Moga, C.N. & Fluharty, S.J. (1994) Intracerebroventricular administration of AT1 receptor antisense oligonucleotides inhibits the behavioral actions of angiotensin II. *Neurochem.* **62**, 2053–2056.

Schlingensiepen, K.H. & Brysch, W. (1992) Phosphorothioate oligomers: Inhibitors of oncogene expression in tumor cells and tools for gene function analysis. In *Gene Regulation: Biology of Antisense RNA and DNA* (eds Erickson, R. & Izan, J.), pp. 317–328. New York, Raven Press.

Schlingensiepen, K.H., Schlingensiepen, R., Kunst, M., Klinger, I., Gerdes, W., Seifert, W. & Brysch, W. (1993) Opposite functions of jun-B and c-jun in growth regulation and neuronal differentiation. *Devel. Genet.* **14**, 305–312.

Schlingensiepen, K.H., Wollnik, F., Kunst, M., Schlingensiepen, R., Herdegen, T. & Brysch, W. (1994) The role of Jun transcription factor expression and phosphorylation in neuronal differentiation, neuronal cell death and in plastic adaptations *in vivo. Cell Mol Neurobiol* **14**, 487–505.

Schütte, J., Viallet, J., Nau, M., Segal, S., Fedorko, J. & Minna, J. (1989) *JunB* inhibits and c-*fos* stimulates the transforming and trans-activating activities of c-jun. *Cell* **59**, 987–997.

Simons, M., Edelman, E.R., DeKeyser, J.L., Langer, R. & Rosenberg, R.D. (1992) Antisense c-*myb* oligonucleotides inhibit intimal arterial smooth muscle cell accumulation *in vivo. Nature* 359, 67–70.

Skutella, T., Criswell, H., Probst, J.C., Breese, G.R., Jirikowski, G.F. & Holsboer, F. (1994a) Corticotropin-releasing hormone (CRH) antisense oligodeoxynucleotide induces anxiolytic effects in rat. *NeuroReport* **5**, 2181–2185.

Skutella, T., Probst, J.C., Engelmann, M. Wotjak, C.T., Landgraf, R. & Jirikowski, G.F. (1994b) Vasopressin antisense oligonucleotide induces temporary diabetes insipidus in rats. *J. Neuroendocrinol.* **6**, 121–125.

Smith, R.C., Dworkin, M.B. & Dworkin-Rastl, E. (1988) Destruction of a translationally controlled mRNA in *Xenopus* oocytes delays progesterone-induced maturation. *Genes Devel.* **2**, 1296–1306.

Smith, R.C., Bement, W.M., Dersch, M.A., Dworkin-Rastl, E., Dworkin, M.B. & Capco, D.G. (1990) Nonspecific effects of oligodeoxynucleotide injection in *Xenopus* oocytes: a reevaluation of previous D7 mRNA ablation experiments. *Development* **110**, 769–779.

Sokoloff, P., Giros, B., Martres, M.P., Bouthenet, M.L. & Schwartz, J.C. (1990) Molecular cloning and characterization of a novel dopamine receptor (D3) as a target for neuroleptics. *Nature* **347**, 146–151.

Sommer, W., Bjelke, B., Ganten, D. & Fuxe, K. (1993) Antisense oligonucleotide to c-*fos* induces ipsilateral rotational behaviour to *d*-amphetamine. *NeuroReport* **5**, 277–280.

Spampinato, S., Canossa, M., Carboni, L., Campana, G., Leanza, G. & Ferri, S. (1994) Inhibition of proopiomelancortin expression by an oligodeoxynucleotide complementary to beta-endorphin mRNA. *Proc. Natl. Acad. Sci. USA* **91**, 8072–8076.

Standifer, K.M., Chien, C.C., Wahlestedt, C., Brown, G.P. & Pasternak, G.W. (1994) Selective loss of delta opioid analgesia and binding by antisense oligodeoxynucleotides to a delta opioid receptor. *Neuron* **12**, 805–810.

Stanley, B.G. (1993) Neuropeptide Y in multiple hypothalamic sites controls eating behavior, endocrine and autonomic systems for body energy balance. In *The Biology of Neuropeptide Y and Related Peptides* (ed. Wahlestedt, C. & Colmers, W.F.), pp. 457–509. Humana Press, Totowa.

Stein, C.A. & Cheng, Y.C. (1993) Antisense oligonucleotides as therapeutic agents – is the bullet really magical? *Science* **261**, 1004–1012.

Stein, C.A. & Krieg, A.M. (1994) Problems in interpretation of data derived from *in vitro* and *in vivo* use of antisense oligodeoxynucleotides. *Antisense Res. Devel.* **4**, 67–69

Stein, C.A., Mori, K., Loke, S.L., Subasinghe, C., Shinozuka, K., Cohen, J.S. & Neckers, L.M. (1988) Phosphorothioate and normal oligodeoxyribonucleotides with 5′-linked acridine: characterization and preliminary kinetics of cellular uptake. *Gene* **72**, 333–341.

Stephenson, M.L. & Zamecnik, P.C. (1978) Inhibition of Rous sarcoma viral RNA translation by a specific oligodeoxyribonucleotide. *Proc. Natl. Acad. Sci. USA* **75**, 285–288.

Strauss, M., Hering, S., Lieber, A., Herrmann, G., Griffin, B.E. & Arnold, W. (1992) Stimulation of cell division and fibroblast focus formation by antisense repression of retinoblastoma protein synthesis. *Oncogene* **7**, 769–773.

Tischmeyer, W., Grimm, R., Schicknick, H., Brysch, W. & Schlingensiepen, K.H. (1994)

Sequence specific impairment of learning by c-*jun* antisense oligonucleotides. *NeuroReport* **5**, 1501–1504.

Vickers, T., Baker, B.F., Cook, P.D., Zounes, M., Buckheit, R.W., Jr., Germany, J. & Ecker, D.J. (1991) Inhibition of HIV-LTR gene expression by oligonucleotides targeted to the TAR element. *Nucleic. Acids Res.* **19**, 3359–3368.

Wahlestedt, C., Pich, E.M., Koob, G.F., Yee, F. & Heilig, M. (1993a) Modulation of anxiety and neuropeptide Y-Y1 receptors by antisense oligodeoxynucleotides. *Science* **259**, 528–531.

Wahlestedt, C., Golanov, E., Yamamoto, S., Yee, F., Ericson, H., Yoo, H., Inturrisi, C.E. & Reis, D.J. (1993b) Antisense oligodeoxynucleotides to NMDA-R1 receptor channel protect cortical neurons from excitotoxicity and reduce focal ischaemic infarctions. *Nature* **363**, 260–263.

Weiss, B., Zhou, L.W., Zhang, S.P. & Qin, Z.H. (1993) Antisense oligodeoxynucleotide inhibits D2 dopamine receptor-mediated behavior and D2 messenger RNA. *Neuroscience* **55**, 607–612.

West, A.P. & Cooke, B.A. (1991) A novel method to modulate desensitization and truncation of luteinizing hormone receptors using antisense oligodeoxynucleotides. *Mol. Cell Endocrinol.* **79**, R9–14.

Whitesell, L., Geselowitz, D., Chavany, C., Fahmy, B., Walbridge, S., Alger, J.R. & Neckers, L.M. (1993) Stability, clearance, and disposition of intraventricularly administered oligodeoxynucleotides: implications for therapeutic application within the central nervous system. *Proc. Natl. Acad. Sci. USA* **90**, 4665–4669.

Wickstrom, E. (1986) Oligodeoxynucleotide stability in subcellular extracts and culture media. *J. Biochem. Biophys. Methods* **13**, 97–102.

Wollnik, F. (1992) Neural control of circadian rhythm in mammals. *Verh. Dtsch. Zool. Ges.* **85**, 231–246.

Wollnik, F., Brysch, W., Uhlmann, E., Gillardon, F., Bravo, R., Zimmerman, M., Schlingensiepen, K.H. & Herdegen, T. (1995) Block of c-*fos* and *junB* expression by antisense-oligonucleotides inhibits light induced phase shifts of the mammalian circadian clock. *Eur. J. Neurosci.* **7**, 388–393.

Woolf, T.M., Melton, D.A. & Jennings, C.G. (1992) Specificity of antisense oligonucleotides *in vivo*. *Proc. Natl. Acad. Sci. USA* **89**, 7305–7309.

Yakubov, L.A., Deeva, E.A., Zarytova, V.F., Ivanova, E.M., Ryte, A.S., Yurchenko, L.V. & Vlassov, V.V. (1989) Mechanism of oligonucleotide uptake by cells: involvement of specific receptors? *Proc. Natl. Acad. Sci. USA* **86**, 6454–6458.

Zamecnik, P.C. & Stephenson, M.L. (1978) Inhibition of Rous sarcoma virus replication and cell transformation by a specific oligodeoxynucleotide. *Proc. Natl. Acad. Sci. USA* **75**, 280–284.

Zhang, M. & Creese, I. (1993) Antisense oligodeoxynucleotide reduces brain dopamine D2 receptors: behavioral correlates. *Neurosci. Lett.* **161**, 223–226.

Zheng, H., Sahai, B.M., Kilgannon, P., Fotedar, A. & Green, D.R. (1989) Specific inhibition of cell-surface T-cell receptor expression by antisense oligodeoxynucleotides and its effect on the production of an antigen-specific regulatory T-cell factor. *Proc. Natl. Acad. Sci. USA* **86**, 3758–3762.

Zhou, L.W., Zhang, S.P., Qin, Z.H. & Weiss, B. (1994) *In vivo* administration of an oligodeoxynucleotide antisense to the D2 dopamine receptor messenger RNA inhibits D2 dopamine receptor-mediated behavior and the expression of D2 dopamine receptors in mouse striatum. *J. Pharmacol. Exp. Ther.* **268**, 1015–1023.

Zhu, N., Liggitt, D., Liu, Y. & Debs, R. (1993) Systemic gene expression after intravenous DNA delivery into adult mice. *Science* **261**, 209–211.

Index

Printed and bound by CPI Group (UK) Ltd, Croydon, CR0 4YY

08/05/2025

01865022-0001